Theoretical Nuclear Physics In Italy

Proceedings of the 11th Conference on Problems in Theoretical Nuclear Physics

Theoretical Nuclear Physics in Italy

Cortona, Italy 11–14 October 2006

edited by

A. Covello
Università di Napoli Federico II, Italy

L. E. Marcucci
Università di Pisa, Italy

S. Rosati
Università di Pisa, Italy & INFN, Italy

I. Bombaci
Università di Pisa, Italy

World Scientific

NEW JERSEY • LONDON • SINGAPORE • BEIJING • SHANGHAI • HONG KONG • TAIPEI • CHENNAI

Published by

World Scientific Publishing Co. Pte. Ltd.
5 Toh Tuck Link, Singapore 596224
USA office: 27 Warren Street, Suite 401-402, Hackensack, NJ 07601
UK office: 57 Shelton Street, Covent Garden, London WC2H 9HE

British Library Cataloguing-in-Publication Data
A catalogue record for this book is available from the British Library.

THEORETICAL NUCLEAR PHYSICS IN ITALY
Proceedings of the 11th Conference on Problems in Theoretical Nuclear Physics

ISBN-13 978-981-270-770-3
ISBN-10 981-270-770-0

Printed in Singapore by World Scientific Printers (S) Pte Ltd

PREFACE

These Proceedings contain the invited and contributed papers presented at the 11th Conference on Problems in Theoretical Nuclear Physics held in Cortona, Italy, from 11th to 14th October, 2006. As usual, the Conference was held at *il Palazzone*, a 16th century castle owned by the Scuola Normale Superiore of Pisa. The aim of this biennal conference is to bring together Italian theorists working in various fields of Nuclear Physics to discuss their latest results and confront their points of view in a lively and informal way. This offers the opportunity to promote collaborations between different groups. A part of the scientific program of this 11th meeting is the mid-term review of the Research Project of National Interest (PRIN) entitled "Theory of Nuclear Structure and Nuclear Matter", which is financially supported through the years 2006-2007 by the Italian Ministero dell'Istruzione, dell'Università e della Ricerca (MIUR).

There were about 60 participants at the conference, coming from 13 Italian Universities (Catania, Ferrara, Firenze, Genova, Lecce, Milano, Napoli, Padova, Pavia, Pisa, Roma, Torino, Trento). The program of the conference, prepared by the Organizing Committee (S. Boffi, A. Covello, L. E. Marcucci and S. Rosati) focused on seven main topics: Few-Nucleon Systems, Nuclear Matter and Nuclear Dynamics, Nuclear Structure, Nuclear Astrophysics, Nuclear Physics with Electroweak Probes, Structure of Hadrons and Hadronic Matter, Quark-Gluon Plasma and Relativistic Heavy Ion Collisions. G. Salmè, M. Baldo, F. Andreozzi, E. Vigezzi, C. Giusti, M. Giannini and M. Nardi took the responsibility of giving general talks on these topics and reviewing the research activities of the various Italian groups. In addition, 24 contributed papers were presented, most of them by young participants. The last session of the conference was devoted to an overview of experimental nuclear physics in Italy. R. Alba from the Laboratori Nazionali del Sud of Catania, G. de Angelis from the Laboratori Nazionali di Legnaro and A. Bracco, who is the Chairperson of the Nuclear Physics group of the Istituto Nazionale di Fisica Nucleare, kindly agreed to join our theoretical meeting and spoke about the present status and perspectives of experimental nuclear physics.

All the talks included in these proceedings are for the purpose of presentation. We would like to thank the authors of the general reports for their hard work in reviewing the main achievements in the various fields as well as our experimental colleagues whose talks bear witness to the vitality of experimental nuclear physics research in Italy.

This preface ends with a profound note of sorrow. Last July, our friend and colleague Adelchi Fabrocini passed away. His remarkable contributions to Nuclear Physics are well known and will remain as lasting proof of his scientific stature. This conference started with a session in memory of Adelchi, in which Sergio Rosati and Artur Polls highlighted his human qualities and scientific achievements. We are very pleased that the written text of this session is included in these proceedings. We would like to mention here that Adelchi has been the driving force behind the Cortona meetings for some 20 years. The death of Adelchi Fabrocini is a great loss to our community and it is our wish that these proceedings be dedicated to his memory.

Aldo Covello
on behalf of the Organizing Committee

CONTENTS

In Memory of Adelchi Fabrocini

REMEMBERING PROFESSOR ADELCHI FABROCINI

Adelchi Fabrocini was born on August 27, 1951 in Sepino (Campobasso) in the Molise region of Italy.

He completed his University studies at the Institute of Physics of Pisa University which was still housed in Piazza Torricelli where he obtained his degree "summa cum laude" with a thesis on "Variational calculations of bosonic systems wih an infinite number of particles".

In 1977 he was associated to the Pisa section of the Italian National

Institute of Nuclear Physics.

From September 1978 he was "Assistente incaricato" in the Engineering Faculty of Pisa University, and he obtained tenure as a Research Associate in October 1981.

In 1985 he was awarded a prize by the "Angelo della Riccia Foundation" for young research physics theoreticians.

From June 1985 until July 1987 he was Post Doctoral Research Associate in the University of Urbana-Champaign in Illinois (USA), and during July and August 1989 he was Visiting Professor there.

In November 1992 he won a competition to become Associate Professor of Nuclear Physics in the Science Faculty of Pisa University.

Between 1990 and 1993 he was responsible for the Pisa unit of the MURST national project on the Physical Theory of the Nucleus and the Multiparticle Systems and in 1994 he was nominated responsible for Italy of a MURST program in a collaboration involving Pisa University and the University of Barcelona (Spain).

From December 1995 until February 1996 he was a Visiting Scientist at the CEBAF laboratory in the United States.

In 2000 he became Full Professor of Nuclear Physics in the Department of Physics of Pisa University and subsequently achieved tenure.

From 2003 until 2005 he was the local coordinator of the Theoretical Physics group of INFN.

He was extensively involved with teaching the Physics I course of the Engineering Faculty of Pisa University and later on the Nuclear Physics and Foundations of Nuclear Physics course for the degree in Physics. As a University teacher, he interacted with a very large number of students, always establishing excellent relations with them. His teaching activities were highly appreciated by all of his students and recognized by his colleagues. He also supervised numerous degree theses, devoted mainly to studying the description of many-body strongly interacting systems. He was also examiner for a host of theses in the fields of Theoretical and Condensed State Physics.

His main scientific activity was devoted to research on the microscopic behaviour of many-body strongly interacting nuclear systems and on the development of techniques for studying such systems realistically. Much of that research work was dedicated to applying the theory of correlated basis functions (CBF) to nuclear and neutronic matter and to doubly magic nuclei, both for the ground state and for reaction processes, essentially inclusive, starting from hamiltonians with two and three-body realistic interaction potentials.

Among the results obtained during this research activity we may mention:

- the derivation of a microscopic equation of state for nuclear matter and its application to study the diffusion of relativistic electrons from heavy nuclei to nuclear matter;
- the study of the inclusive longitudinal and transverse response functions (electromagnetic and spin) for nuclear and neutron matter;
- the application of the CBF theory, with realistic potentials, to the study of the ground state of doubly magic nuclei;
- application and generalization of the methods developed for nuclear systems to the study of liquid helium;
- application of the CBF theory to the microscopic study of Bose-Einstein condensates of alkaline atoms in harmonic traps.

He published about a hundred articles in international scientific journals and conference proceedings.

His activity as organizer of Conferences and Workshops was also notorious:

- ten editions of the "Convegno su problemi di Fisica Nucleare Teorica" in Cortona;
- he was one of the promotors of the "Elba International Physics Center" in Marciana Marina where he organized several Workshops on "Electron-nucleus scattering" and "Two-nucleon emission reactions";
- "Workshop on Monte Carlo methods in theoretical physics", that was held in the Elba International Physics Center;
- summer school on "Microscopic Quantum Many-Body Theories and their Applications", that was hosted by the Abdus Salam Center for Theoretical Physics in Trieste.

He was member of the International Advisory Committee for the Series of International Conferences on Recent Progress in Many-Body Theories, very active in organizing activities to promote the discipline of Quantum

Many-Body Physics among the young generations.

He participated as invited speaker in numerous conventions, workshops and international schools and gave many lectures and seminars in Italian institutes and abroad.

He was frequently consulted as a referee by international physics journals, such as Journal of Low Temperature Physics, Journal of Physics, Nuclear Physics, Physics Letters, Physical Review, etc. .

He had numerous and profitable scientific collaborations with Italian institutions (university of Lecce, Sissa, the Genoa and Rome I sections of INFN) and abroad (the universities of Illinois, Basel, Barcelona, Granada, Athens, Tempe, the Argonne Nationale Laboratories and the Thomas Jefferson Laboratory National Accelerator Facility).

We would again like to recall the enormous capacity of Adelchi Fabrocini in contributing to the diverse teaching, scientific and organisational activities generally required in the everyday life of the Physics Department in the University of Pisa combined with the simplicity and cordiality always demonstrated in his relations with other people.

His death will prove to be a serious loss for the University of Pisa, and for the International Many-Body Physics community. He will certainly be remembered with unbounded gratitude and affection by all those who knew him and had the fortune to have his friendship and collaboration.

On behalf of the members of the Pisa Nuclear Theoretical Physics group and all the friends of Adelchi Fabrocini, S. ROSATI, *Dipartimento di Fisica, Università di Pisa, Largo B. Pontecorvo 3, 56127 Pisa, Italy. Email: sergio.rosati@df.unipi.it.*

LEARNING MANY-BODY PHYSICS WITH ADELCHI

O. BENHAR

INFN Sezione di Roma, and Dipartimento di Fisica
Università "La Sapienza", I-00185 Roma, Italy

G. CO'

Dipartimento di Fisica, Università di Lecce,
and INFN Sezione di Lecce, I-73100 Lecce, Italy

A. POLLS

Departament d'Estructura i Constituents de la Matèria,
Universitat de Barcelona, E-08028 Barcelona, Spain

We present an overview of the contributions that Adelchi Fabrocini gave to the field of many-body physics during the last thirty years. He has left us while he was still in full activity, and his work, which is certainly a reference for all of us, will motivate and guide the work of future research in many-body physics for a long time.

1. Introduction

It is already five months that Adelchi has left us, and it becomes evident that we will not get used to his absence. We often talk as he is still present, or we think of common future projects. More than once, during these months, we found ourselves trying to make impossible phone calls. Both, from the personal and scientific points of view, we feel an emptiness hard to fill. In this short article, we want to remember his contributions to science, and give a feeling of his attitude with respect to physics.

For Adelchi, physics, or science in general, was not just a job, but a way of life, a way of understanding the world. In some sense, the rigor in the analysis, the imagination and the creativity in the invention of new formalisms, the curiosity and the motivations in asking "why" and "how", were characteristics of his scientific personality, and they were projected in his daily life. We are of similar ages, and from the very beginning we have made our careers together, sharing and enjoying common scientific interests. In fact, the many-body problem in several fields have occupied us for about thirty years.

Adelchi remembered often to us his first participation to a Many-Body Conference, in Trieste, in 1978. At that time, the discrepancies between the results of the Brueckner-Hartree-Fock theories and those of the variational calculations for nuclear matter were driving the main discussions at the Conference. At that time, the many-body problems were centered around nuclear physics, although the first Monte Carlo calculations for bosons systems were also presented. But what remained in Adelchi's memory was the special seminar that P.A.M. Dirac gave about the time dependence of universal constants.

After that conference, Adelchi was convinced that Many-Body Physics was a good subject of study and he dedicated his efforts to it. The guidance of Sergio Rosati and Stefano Fantoni was crucial to make the first steps in this direction. The reasons for this motivations are well summarized at the beginning of one of the review articles that Adelchi wrote together with Stefano Fantoni [1]: " *Nearly all of physics is many-body physics at the most microscopic level of understanding, appropriate to the energy scale of the particular branch of physics under consideration. Thus, the subject of quantum many-body theory (QMBT) can fairly be said to virtually strengthen from beneath all of modern physics. The fundamentally many-particle nature of nuclei, atoms, molecules, solids, and fluids are all manifestly apparent, but even the single nucleon problem is itself becoming a multiparticle problem at the deepest level of understanding"* .

The QMBT was applied by Adelchi to describe different many-body systems, from cold atoms to neutron stars. In spite of the different scales of length and energy of these systems, many of the driving physics phenomena can be described by using similar strategies. The microscopic description of these systems is not only an intellectual goal, but it is the appropriate approach to put in evidence the similarities which define a unified view of different physical phenomena.

The microscopic description of the many-body systems, requires first the identification of their appropriate constituents. The proper choice depends on the energy scale used to probe the system. The determination of the interaction between these constituents is the following step. The knowledge of the masses of the constituents and of their interactions is a sufficient information to define the hamiltonian of the systems. At this point, the goal is the solution of the corresponding Schrödinger equation to evaluate the binding and excitation energies of the system, and to describe its dynamics.

The many-body systems Adelchi and us have studied, are characterized by strong interactions between their components, producing large correlations which strongly modify the independent particle model picture. These strong interactions are the origin of serious difficulties in the application of traditional perturbative approaches. A way out of this difficulty is to produce, and use, effective interactions, weak enough to allow a rapid convergence of the traditional perturbation expansion. An alternative approach, is that of the Correlated Basis Function (CBF) theory, that Adelchi has significantly contributed to improve and develop in the last thirty years.

The core of CBF theory, is to incorporate the correlations from the very beginning into a trial wave function Ψ_T of the form:

$$\Psi_T(1,...,N) = F(1,...,N)\Phi(1,...,N) \; . \tag{1}$$

In the above equation $\Phi(1,...,N)$ describes the system in the Independent Particle Model (IPM) picture, i.e. in absence of interaction between the components, and, evidently, of correlations. We have indicated with $F(1,...,N)$ a correlation operator which takes into account in a direct, but also not exhaustive, way some of the correlations induced by the interactions.

The quantum statistic of the system is taken into account by the IPM function $\Phi(1,...,N)$. For instance, for a homogeneous bosonic liquid, $\Phi(1,...,N)$ is a constant, i.e. all the particles are assumed to occupied the zero momentum state. In the case of an infinite Fermi system, Φ is a Slater determinant of plane waves, with all the momenta occupied up to the Fermi level. For finite systems, $\Phi(1,...,N)$ is built on single particle wave functions generated by a mean-field potential.

The variational principle,

$$\frac{\langle \Psi_T \mid H \mid \Psi_T \rangle}{\langle \Psi_T \mid \Psi_T \rangle} = E_T \geq E_0 \; , \tag{2}$$

provides the first estimation of the ground state energy (E_0).

The calculation of the expectation value (2) is rather involved, and different methods, for example Monte Carlo techniques, have been devised to evaluate it. In the second half of the seventies, important progresses have been done on methods based on integral equations as hypernetted chain and its generalization to the Fermi systems (Fermi-Hypernetted Chain, FHNC).

The CBF theory provides a way of systematically improve on a trial wave function (1) by doing perturbation theory on correlated functions. Since many correlations are already contained in the variational wave function (1), one expects that a satisfactory description of the system can be obtained by the lowest order perturbation term of the CBF theory. Obviously, the perturbation theory done with correlated functions is more complicated than the standard one. This is the field in which Adelchi has given important contributions, defining also paths to be followed in the future. In the next sections, we briefly describe these contributions in different fields.

We want to end this introduction by pointing out the important role of Adelchi as author of several review articles which, right now, are the best places to find the latest developments of the CBF theory. We want to remember the great amount of energy he devoted to coordinate and organize activities aimed to catalyze the Italian Theoretical Nuclear Physics community, among other things, through the organization of the Cortona meetings which have been very useful, especially for the young generations of nuclear physicists in Italy.

Finally, we want to say that we have been very fortunate to meet Adelchi, to enjoy his friendship, to share projects with him, and especially, to enjoy his sweet and clever irony which helped us a lot to look at the problems, in both life and physics, under unusual perspectives.

2. Quantum liquids and cold atoms

The first article of Adelchi [2], published in *Il Nuovo Cimento* in March 1980, was devoted to study an infinite-hard-sphere Fermi system by using Jastrow variational wave functions. The FHNC equations were used to study its energy and the momentum distribution. Special attention was devoted also to the evaluation of the different types of elementary diagrams, which appear in the diagrammatic representation of the cluster expansion of the two-body distribution function. The elementary diagrams cannot be calculated in a closed form, and their evaluation is a recurrent problem of all the integral equations methods [3]. The existence of low-density expansions and of the Brueckner-Hartree-Fock calculations for this system gave the possibility of a critical comparison between the results obtained with different many-body methods. Recently, this system has renewed the interest of the community in connection to the description of, both fermionic and bosonic, cold atoms.

Immediately after that, Adelchi started to study more realistic systems, and paid attention to the ^{3}He-^{4}He mixtures. At that time, there was great interest in the microscopic description of both ^{4}He and ^{3}He quantum liquids which are of bosonic and fermionic nature respectively. One of the reasons of this interest, was that the interaction between the basic constituents, i.e. the helium atoms, is rather simple and depends only on their relative distance. The interaction shows a repulsive hard-core at short distances and a weak attraction at larger distances. Furthermore, the interaction, which is of electromagnetic origin, is the same for both isotopes. The fact that the interaction is so simple in comparison with nuclear interaction is one of the reasons to consider quantum liquids as an excellent laboratory to test the accuracy of quantum many-body theories.

These systems remain liquid at zero temperature. In the case of the mixtures, there is an incomplete phase separation, being the maximum solubility of ^{3}He in liquid ^{4}He of about 6.5% at zero pressure. The coexistence of fermionic and bosonic statistics, and the fact that the concentration of the fermionic component is small, make the system very interesting. To study the ground state of these mixtures, we derived in 1982 [4], a system of seven non-linear HNC/FHNC coupled integral equations. Soon after, we studied the momentum distributions of the mixture and how the condensate fraction of ^{4}He, i.e. the fraction of ^{4}He atoms in the zero momentum state, changes with the concentration [5]. We predicted a small increase of the condensate fraction, based on the fact that the total density of the mixture decreases, when the concentration of ^{3}He increases, due to its larger mobility. The discrepancies with the results of a later deep inelastic neutron scattering experiment, forced us to revise our calculations with much more sophisticated wave functions, including two- and three-body correlations. We finally obtained results [6] which are in agreement with recent, new, experimental data.

The limit of ^{3}He zero concentration, defines the impurity problem. The experimental chemical potential of the ^{3}He impurity, at the saturation density of liquid ^{4}He, turns out to be -2.785 K, to be compared with the binding energy per particle of liquid ^{3}He, which is -2.5K. The excitation spectrum of the impurity is also important, and it is characterized by its effective mass, which, at zero momentum, is $m^*/m_3 = 2.3$. In 1986 we studied all these observables at the variational level, by using back-flow correlations, and we obtained an effective mass of $m^*/m_3 = 1.7$ [7], clearly far away from the experimental value. This result forced us to make an exhaustive analysis of the correlated-basis perturbation theory. We improved the wave function by including backflow correlations, not only around the impurity, but also around the ^{4}He atoms in the medium. These additional correlations are described by means of a basis, were the momentum of the excitation is shared between the Feynman phonons that can be excited in the medium. Finally, by including all the perturbative diagrams up to two independent phonons, we obtained an effective mass of $m^*/m_3 = 2.2$. This analysis was extended to the full range of momentum, to study the momentum dependence of the effective mass [8], which is an important ingredient to understand the two branches of the low energy spectrum of the mixtures: the phonon and roton excitations of the ^{4}He medium which are very little affected by the presence of ^{3}He atoms, and the ^{3}He quasiparticle-excitations characterized by the effective mass studied in the impurity problem. All these ingredient brought us to investigate the response of ^{3}He-^{4}He mixtures [9], in the range of low momentum transfer were the two branches of the response appear well separated in energy.

At the same time, we have done an analysis of the variational contents of the Average Correlation Approximation, i.e. the approximation which considers the same correlations for all pair of particles, allowed us to use the impurity as a probe in the liquid ^{4}He and to obtain a lower bound to the kinetic energy per particle of liquid ^{4}He:

$$t_4(\rho) \geq \left(\frac{m_4}{m_3} - 1\right)^{-1} (\mu_3^{exp}(\rho) - \mu_4^{exp}(\rho)) \ . \tag{3}$$

This lower bound was useful to constrain the value of the kinetic energy of liquid ^{4}He, extracted from deep-inelastic neutron scattering data [10]. The study of the analogies between the deep-inelastic regime for liquid ^{3}He and the inelastic electron scattering off nuclei was also an important piece of work [11,12]. The ^{4}He impurity problem in ^{3}He liquid was also calling our attention. In this case, the microscopic calculation of the binding energy and the effective mass of the ^{4}He impurity, done by using an extended Jastrow-Slater wave function, and by including two- and three-body correlations and also back-flow correlations between the ^{4}He atom and the particles in the medium, was providing the result $m^*/m_4 = 1.21$, at the ^{3}He saturation density, in very good agreement with the experimental value [13].

After the experimental realization of the Bose-Einstein condensation (BEC) with magnetically trapped alkali atoms, with all the system lying in the condensate, in contrast to the liquid ^{4}He, where the condensate fraction is of the order of 8%, we dedicated some efforts to the description of these dilute systems.

We have studied the ground state of Bose hard-spheres trapped in a harmonic trap, to investigate the effects of the interatomic correlations, and the accuracy of the mean field Gross-Pitaevskii equation. We have proposed a modified Gross-Piatevskii equation [14–16], based on a local density approximation, which being still a mean field, and having therefore all the atoms in the condensate, incorporates additional terms of the low density expansion of the energy of a homogeneous hard-sphere system. The existence of the Fesbach resonances opens the possibility of tuning the interaction between the atoms, and therefore it becomes necessary the study of correlation effects in the field of cold atoms.

Finally, we want to point out one of the last publications of Adelchi, which faces an old problem which presently is very important both in fermionic cold atoms and also in the context of nuclear physics. We are referring to the pairing problem. In this case, one should use a trial wave function, with Jastrow correlations multiplying a Slater determinant built with pairs of particles of opposite momenta. At present, the results do not show significant differences with the standard BCS. Certainly, more efforts will be devoted in this new, and promising, line of research.

3. Correlated basis function theory of nuclear matter

Nuclear matter can be thought of as a giant nucleus, with given numbers of protons and neutrons interacting through nuclear forces only. The calculation of the binding energy of such a system, whose equilibrium value can be inferred from nuclear systematics, is greatly simplified by translational invariance. A quantitative understanding of the properties of nuclear matter, besides being a necessary intermediate step towards the description of real nuclei, is needed to develop realistic models of neutron star matter.

At the end of the 1970s, as Adelchi was beginning his scientific life, the study of nuclear matter was regarded as the hottest topic in Many-Body theory. The accuracy of calculations carried out using the well established formalism of G-matrix perturbation theory and the hole-line expansion was being questioned by the results of new variational approaches, based on correlated wave functions and cluster expansion techniques. In the spring of 1977 a number of outstanding physicists, including the Nobel laureate Hans Bethe, had gathered at the University of Illinois at Urbana-Champaign to attend a Workshop on Nuclear and Dense Matter, aimed at assessing the status of the field and pin down the sources of the stricking disagreement between the results of different approaches. As mentioned in Section 1, the Conference on Recent Progress in Many-body Theory, held the following year at ICTP, Trieste, was also largely devoted to nuclear matter.

Adelchi played an important role in the development of CBF theory of nuclear matter. In the 1980s, in collaboration with Bob Wiringa, he carried out a detailed study of the equation of state of charge neutral nucleon matter in weak equilibrium, whose results set the standard of the field for over a decade and have been employed in a number of calculations of neutron star properties.

At about the same time, the large body of electron scattering data flowing from the new facilities, operating in both Europe and the United States, provided previoulsly unavailable information on nuclear dynamics, exposing the limitations of the nuclear shell model and the importance of nucleon-nucleon correlations. Extrapolation of the available data to the limit of infinite target allowed one to extract empirical information on the linear response and Green's function of nuclear matter at equilibrium density.

The CBF formalism is ideally suited to carry out theoretical studies of correlation effects on electron scattering observables. Adelchi was quick to realize this potential and engaged in a number of projects aimed at providing quantitative predictions to be compared to the data.

Calculations at moderate momentum transfers ($|\mathbf{q}| < 0.5\,\text{GeV}$) can be carried out using nonrelativistic wave functions to describe both the initial and final nuclear states and expanding the nuclear current operator in powers of $|\mathbf{q}|/m$, m being the nucleon mass. Within this approach Adelchi and Stefano Fantoni carried out a CBF calculation of the response of nuclear matter to longitudinally polarized photons [17], measured in inclusive electron-nucleus scattering. Their results clearly demonstrated that correlation effects dominate the nuclear cross section in the regions of both low and high electron energy loss.

At higher values of $|\mathbf{q}|$, corresponding to beam energies larger than ~ 1 GeV, the description of the final states in terms of nonrelativistic nucleons is no longer possible. Calculations of the nuclear cross section in this regime require a set of simplifying assumptions, allowing one to take into account the relativistic motion of final state particles carrying momenta $\sim \mathbf{q}$, as well as the occurrence of inelastic processes leading to the appearance of hadrons other than protons and neutrons.

The impulse approximation scheme is based on the assumption that, as the spatial resolution of a probe delivering momentum $\mathbf{q}$ is $\sim 1/|\mathbf{q}|$, at large enough $|\mathbf{q}|$ the scattering process off a nuclear target reduces to the incoherent sum of elementary processes involving individual nucleons. As a consequence, the nuclear cross section can be written in terms of the spectral function, i.e. the Green's function yielding the energy and momentum distribution of the target nucleons.

Adelchi gave relevant contributions to both the analysis of the limits of the impulse approximation [12] and the derivation of the nuclear matter Green's functions [18,19] within CBF. The results of these calculations are still routinely used in the data analysis of many electron scattering experiments.

Further studies of the analytic structure of the nuclear matter Green's function also led to generalize Migdal's theorem to momenta different from the Fermi momentum, thus providing a clearcut identification of correlation effects in the spectroscopic strengths measured in high resolution proton knock out experiments [20].

While being very fond of the beauty of the mathematical formalism of Many-Body theory, Adelchi was also interested in phenomenology, and managed to interact with experimentalists in a remarkably productive fashion. Starting in the early 1990s, he was part of a collaboration including Ingo Sick, that led to the development of a formalism to describe final state interactions in electron-nucleus scattering within CBF [21,22]. During this period Adelchi coauthored a very peculiar paper for a theorist: a letter on the interpretation of the ratios of inclusive nuclear cross section that contained no equations [23].

Perhaps, the most amazing feature of Adelchi's personality was his ability of carrying out outstanding research work, that led to important and lasting contributions to the theory of nuclear matter, keeping at the same time an ironic and almost self-mocking attitude. He certainly believed that, as stated by the latin poet Horace, "humor does not prevent one from speaking the truth" ("ridentem dicere verum, quid vetat ?" (Horace, Sat I.1.24)).

4. Correlated basis functions for finite nuclei

At beginning of the '90s, we started a project aiming to apply the FHNC/SOC computational scheme to finite nuclear systems. The idea of the project was triggered by S. Fantoni that, together with S. Rosati, proposed in the late '70s a formal extension of the FHNC theory to finite Fermi systems. The traditional FHNC equations should be reformulated to consider the effects of the not, any more, constant density of these systems, by means of the so-called vertex corrections. The new set of integral equations are known as Renormalized Fermi Hypernetted Chain (RFHNC) equations.

The results of the first numerical application of the RFHNC equations to finite nuclear systems, have been presented in Ref. [24]. In that article, model nuclei have been described. Protons and neutrons wave functions were produced by a unique mean field potential, and in a ls coupling scheme. The nucleon-nucleon interactions considered, had only central terms, and the correlations were scalar functions. This simplified situation was used to test the theoretical, and numerical, feasibility of the approach. Results for binding energies of ^{16}O and ^{40}Ca model nuclei, were presented in [24], while the momentum distributions have been shown in a following article [25].

A more realistic description of doubly closed shell nuclei was given in [26]. Protons and neutrons were separately treated, and the single particle wave functions were written in a jj coupling scheme. The RFHNC equations required a non trivial reformulation. Binding energies, matter densities and momentum distributions, have been calculated for various doubly magic nuclei up ^{208}Pb. Also in this case, however, simple central interactions and scalar correlations were used. In this framework, we described hypernuclei with a single Λ, by considering the hyperon as an impurity in the nucleonic system [27].

In the following step, the RFHNC equations were extended to treat operator dependent correlations, which do not commute with the hamiltonian, and also among themselves. This implied the use of the single operator chain (SOC) approximation. Because of the technical difficulties, the RFHNC/SOC equations have been first formulated to deal with spin and isospin saturated nuclei, and the single particle wave functions were described in ls coupling scheme. Again, only ^{16}O and ^{40}Ca nuclei could be treated. The results of these calculations have been presented in Refs. [28]. In this phase of the project, the great experience of Adelchi in FHNC nuclear matter calculations has been exploited at its best. The calculations of Ref. [28] where done with two-body nucleon interactions only. The results of fully realistic calculations, where the two-nucleon interactions of the Argonne-Urbana family have been implemented with the appropriated three-body forces, have been presented in Refs. [29,30].

A formulation of the RFHNC/SOC equations general enough to handle separately protons and neutrons in the more realistic jj coupling scheme, was finally achieved. In Ref. [31] binding energies and density distributions are shown for the ^{12}C, ^{16}O, ^{40}Ca, ^{48}Ca and ^{208}Pb nuclei. To the best of our knowledge these are the first calculations of medium-heavy nuclei, done with fully realistic interactions containing both two- and three-body forces.

We feel now as we have reached the top of the mountain. The hard work has finished, and we have in front of us a wide range of applications of the RFHNC/SOC computational scheme. We feel the absence of Adelchi in this new phase of our project. His talent, his guide, his experience and also, last and not least, his subtle sense of humor, would have given different perspectives to our future work.

References

1. S. Fantoni and A. Fabrocini, in Microscopic Quantum Many-Body Theories and Their Applications, Lectures Notes in Physics, Vol. 510, J. Navarro and

A. Polls eds., Springer Verlag, 1998.
2. A. Fabrocini, S. Fantoni, A. Polls, and S. Rosati, Nuovo Cimento **A56**, 33 (1980).
3. A. Fabrocini and S. Rosati, Nuovo Cimento **D1**, 567 (1982); **D1**, 615 (1982).
4. A. Fabrocini and A. Polls, Phys. Rev. B **25**, 4533 (1982).
5. A. Fabrocini and A. Polls, Phys Rev. B **26**,1438 (1982)
6. J. Boronat, A. Polls, and A. Fabrocini, Phys. Rev. B **56**, 11854 (1997).
7. A. Fabrocini, S. Fantoni, A. Polls, and S. Rosati, Phys. Rev. B **33**, 6057 (1986).
8. A. Fabrocini and A. Polls, Phys. Rev. B **58**, 5209 (1998).
9. A. Fabrocini, L. Vichi, F. Mazzanti, and A. Polls, Phys. Rev. B **54**, 1035 (1996).
10. J. Boronat, A. Fabrocini, and A. Polls, Phys. Rev. B **39**, 2700 (1989).
11. S. Moroni, S. Fantoni, and A. Fabrocini, Phys. Rev. B **58**, 11607 (1998).
12. O. Benhar, A. Fabrocini, and S. Fantoni, Phys. Rev. Lett. **87**, 052501 (2001).
13. F. Arias de Saavedra, J. Boronat, A. Polls, and A. Fabrocini, Phys. Rev. B **50**, R4248 (1994).
14. A. Fabrocini and A. Polls, Phys. Rev. A **60**, 2319 (1999).
15. A. Fabrocini and A. Polls, Phys. Rev. A **64**, 063610 (2001).
16. F. Mazzanti, A. Polls, and A. Fabrocini, Phys. Rev. A **67**, 063615 (2003).
17. A. Fabrocini and S. Fantoni, Nucl. Phys. **A503**, 375 (1989).
18. O. Benhar, A. Fabrocini, and S. Fantoni, Nucl. Phys. **A505**, 267 (1989).
19. O. Benhar, A. Fabrocini, and S. Fantoni, Nucl. Phys. **A550**, 201 (1992).
20. O. Benhar, A. Fabrocini, and S. Fantoni, Phys. Rev. C **41**, R24 (1990).
21. O. Benhar, A. Fabrocini, S. Fantoni, G.A. Miller, V.R. Pandharipande, and I. Sick, Phys. Rev. C **44**, 2328 (1991).
22. O. Benhar, A. Fabrocini, S. Fantoni, V.R. Pandharipande, S.C. Pieper, and I. Sick, Phys. Lett. **B359**, 8 (1995).
23. O. Benhar, A. Fabrocini, S. Fantoni, and I. Sick, Phys. Lett. **B343**, 47 (1995).
24. G. Co', A. Fabrocini, S. Fantoni, and I. Lagaris, Nucl. Phys. **A549**, 439 (1992).
25. G. Co', A. Fabrocini, and S. Fantoni, Nucl. Phys. **A568**, 73 (1994).
26. F.Arias de Saavedra, G.Co', A. Fabrocini, and S. Fantoni, Nucl. Phys. **A605**, 219 (1996).
27. F.Arias de Saavedra, G. Co', and A. Fabrocini, Phys. Rev. C **63**, 064308 (2001).
28. A. Fabrocini, F.Arias de Saavedra, G. Co', and P. Folgarait, Phys. Rev. C **57**, 1668 (1998).
29. A. Fabrocini, F.Arias de Saavedra, and G. Co', Phys. Rev. C **61**, 044302 (2000).
30. A. Fabrocini and G. Co', Phys. Rev. C **63**, 044319 (2001).
31. C. Bisconti, F.Arias de Saavedra, G. Co', and A. Fabrocini, Phys. Rev. C **73**, 054304 (2006).

Few-Nucleon Systems

FEW-NUCLEON SYSTEMS

G. SALMÈ

Istituto Nazionale di Fisica Nucleare, Sezione di Roma , P.le A. Moro 2, I-00185 Roma, Italy

Highlights on the recent research activity, carried out by the Italian Community involved in the Few-Nucleon Systems field, will be presented.

1. Introduction

The research in the Few-Nucleon Systems field is a very active one in Italy, and it is not an easy task to report the whole amount of studies that were performed in the last two years. I will present only a partial overview of this activity, attempting to highlight major contributions from the different Groups participating to this Conference.

The introduction of new techniques i) for obtaining solution of the Schrödinger equations for three- and four-nucleon systems (see, e.g., Refs. [1,2]) and ii) for directly evaluating the electromagnetic responses for the Few-Nucleon systems (see, e.g., Refs. [3,4]), has allowed to greatly extend the accuracy of the description of both the ground-state properties and the key issue represented by the Final-State-Interaction (FSI) effects, so important for elucidating/extracting many nuclear features, like, e.g., the role of the three-nucleon forces [5].

Relativity is an emerging issue, given the accuracy reached by present-day experiments at momentum transfer $\geq$ 300 MeV/c (see, e.g., the extraction of the neutron magnetic form factor from the asymmetry for the reaction $^3\vec{\mathrm{H}}e(\vec{e}, e')$ [6]). Efforts are going on in order to construct a relativistic framework for studying Few-Nucleon systems (see below and Contributions in this Conference Session), and this activity appears attractive since it could open the possibility to gain new insights in the many-body interaction and in the short range correlations.

Finally, it should be pointed out that relevant outcomes will be discussed in other Sessions as well.

2. Relativistic Hamiltonian Dynamics

Some Groups are becoming involved in applying the Relativistic Hamiltonian Dynamics (RHD) models to Few-Nucleon systems, therefore it could be useful to briefly introduce some of the issues typical of this approach, proposed by Dirac [7] in 1949.

RHD could be seen as an intermediate step between the non-relativistic approach and the full-glory field theory. Within RHD, one can embed the successful few-nucleon phenomenology, developed within non relativistic approaches, in a Poincaré covariant theory [8].

Dirac aim was to merge the principles from Poincaré Covariance and the dynamical description of an interacting system He exploited the Hamiltonian framework, the most suitable one for dealing with interacting system in a non perturbative way. In non relativistic Quantum Mechanics, once the quantum state at some instant $t = t_0$, $\psi(x, t_0)$, is known, then the time (dynamical) evolution can be obtained by applying the proper operator, i.e. $\psi(x,t) = exp[-\imath H(t - t_0)]\psi(x, t_0)$, where H is the generator of the time translations. Differently from the non relativistic case, in a relativistic framework there are various possible choices for the "initial" hypersurface where the state could be known, due to the finite value of the light speed. In particular, Dirac listed i) the hypersurface defined by $t = 0$ and any values for $\vec{r}$, leading to the *Instant form* of the RHD, ii) the hypersurface defined by $z + t = 0$, leading to the *Front-form* RHD, and iii) the hypersurface defined by $t^2 - r^2 = a > 0$, leading to the *Point-form* RHD. It is worth noting that the variable labeling the dynamical evolution of the system, or "time" in the standard language, is given by different combinations of time and spatial coordinates, depending on the RHD one adopts.

The properties of RHD's are related to the generators of the Poincaré Algebra (i.e. the four-momentum, P^μ, the 3 boosts, B^i, and the three rotations, J^i) that allow one to accomplish all the possible transformations on a quantum state in the Minkowski space. In particular, depending on the chosen RHD, some generators contain the interaction and are called dynamical, while others do not and they are called kinematical. These features of the generators are dictated by the symmetry properties of the "initial" hypersurfaces. For instance, in non relativistic QM, the Hamiltonian (generator of the time translation) contains the interaction, while three-momenta and rotations do not (given the translational and rotational invariance of the 3D Euclidean space). For each RHD introduced by Dirac, the Poincaré generators that have kinematical or dynamical nature are in order

- Instant form: space translations, $\vec{P}$, and space rotations, $\vec{J}$, are kinematical, while $P^0 = H$ and the boosts are dynamical, in order to allow the evolution of the quantum state.
- Front form (or Light-front form): $\vec{P}_\perp$, $P^0 + P_z$, J_z, B_z and the other two Light-front (LF) boosts are kinematical, while $P^0 - P_z$ and two LF rotations are dynamical.
- Point form: boost $\vec{B}$ and rotations $\vec{J}$ are kinematical, while P^μ is dynamical.

A simple consequence for an interacting system (consider a two-body case for the sake of simplicity) is that $p_1^\mu + p_2^\mu \neq P^\mu$, where p_i^μ are the momenta of free constituents and P^μ the 4-momentum of the interacting system. Such a disequality means that at least one component of P^μ must contain interaction. In particular, i) for the Instant form, where the standard t is the evolution variable one has $p_1^0 + p_2^0 \neq P^0$, ii) for the Front form, where the combination $x^+ = t + z$ plays the role of evolution variable, one has $p_1^- + p_2^- \neq P^-$, iii) for the Point form, where t can be used as evolution variable, but x^+ as well, all the four components of $P^\mu \neq p_1^\mu + p_2^\mu$, viz :

$$P^\mu = M_{int} \frac{p_1^\mu + p_2^\mu}{M_{free}} \qquad \text{with} \quad M_{int} \neq M_{free} = |p_1^\mu + p_2^\mu| \tag{1}$$

This extremely introductory list of features of the RHD approach could give a snapshot on some advantages and drawbacks in the application of RHD to interacting Few-Body systems (see, e.g., Ref. [8]).

3. Relativistic approaches for few-body systems in Padoa

It appears far-reaching the attempt carried out by the collaboration Padoa-Kharkov [9] of including, in a consistent way, the pion in elementary processes like $NN \to NN$, $\pi N \to \pi N$ and $NN \to \pi NN$. For accomplishing this, one has to devise a suitable relativistic treatment, since i) the absorption/emission of particles forces to enlarge the energy range of the theoretical investigation and ii) the virtual processes in the π-NN vertex have to be described consistently. In particular, the Instant-form framework is adopted. The aim is to rearrange the Hamiltonian, $\mathbf{H}(g)$ (where g is the coupling constant) by applying a suitable Unitary Transformation (UT), in order to have, at a given order of g, a $\tilde{\mathbf{H}}(g)$ such that

$$\tilde{\mathbf{H}}(g)\ |Vacuum\rangle = \mathbf{H}_{free}\ |Vacuum\rangle \qquad \tilde{\mathbf{H}}(g)\ |1\rangle = \mathbf{H}_{free}\ |1\rangle \tag{2}$$

In this way, at a given order of g, no virtual processes, dressing the vacuum and the one-particle state, are present, namely they are summarized in the creation/destruction operators *clothed* by the Unitary Transformation.

For the sake of concreteness the clothing procedure is applied to a model Hamiltonian, that contains a simple Yukawa interaction with a pseudoscalar meson-nucleon coupling, i.e. $V = ig\bar{\psi}\gamma_5\psi\phi$.

After applying a proper mass-changing UT to the Hamiltonian, in order to dress the bare masses of the particles, one has to look for a *Unitary Clothing Transformation*, that makes the eigenstates of the free Hamiltonian, with 0 and 1 particle, be eigenstates of the total one, viz

$$\begin{aligned} H = H(\alpha) = H_{free}(\alpha) + H_{int}(\alpha) = H\left(W\left(\alpha_c\right)\alpha_c W^\dagger\left(\alpha_c\right)\right) = \\ = W\left(\alpha_c\right) H\left(\alpha_c\right) W^\dagger\left(\alpha_c\right) = K\left(\alpha_c\right) = K_{free}(\alpha_c) + K_{int}(\alpha_c) \end{aligned} \tag{3}$$

where α indicates the set of creation/destruction operators and α_c the new, clothed ones, $W(\alpha_c) = \exp R(\alpha_c)$ is a UT with $R(\alpha_c)$ an operator to be determined, and the operator $K_{int}(\alpha_c)$ must have the property: $K_{int}(\alpha_c)|1\rangle_c = 0$. Note that $K(\alpha_c)$ is $H(\alpha)$ expressed in terms of α_c.

It is found that the explicit expression obtained for $R(1)$ (at the first order of g) eliminates all the terms in the Hamiltonian that prevent the vacuum and the one-particle state to be eigenstate of the total Hamiltonian. Some bad terms of order g^2 can be eliminated with a further transformation, $R(2)$. In this way, one restricts the presence of bad terms to the g^3 order.

Along these guidelines, one can decompose the Hamiltonian, as follows

$$\begin{aligned} K(\alpha_c) &= K_{free}(\alpha_c) + K_{(NN\to NN)} + K_{(\bar{N}\bar{N}\to\bar{N}\bar{N})} + K_{(N\bar{N}\to N\bar{N})} \\ &+K_{(\pi N\to\pi N)} + K_{(\pi\bar{N}\to\pi\bar{N})} + K_{(\pi\pi\leftrightarrow N\bar{N})} + K_{(NN\leftrightarrow\pi NN)} + K_{(\bar{N}\bar{N}\leftrightarrow\pi\bar{N}\bar{N})} \\ &+K_{(N\bar{N}\leftrightarrow\pi N\bar{N})} + K_{(N\bar{N}\leftrightarrow\pi\pi\pi)} + K_{(\pi N\leftrightarrow\pi\pi N)} + K_{(\pi\bar{N}\leftrightarrow\pi\pi\bar{N})} + \ldots \end{aligned} \tag{4}$$

where the interactions between the clothed nucleons (N), antinucleons ($\bar{N}$) and pions (π) have been separated out. For instance, the $NN \to NN$ interaction operator is given by

$$K_{(NN\to NN)} = \int d\mathbf{p}_1 d\mathbf{p}_2 d\mathbf{p}_1' d\,\mathbf{p}_2' V_{NN}(\mathbf{p}_i', \mathbf{p}_i) b_c^\dagger(\mathbf{p}_1') b_c^\dagger(\mathbf{p}_2') b_c(\mathbf{p}_1) b_c(\mathbf{p}_2)$$

where (ω is the energy of the exchanged pion)

$$\begin{aligned} V_{NN}(\mathbf{p}_i', \mathbf{p}_i) = -\frac{1}{2}\frac{g^2}{(2\pi)^3}\frac{1}{2\omega_{\mathbf{p}_1'-\mathbf{p}_1}} \times \\ \frac{m^2}{\sqrt{E_{\mathbf{p}_1}E_{\mathbf{p}_2}E_{\mathbf{p}_1'}E_{\mathbf{p}_2'}}}\delta(\mathbf{p}_1' + \mathbf{p}_2' - \mathbf{p}_1 - \mathbf{p}_2)\bar{u}(\mathbf{p}_1')\gamma_5 u(\mathbf{p}_1) \times \\ \left\{\frac{1}{E_{\mathbf{p}_1} - E_{\mathbf{p}_1'} - \omega_{\mathbf{p}_1'-\mathbf{p}_1}} + \frac{1}{E_{\mathbf{p}_1'} - E_{\mathbf{p}_1} - \omega_{\mathbf{p}_1'-\mathbf{p}_1}}\right\}\bar{u}(\mathbf{p}_2')\gamma_5 u(\mathbf{p}_2) \end{aligned} \tag{5}$$

A relevant issue for the possible extension to Few-Nucleon systems of a Point-Form approach (see the Contribution by Luca Girlanda in these Proceedings), could be the analysis of the normalization of the wave function of an interacting system, performed [10] in detail for baryonic systems, seen as bound three quark system. In Ref. [10], the ambiguity related to the choice of the normalization factor of the baryon wave function has been elucidated within the so-called Point-Form Spectator model (PFSM) (see Ref. [11]), pointing to the simplifying assumption of a spectator pair with an unchanged three-momentum, before and after the reaction process. The 3-momentum conservation for an electromagnetic process involving a baryon reads

$$\vec{P}\,' - \vec{P} = \vec{q} = M'\frac{\vec{p}_1\,' + \vec{p}_2\,' + \vec{p}_3\,'}{M'_{free}} - M\frac{\vec{p}_1 + \vec{p}_2 + \vec{p}_3}{M_{free}} \tag{6}$$

where $P'(P)$ is the final (initial) baryon 3-momentum, $\vec{q}$ the 3-momentum transfer. Note that it is different from the 3-momentum absorbed by the struck quark (labeled by 1, for concreteness), namely $\vec{q} \neq \vec{p}_1\,' - \vec{p}_1$, but such a quantity has to be determined in order to perform actual calculations.

In the initial-baryon CM frame one has: $\vec{p}_1 + \vec{p}_2 + \vec{p}_3 = 0$, and

$$\vec{q} = M'\frac{\vec{p}_1\,' + \vec{p}_2\,' + \vec{p}_3\,'}{M'_{free}} \tag{7}$$

where M' is the mass of the final baryon and M'_{free} the corresponding free mass. *If* we assume that the spectator quarks do not change their-own 3-momenta one has

$$\vec{q} = M'\frac{\vec{p}_1\,' + \vec{p}_2 + \vec{p}_3}{M'_{free}} = M'\frac{\vec{p}_1\,' - \vec{p}_1}{M'_{free}} \tag{8}$$

Then, the normalization factor to be used is [10]

$$\mathcal{N}_{in} = \left(\frac{M'}{M'_{free}}\right)^3 \tag{9}$$

In the final-baryon CM frame one ends up with the following normalization factor

$$\mathcal{N}_{fin} = \left(\frac{M}{M_{free}}\right)^3 \tag{10}$$

A workable rule for solving the mentioned ambiguity has been found from the comparison between the experimental values of the mesonic decays of baryon resonances and the theoretical calculations obtained by using different choices of the normalization factor (between the two extrema given by Eqs. (9) and (10)) and the baryon wave functions evaluated within a Goldstone-boson exchange model [11]. It is worth noting that the normalization factor, given the dependence upon the mass of the interacting system, introduces in the evaluation of matrix elements of one-body operators many-body effects.

See also the Contribution by Thomas Melde in these Proceedings.

4. The Hyperspherical Harmonics approach and its applications: the research activity in Pisa

The first important topic, thoroughly investigated by the Pisa Group, is the effects of the Final State interaction in states with three and four nucleons. The novelty of their approach consists in the successful merging of the asymptotic behavior, typical of the continuum states, and a variational technique [12], where the key ingredient is the expansion of the nuclear wave function onto a basis, given by the spin-isospin states ⊗ the Hyperspherical Harmonics (HH) functions (that depend upon the polar and azimuthal angles for each of the 3 Jacobi coordinates, and the two hyperspherical angles). Note that the remaining dependence upon the hyperradius is contained in the coefficients (better to say functions) of the expansion. Let us briefly illustrate such a powerful approach in the case of the N-deuteron scattering, for the sake of simplicity.

In HH methods, the fully-interacting scattering state, for a 3N system, can be decomposed [12] as follows

$$\begin{aligned}\Phi^{jj_zTT_z\pi} &= \Psi_A^{jj_zTT_z\pi} + \Psi_C^{jj_zTT_z\pi} = \\ &= \sum_{i=3}\left[\psi_A^{jj_zTT_z\pi}(i) + \psi_C^{jj_zTT_z\pi}(i)\right] \qquad (11)\end{aligned}$$

where the intrinsic energy is understood; $\Psi_A^{jj_zTT_z\pi}$ is the solution of the Schrödinger equation in the asymptotic region, $\Psi_C^{jj_zTT_z\pi}$ describes the system when the three nucleons are close each other. The functions $\psi_A^{jj_zTT_z\pi}(i)$ and $\psi_C^{jj_zTT_z\pi}(i)$ are Faddeev-like amplitudes, corresponding to the three permutations of the coordinates ($\equiv \{\mathbf{r_1},\ \mathbf{r_2},\ \mathbf{r_3}\}$).

The asymptotic component, $\Psi_A^{jj_zTT_z\pi}$, can be recast in a different way, in order to emphasize its physical content; then one has

$$\begin{aligned}\Psi_A^{LXjj_zTT_z\pi} &= \Omega^R_{LXj}(\mathbf{x}_1,\mathbf{y}_1) + \\ &+ \sum_{L'X'} {}^j\mathcal{L}^{XX'}_{LL'}\left[\imath\, \Omega^R_{L'X'j}(\mathbf{x}_1,\mathbf{y}_1) + \Omega^I_{L'X'j}(\mathbf{x}_1,\mathbf{y}_1)\right] + \\ &+ \left[\psi_A^{LXjj_zTT_z}(2) + \psi_A^{LXjj_zTT_z}(3)\right] \qquad (12)\end{aligned}$$

where L is the relative orbital angular momentum of N with respect to the deuteron, X is the intermediate coupling of the spin of the nucleon with the total angular momentum of the deuteron, and the intrinsic coordinates $\{\mathbf{x}_1,\ \mathbf{y}_1\}$ are defined as follows

$$\mathbf{x}_1 = \mathbf{r}_2 - \mathbf{r}_3 \qquad \mathbf{y}_1 = \frac{1}{\sqrt{3}}\left[\mathbf{r}_2 + \mathbf{r}_3 - 2\mathbf{r}_1\right]. \tag{13}$$

In Eq. (12), $\Omega^{R(I)}_{LXJ}$ represents the regular ("irregular", but properly regularized at small distances [12,13]) solution describing the free scattering of a nucleon by an interacting pair (in this case a deuteron); the matrix $\mathcal{L}$ is

$$\mathcal{L} = \frac{\mathcal{S} - 1}{2\imath} = -\pi\mathcal{T} \tag{14}$$

with $\mathcal{S}$ and $\mathcal{T}$ the S-matrix and the T-matrix, respectively. Similar expressions hold for $\psi_A^{LXjj_zTT_z}(2)$ and $\psi_A^{LXjj_zTT_z}(3)$. Three terms can be recognized in Eq. (12): i) the first one produces the Plane Wave Impulse Approximation (PWIA), i.e. it contains a fully-interacting pair and a free particle; ii) the second term describes the rescattering between the interacting pair and the asymptotically free particle; iii) the third term, $\left[\psi_A^{LXjj_zTT_z}(2) + \psi_A^{LXjj_zTT_z}(3)\right]$, takes care of the correct antisymmetrization of $\Psi_A^{jj_zTT_z\pi}$.

The core component, $\Psi_C^{jj_zTT_z\pi}$, goes to zero for large interparticle distances and energies below the deuteron breakup threshold, while for higher energies, must reproduce a three outgoing particle state. In the approach developed in Ref. [12], $\Psi_C^{jj_zTT_z\pi}$ is explicitly written as an expansion on a basis of HH Polynomials, with the inclusion of pair-correlation functions, to be determined along with the elements of the S-matrix (see Eq. (14)), through a variational procedure (complex Kohn variational principle [12]).

The proton-deuteron elastic scattering in the energy region between 3-65 MeV has been investigated in detail in Ref. [14], where a benchmark calculation has been carried out for comparing the predictions obtained by using the Correlated HH approach (in coordinate space) and the Alt-Grassberger-Sandhas equation (in momentum space). The NN interaction was the charge-dependent Argonne V18 (AV18) potential [15], and different approaches were used for including the repulsive Coulomb interaction. In general, the methods produce essentially the same results for a large set of elastic observables, but at energies below the 3-body breakup threshold the CHH works better, while for energies higher than 65 MeV, the AGS method seems more efficient. It should be pointed out that the present techniques are able to master the Coulomb effects, clearly identifying the kinematical region where they become negligible.

The extended theoretical analysis of the polarization observables, extracted from the reactions $d(\vec{p},\vec{p})d$ and $d(\vec{p},\vec{d})p$ at $E_p^{lab} = 22.7$ MeV [5], was based on solutions of the scattering states, obtained by using both the Correlated HH method and the Faddeev equations in momentum space. Different interactions were tested: i) realistic 2N (AV18 [15], CD Bonn [16], Nijm I and Nijm II [17]) and 3N (Tucson-Melboune [18] and Urbana IX [19]) potentials and ii) 2N interaction based on a chiral perturbation theory (NLO and NNLO [20]). Such an investigation has yielded another piece of evidence of the relevance of 3N forces, when the energy of the 3N system become larger and larger. This immediately becomes a practical suggestion for calling new accurate experiments at higher energies.

A great achievement within the HH approach, has been the calculation for the first time of solutions of the Schrödinger equation for the four-nucleon ground state [2]. In particular, both i) model interactions (like the MT-V, MT-I/III, Volkov etc.) and ii) realistic 2N (AV18 [15], Nijm II [17]) and t3N (Urbana IX [19], Tucson-Melbourne [18]) potentials have been adopted.

A careful analysis has been devoted to the convergence of different classes of HH functions, since, in presence of a huge amount of basis functions, some criteria have to be introduced in order to reduce the complexity of the numerical task. The results, essential for sound theoretical evaluations of the ^{4}He properties, were very successful as shown in Table 1.

It is worth noting that, in ^{4}He, the isospin components with $T > 0$ are very small, but nonetheless important for the key-issue of the parity violation in experiments involving ^{4}He. The presence of a $T = 1$ component could be relevant in extracting information on the $s\bar{s}$ pair in the nucleon (see the Contribution by Michele Viviani in these Proceedings).

As to the 4N system, besides the above mentioned careful study of the ground state, an analysis of the states in the continuum, at low energy, has been undertaken [24]. The low energy elastic n-^{3}H scattering appears on one side a simple reaction channel (no Coulomb interaction, and to a very good approximation a pure T=1 state), on the other side has a rich dynamics, since a resonance structure is present around $E_{cm} \sim 3$ MeV. Its investigation can be considered highly complementary for studying 4N ground state, and allows one to more deeply test both the existing models for 2N and 3N interactions and the new ones. The thorough comparison of the outcomes of present-day techniques (Alt-Gassberger-Sandhas, Faddeev-

Table 1. The α-particle binding energies B (MeV), the rms radii (fm), the expectation values of the kinetic energy operator $\langle K \rangle$ (MeV), and the P and D probabilities (%) for various realistic interaction models as computed by means of the HH expansion are compared with the results obtained by the Faddeev-Yakubosky (FY) method and the Green-Function Montecarlo (GFMC) method (the binding energies obtained by using an extrapolation technique are enclosed in parentheses).

Interaction	Meth.	B	$\langle K \rangle$	$\langle r^2 \rangle^{1/2}$	P_P	P_D
AV18	HH	24.210(24.222)	97.84	1.512	0.347	13.74
	FY [21]	24.25	97.80		0.35	13.78
	FY [22]	24.223	97.77	1.516		
Nijm II	HH	24.419(24.432)	100.27	1.504	0.334	13.37
	FY [21]	24.56	100.31			
AV18+UIX	HH	28.462(28.474)	113.30	1.428	0.73	16.03
	FY [21]	28.50	113.21		0.75	16.03
	GFMC [23]	28.34(4)	110.7(7)	1.44		
AV18+TM'	HH	28.301(28.313)	110.27	1.435	0.73	15.63
	FY [21]	28.36	110.14		0.75	15.67

Yakubosky and Hyperspherical Harmonics), with a particular care to the issue of the convergence by increasing the angular momenta taken into account (in particular for FY and HH) has led to recognize a failure of the existing nuclear interactions to reproduce the n-^{3}H total cross section. It should be pointed out that the results obtained within FY and HH methods

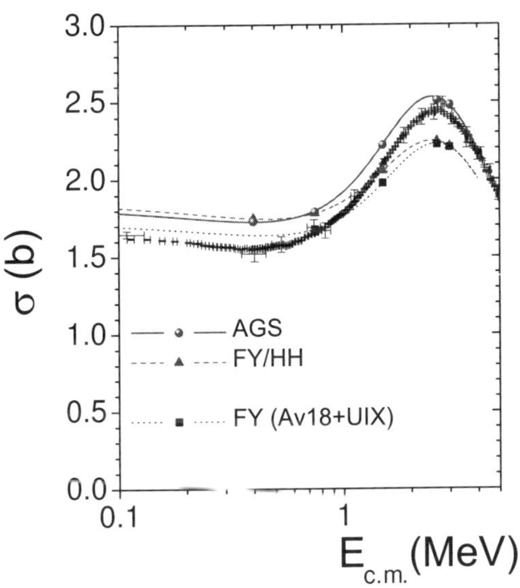

Fig. 1. Comparison among experimental and theoretical n-^{3}H total cross sections calculated using different methods (AGS, FY and HH) and with the AV18 potential. Solid line: AGS method; short-dashed line: overlapping HH and FY results; dotted curve presents AV18 [15] + Urbana IX [19] results of Ref. [22]. (After Ref. [24]).

nicely agree, but underestimate the data at the peak, as shown in Fig. 1. Indeed AGS yields a better description of the peak, but it is based on a simple rank-one approximation of the two-body T-matrix.

In collaboration with the experimental group, the detailed analysis, of the proton-^{3}He elastic scattering [25], at low energy (1.6 MeV$\leq E_p \leq$ 4.05 MeV) has shown the analogous of the "A_y puzzle" known for the past 20 years in the nucleon-deuteron elastic channel. The unpolarized cross section is very well described by the variational wave functions for 3N and 4N states, when the 3N interaction (Urbana IX [19]) is included. The puzzle, see Fig. 2, arises when the comparison with the measured A_y is carried out. A discrepancy of about 50% at the maximum of the proton analyzing power,

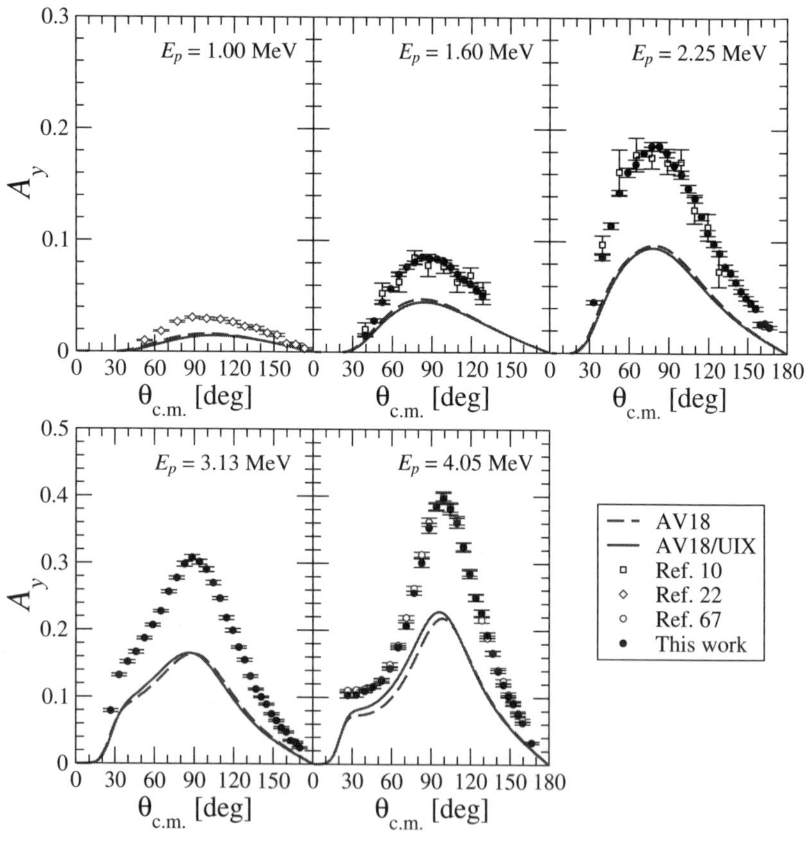

Fig. 2. Measured p-^{3}He proton analyzing power A_y at five different energies. Curves show the results of theoretical calculations for the AV18 (dashed lines) and AV18 [15]+Urbana IX [19] (solid lines) potential models. (After Ref. [25]).

for each values of E_p measured, is found. The challenge is to understand the source of such a discrepancy: a possible failure in our understanding of 3N forces or a subtle role of higher angular momenta (P-waves?).

In the last few years, a new class of realistic NN interactions (like the Charge-Dependent Bonn 2000 or N3LO [26], based on chiral perturbation theory) has been proposed. These interactions have a χ^2/datum $\simeq$ 1 and are given in momentum space, namely they are non local. They appear a profitable tool that could shed light on still-unclear topics, like the structure and the role played by 3N forces. Therefore, accurate techniques for solving the corresponding Schrödinger equation for A=3,4 are very important. The Pisa Group [1] has applied their HH variational approach (without correlation functions) for solving the Schrödinger equation with those momentum-dependent potential. This approach is very effective, since the matrix elements of the various terms of the interactions, can be evaluated in the coordinate space or in momentum space, depending upon convenience (local or non local terms), in almost a straightforward way (but, yet, a non trivial numerical task!). The convergence issue for these non-local potentials is less critical, since the short-range repulsion is rather soft compared to the AV18 one. The impressive accuracy that can be reached for the ground state properties of A=3 and A=4 nuclei is shown in the Tables 2 and 3, respectively.

Table 2. The triton binding energies B (MeV), the mean square radii $\sqrt{\langle r^2\rangle}$ (fm), the expectation values of the kinetic energy operator $\langle T\rangle$ (MeV), and the mixed-symmetry S', P, D, and isospin T=3/2 (largely dominated by the charge symmetry breaking!) probabilities (all in %), calculated with the CD Bonn 2000 and N3LO potentials, are compared with the results obtained within the Faddeev equations approach (FE) and within the No Core Shell Model (NCSM) approach. The results within HH, CHH and FE approaches for the AV18 potential have been also reported for sake of comparison. These last HH results do not include the T=3/2 states.

Inter.	Meth.	B	$\langle T\rangle$	$\sqrt{\langle r^2\rangle}$	$P_{S'}$	P_P	P_D	$P_{T=3/2}$
CDB	HH	7.998	37.630	1.721	1.31	0.047	7.02	0.0049
2000	FE	7.997	37.620	-	1.31	0.047	7.02	0.0048
	FE	7.998	37.627	-	1.31	0.047	7.02	0.0048
	NCSM	7.99(1)	-	-	-	-	-	-
N3LO	HH	7.854	34.555	1.758	1.36	0.037	6.31	0.0009
	FE	7.854	34.546	-	1.37	0.037	6.32	0.0009
	FE	7.854	34.547	-	1.37	0.037	6.32	0.0009
	NCSM	7.85(1)	-	-	-	-	-	-
AV18	CHH	7.624	46.727	-	1.293	0.066	8.510	0.0025
	HH	7.618	46.707	1.770	1.294	0.066	8.511	-
	FE	7.621	46.73	-	1.291	0.066	8.510	0.0025

Table 3. The α–particle binding energies B (MeV), the mean square radii $\sqrt{\langle r^2\rangle}$ (fm), the expectation values of the kinetic energy operator $\langle T\rangle$ (MeV), and the P, D, $T=1$ and $T=2$ probabilities (%) for the two non-local potentials considered. The results obtained for the AV18 potential have been also reported for sake of comparison. The results obtained by other techniques are also listed.

Inter.	Meth.	B	$\langle T\rangle$	$\sqrt{\langle r^2\rangle}$	P_P	P_D	$P_{T=1}$	$P_{T=2}$
CDB	HH	26.13	77.58	1.454	0.223	10.74	0.0029	0.0108
2000	FY	26.16	77.59	-	0.225	10.77	0.0030	0.0108
N3LO	HH	25.38	69.24	1.516	0.172	9.289	0.0035	0.0024
	FY	25.37	69.20	-	0.172	9.293	0.0033	0.0024
	NCSM	25.36(4)	-	-	-	-	-	-
AV18	HH	24.210	97.84	1.512	0.347	13.74	0.0028	0.0052
	FY	24.25	97.80	-	0.35	13.78	0.003	0.005
	FY	24.223	97.77	1.516	-	-	-	-

5. A=3,4 nuclei and the Lorentz Integral Transform in Trento

The Lorentz Integral Transform (LIT) approach allows microscopic calculations of the electromagnetic scattering by light nuclei, without an explicit knowledge of the nuclear wave functions in the final state, namely one can fully include the FSI effects, without solving an infinite set of equations.

On the other side, the inversion of the Lorentz Transform is a delicate step, since it constitutes a so-called ill-posed problem, due to instabilities in the solution, for a given interval of the involved variables. The Trento Group, that has applied for the first time such an approach to the electromagnetic responses of few-nucleon systems, has devoted a careful analysis of this issue, introducing consequently new inversion techniques, that contain suitable regularization schemes, necessary to overcome the ill-posed problem. Moreover, the new techniques enlarge the range of applicability of LIT. Let us briefly illustrate the main ingredients of the LIT method (that can be seen as a two-step method) for the electromagnetic response of a nucleus, $R(\omega)$, given by

$$R(\omega) = \int d\Psi_f \ |\langle\Psi_f|\mathbf{O}|\Psi_0\rangle|^2 \ \delta(E_f - E_0 - \omega) \tag{15}$$

where $\mathbf{O}$ is the suitable electromagnetic operator, $|\Psi_0\rangle$ is the ground state of the target nucleus ($\mathbf{H}|\Psi_0\rangle = E_0|\Psi_0\rangle$), and $|\Psi_f\rangle$ one of the possible final states reachable after interacting with the probe, ($\mathbf{H}|\Psi_f\rangle = E_f|\Psi_f\rangle$).

Then the Lorentz Transform reads

$$L(\sigma_R, \sigma_I) = \int d\omega \; \frac{1}{(\omega - \sigma_R)^2 + \sigma_I^2} \; R(\omega)$$
$$= \langle \Psi_0 | \mathbf{O}^\dagger \; \frac{1}{(\mathbf{H} - E_0 - \sigma_R)^2 + \sigma_I^2} \; \mathbf{O} | \Psi_0 \rangle = \langle \tilde{\Psi} | \tilde{\Psi} \rangle \tag{16}$$

where σ_R and $\sigma_I > 0$, $(\mathbf{H} - E_0 - \sigma_R - \imath\sigma_I) \, |\tilde{\Psi}\rangle = \mathbf{O}|\Psi_0\rangle$.

The standard inversion method is constructed by expanding $R(\omega)$ in terms of a given basis, and the coefficients are determined through a fit of the Lorentz Transform, previously calculated by solving the differential equation for $|\tilde{\Psi}\rangle$. The number of functions in the expansion plays the role of regularization parameter, that allows one to damp out high frequency oscillations. These unphysical oscillations make unstable the inversion, and unfortunately, cannot be simply separated from the solution: this forces one to introduce a regularization scheme. Indeed, such a procedure turns out to be a very reliable one and quickly converging for $R(\omega)$, without too much structure. In the case of a more complicate structure an extra attention should be paid, for instance isolating the different contributions to $R(\omega)$ and applying a LIT to each structure. This has been applied to the inclusive longitudinal response of the deuteron, at a constant momentum transfer, where a more complicated structure than the simple one due to the quasi-elastic peak [27] appears. One has to separate out the Coulomb monopole and quadrupole transitions, which lead to a shoulder in the corresponding response, at the break-up threshold, while the remaining part of the response has the typical quasi-elastic peak structure. Then one can apply LIT to the Coulomb monopole and quadrupole contribution separately. The possibility to test LIT by using different regularization methods, and hopefully to make more simple the application of LIT in presence of a more complicated structure, is the aim of Ref. [28]. Two proposals are explored:

i) the Fridman approach: where the LIT is based on the iterative procedure for solving a Fredholm integral equation of first order

$$L(\sigma_R, \sigma_I) = \int_a^b d\omega \; K(\sigma_R, \sigma_I, \omega) \; R(\omega)$$

with the solution given by

$$R_{n+1}(x_j) = R_n(x_j) + \lambda \left[w_j L(\sigma_I, x_j) - \sum_{\ell=0}^{N} K(\sigma_I, x_j, x_\ell) R_n(x_\ell) \right]$$

where w_j are the proper weights for the chosen quadrature rule for integrating and $0 < \lambda < 2\lambda_1$, (λ_1 is the smallest characteristic number of K).

The regularization scheme is provided by the chosen grid for the variables x_j.

ii) The Banach fix-point approach: where the LIT can be seen as the matrix elements of a Lorentz operator $\mathbf{L}(\sigma_I)$

$$\langle\omega'|\mathbf{L}(\sigma_I)|R\rangle = \int d\omega\, \frac{1}{(\omega'-\omega)^2+\sigma_I^2}\, R(\omega)$$

with $|R\rangle$ belonging to the space of the square-integrable functions. Then, one introduces a mapping, $\mathbf{T}(\sigma_I, L)$, acting on $R(\omega)$, viz

$$\mathbf{T}(\sigma_I, L)\,[R(\omega')] = \frac{\sigma_I}{\pi}\, L(\omega', \sigma_I) - \frac{\sigma_I}{\pi} \int d\omega\, \frac{R(\omega') - R(\omega)}{[(\omega'-\omega)^2+\sigma_I^2]}$$

that has a fixpoint, namely

$$\mathbf{T}(\sigma_I, L)\,[R(\omega')] = R(\omega')$$

By using the *Banach's fixpoint theorem*, one demonstrates the convergence of the iteration procedure, viz

$$R_{n+1}(\omega') \equiv \mathbf{T}(\sigma_I, L)\,[R_n(\omega')]$$
$$\lim_{n\to\infty} R_n(\omega') = R_{fix}(\omega') \equiv R(\omega') \quad \text{for} \;\; 0 < \sigma_I < \epsilon$$

The regularization scheme involves the mesh points and the number of iterations.

The careful analysis of the new methods were carried out considering realistic, analytical responses [28].

As above mentioned, the main field of application of LIT are the electromagnetic processes involving light nuclei. In particular, the reaction

$$^4\mathrm{He} + e \rightarrow ^3\mathrm{H} + p + e$$

has been investigated [29]. Longitudinal responses (extracted experimentally through a Rosenbluth separation method) have been calculated by using LIT and the semirealistic MT-I-III potential. In particular, the full calculations, namely FSI + proper antisymmetrization in the 4N final states, has been compared with calculations in PWIA, where the final state is the product of a plane wave (describing the struck nucleon) and the *fully interacting* wave function of the spectator 3N system. The aim was to investigate the limits of the PWIA (and the fully symmetrized PWIA), in extracting relevant physical quantities, like the nucleon distribution inside the target nucleus and the spectroscopic factor (a shell model quantity!; but one could argue that ^{4}He is "almost" heavy, i.e. $A = 4 \sim A = \infty$). Within PWIA the cross section has a factorized form, and then one can extract the above mentioned quantities. This extraction is expected to be a reliable one in a parallel kinematics, where a direct proton knock-out mechanism is highly likely. The comparison shows that PWIA might be a reasonable approximation for small missing momenta (below 100 MeV/c) and higher momentum transfer (above 400 MeV/c), see Fig. 3. Furthermore, the antiparallel kinematics drastically enhances (as expected) both FSI and antisymmetrization effects [29]. This finding could be used in a reverse mode. This kinematical region could be the most profitable one for studying different approaches for including FSI, given its sensibility to such an effect.

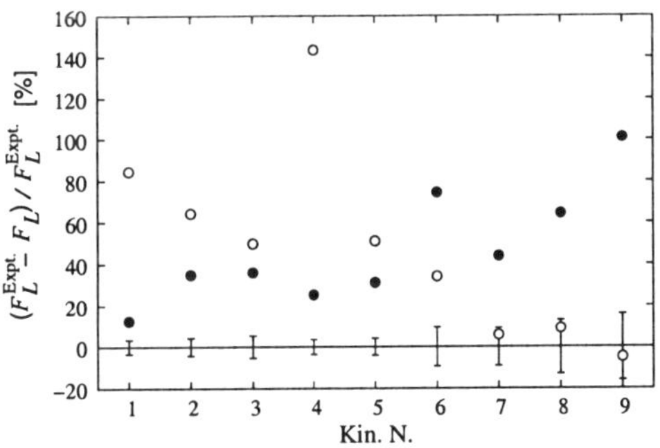

Fig. 3. Percentage deviation from the experimental values: PWIA (open circles), FULL results (full circles). Kin. 1:(q = 299 MeV, ω = 57.68 MeV),→ Kin. 9:(q = 680 MeV, ω = 146.48 MeV). (After Ref. [29]).

Another reaction channel investigated in Trento was $^4He(e, e'd)d$. This analysis [4] has given clear hints about the necessary improvements to be adopted for a better description of the experimental data: the calculation of FSI should be refined going beyond the central interaction, like MT -I/III adopted in this work. The comparison with NIKHEF data, see e.g. Fig. 4 excludes the necessity of a substantial improvement in the description of the ^{4}He state, while the remarkable overestimate of the experimental cross-section at low momentum transfer and the comparison with other models for the ground state suggest to include the tensor force and to adopt other realistic NN interactions in the description of the final states.

A substantial step forward in accurately calculating the total photo-absorption cross section [30] for ^{4}He within the LIT, has been done by considering for the first time a realistic NN interaction, like AV18 [15], and an up-to date 3N force, like Urbana IX [19]. See the Contribution, devoted to this topic, by Sonia Bacca, in these Proceedings.

The ^{3}He longitudinal electromagnetic response has been studied for different values of the momentum transfer q = 500 MeV/c, 600 MeV/c, and 700 MeV/c at the quasi-elastic peak, in order to elucidate the frame de-

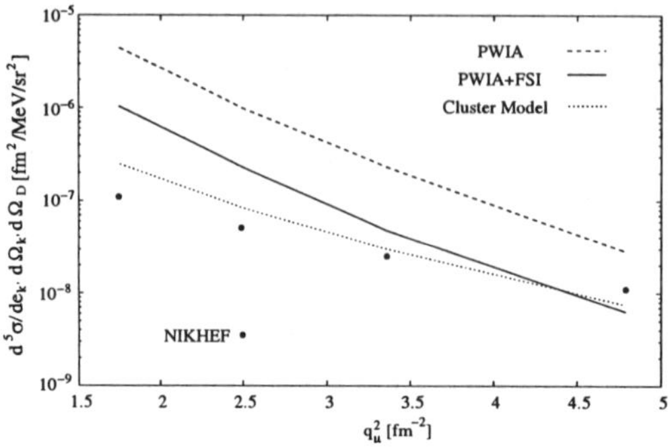

Fig. 4. Differential cross-section at $E_{d,d}$= 35 MeV and averaged over 100 MeV < $|p_m|$ < 150 MeV/c, as function of q_μ^2, with (full curve) and without FSI (dashed curve). Dots: NIKHEF experimental data; also shown results with a HO ^{4}He ground state and FSI in a d-d cluster model for $|p_m|$ = 125 MeV/c (dotted curve). (After Ref. [4]).

pendence, and therefore the relevance of the relativity, of the theoretical calculations [31]. The response in the Lab frame can be put in relation to the responses calculated in other frames (all related by boosts along the direction of $\vec{q}$) through a pure kinematical factor, without considering the effect of the boosts on the 3N wave functions. In particular, three frames are considered:

i) Antilab frame : $\vec{P}_i = -\vec{q}$ and $\vec{P}_f = 0$. Roughly speaking, $\vec{p}_i \sim -\vec{q}/3$, and $\vec{p}_f \sim 2\vec{q}/3$ (where $\vec{p}_{i(f)}$ is the nucleon momentum in the initial (final) 3N state).

ii) Breit frame: $\vec{P}_i + \vec{P}_f = 0$, where $\vec{P}_i = -\vec{q}/2$ and $\vec{P}_f = \vec{q}/2$. Roughly speaking, $\vec{p}_i \sim - - \vec{q}/6$, and $\vec{p}_f \sim 5\vec{q}/6$.

iii) Active nucleon Breit frame: $\vec{P}_i = -3\vec{q}/2$ and $\vec{P}_f = -3\vec{q}/2 + \vec{q} = -\vec{q}/2$. Roughly speaking, $\vec{p}_i \sim -\vec{q}/2$, and $\vec{p}_f \sim \vec{q}/2$.

The nuclear potential adopted was the AV18 (2N) + Urbana IX (3N) [15,19]. The best results can be obtained by using the Active nucleon Breit frame, where nucleons with moderate momenta are present.

Finally, it has to be quoted the impressive taxonomy of responses of a polarized deuteron target in inclusive and exclusive electrodisintegration processes presented in Ref. [32].

6. Trinucleon electromagnetic form factors in the Light-Front Hamiltonian Dynamics: the Rome approach

Following the Light-Front approach already applied to the deuteron [33], the electromagnetic elastic form factors of ^{3}H and ^{3}He have been evaluated in a preliminary calculation [34], retaining only the S and S' waves in the ground state. The main motivation for performing such a calculation is given by the possibility to embed in a Poincaré covariant framework the non relativistic nuclear wave functions (obtained within a Galilean covariant framework). As a matter of fact, the so-called Bakamjian-Thomas construction allows one to settle the general properties to be fulfilled by a mass operator within a Poincaré covariant theory with a fixed-number of interacting constituents: i) independence upon the total momentum and ii) rotational invariance, namely the same properties requested in any non relativistic approach (see, e.g.. Ref. [8] for more details). Furthermore, one can consider the mass operator constructed within the non relativistic framework as an effective approximation (with suitable covariance properties, and reproducing the binding energies) of the "true" relativistic mass operator, i.e. M_{NR} can be seen as an assembly of some operators multiplied by phenomenological functions (depending upon scalars constructed from intrinsic variables of the actual problem) that approximates M_{rel}. Even the dependence of the mass upon the square of the intrinsic 3-momenta of the constituents, that the common wisdom consider a typical signature of non relativistic theories could be acceptable within the scheme outlined before. By the way, for a two-body case such a dependence upon the squared intrinsic 3-momentum is exact, once the Bakamjian -Thomas construction is applied to the square mass operator.

The ingredients of our approach are:

i) a Breit frame where $\mathbf{q}_\perp = 0$ and $q^+ \neq 0$ [33],

ii) a Poincaré covariant current

$$j^\mu(K\vec{e}_z) = \frac{\mathcal{J}^\mu(K\vec{e}_z)}{2} + L^\mu_\nu[r_x(-\pi)]\, e^{\imath\pi S_x} \frac{\mathcal{J}^\nu(K\vec{e}_z)^*}{2} e^{-\imath\pi S_x} \tag{17}$$

with

$$\begin{aligned} \mathcal{J}^+(K\vec{e}_z) &= \mathcal{J}^-(K\vec{e}_z) = \langle \vec{P}_\perp = 0, P^{'+}|\Pi J^+_{free}(0)\Pi|\vec{P}_\perp = 0, P^+\rangle \\ \vec{\mathcal{J}}_\perp(K\vec{e}_z) &= \langle \vec{P}_\perp = 0, P^{'+}|\Pi \vec{J}_{free\perp}(0)\Pi|\vec{P}_\perp = 0, P^+\rangle \end{aligned} \tag{18}$$

where S_x is a spin component, $K = q/2$, Π is a projector onto the subspace of a trinucleon bound state, $|\chi_{\frac{1}{2}}\rangle$, with mass M_T and spin 1/2, the free em current is $J^\mu_{free}(0) = \sum_i J^\mu_{pi}(0)(1+\tau_3)/2 + J^\mu_{ni}(0)(1+\tau_3)/2$, with $J^\mu_{p(n)} = -F_2^{p(n)}(p^\mu + p'^\mu)/2M + \gamma^\mu(F_1^{p(n)} + F_2^{p(n)})$,

iii) a trinucleon bound state, obtained through a variational technique by Kievsky, Rosati, Viviani in Ref. [13] with the AV18 interaction [15].

Preliminary calculation with only S+S' waves (the corresponding probabilities are: $\mathcal{P}_S(Av18) =$ 90.1% and $\mathcal{P}_{S'}(Av18) =$ 1.29%) are shown in Figs. 5 and 6. Note that the difference between the theoretical diffraction pattern and the experimental one in all the form factors should be reduced (hopefully eliminated) by adding the D wave (as it happens in the non relativistic treatment). The calculated magnetic moments of ^{3}H in correspondence to S and S+S' waves are: $\mu_{\rm T}(S)$ =-1.778 (non rel.:-1.723) and $\mu_{\rm T}(S+S')$ =-1.759 (non rel.: -1.707)($\mu_{\rm T}(exp)$ = -2.1276) respectively, while for ^{3}He they are: $\mu_{\rm He}(S)$ =2.597 (non rel.: 2.515) and $\mu_{\rm He}(S+S')$ =2.652 (non rel.: 2.518)($\mu_{\rm He}(exp)$ = 2.9789) respectively.

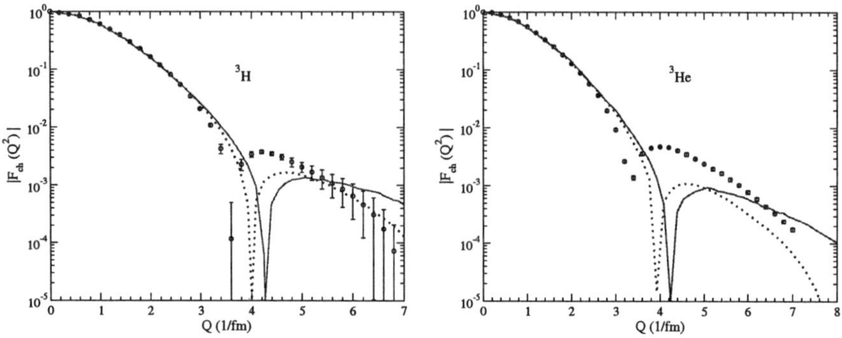

Fig. 5. Charge form factors of ^{3}H (leftmost panel) and ^{3}He (rightmost panel) vs. the square momentum transfer, Q^2, within a LF approach. Solid line: LFD calculation with S+S' wave only,in a Breit frame where $\mathbf{q}_\perp = 0$. Dotted line: non relativistic calculation. The Gari-Krümpelmann N form factors [35] have been adopted. Data after Ref. [36].

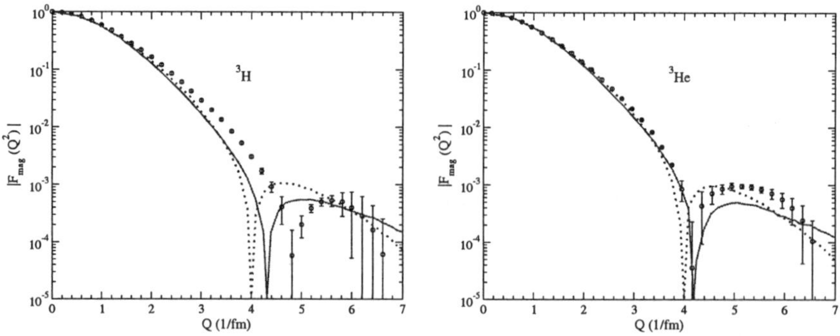

Fig. 6. The same as in Fig. 5, but for the magnetic form factor.

7. Conclusions and perspectives

To summarize, the work done in the last two years appear very fruitful, since the A=3,4 ground state properties have been investigated with a great accuracy, constructing a sound basis for addressing new issues, like the application of new NN interactions based on the chiral perturbation theory or the role of 3N forces. Furthermore, different methods for dealing with the FSI effects allow us to get rid of uncertainties related to the model dependence, when relevant physical quantities are extracted from reactions involving few-nucleon systems, and this fact yields more confidence in the theoretical tools for addressing the subtle effects of 3N forces. Finally, including the general constraints requested by a theory Poincaré covariant for interacting few-nucleon system could be the challenge for the next future calculations of ground state properties.

Acknowledgments

This work is dedicated to the memory of our friend and colleague Adelchi Fabrocini.

References

1. M. Viviani, L.E. Marcucci, S. Rosati, A. Kievsky, and L. Girlanda, Few-Body Sys. **39**, 159 (2006).
2. M. Viviani, A. Kievsky, and S. Rosati, Phys. Rev. C **71**, 024006 (2005).
3. V.D. Efros, W. Leidemann, G. Orlandini, and E.L. Tomusiak, Phys. Rev. C **72**, 011002R1 (2005).
4. D. Andreasi, S. Quaglioni, V.D. Efros, W. Leidemann, and G. Orlandini, Eur. Phys. J. A **27**, 47 (2006).
5. H. Witała, J. Golak, R. Skibinski, W. Gloeckle, A. Nogga, E. Epelbaum, H. Kamada, A. Kievsky, and M. Viviani, Phys. Rev. C **73**, 044004 (2006).
6. W. Xu *et al.*, Phys. Rev. Lett. **85**, 2900 (2000).
7. P.A.M. Dirac, Rev. Mod. Phys. **21**, 392 (1949).
8. B.D. Keister and W.N. Polyzou, Adv. Nucl. Phys. **20**, 225 (1991).
9. V.Yu. Korda , L. Canton, and A.V. Shebeko, nucl-th/0603025.
10. T. Melde, L. Canton, W. Plessas, and R.F. Wagenbrunn, Eur. Phys. J. A **25**, 97 (2005).
11. R.F. Wagenbrunn, S. Boffi, W. Klink, W. Plessas, and M. Radici, Phys. Lett. **B511**, 33 (2001).
12. A. Kievsky, M. Viviani, and S. Rosati, Phys. Rev. **C 64**, 024002 (2001).
13. A. Kievsky, M. Viviani, and S. Rosati, Nucl. Phys. **A577**, 511 (1994).
14. A. Deltuva, A.C. Fonseca, A. Kievsky, S. Rosati, P.U. Sauer, and M. Viviani, Phys. Rev. C **71**, 064003 (2005).
15. R.B. Wiringa, R.A. Smith, and T.A. Ainsworth, Phys. Rev. C **29**, 1207 (1984).
16. R. Machleidt, F. Sammarruca, and Y. Song, Phys. Rev. C **53**, R1483 (1996).
17. V.G.J. Stoks, R.A.M. Klomp, C.P.F. Terheggen, and J.J. de Swart, Phys. Rev. C **49**, 2950 (1994).

18. S.A. Coon and W. Glökle, Phys. Rev. C **23**, 1790 (1981).
19. B.S. Pudliner, V.R. Pandharipande, J. Carlson, and R.B. Wiringa, Phys. Rev. Lett. **74**, 4396 (1995).
20. E. Epelbaum, W. Glöckle, and Ulf-G. Meissner, Nucl. Phys. **A747**, 362 (2005); E. Epelbaum, W. Glöckle, and U.-G. Meissner, Eur. Phys. J. A **19**, 401 (2004).
21. A. Nogga *et al.*, Phys. Rev. C **65**, 054003 (2002).
22. R. Lazauskas and J. Carbonell, Phys. Rev. C **70**, 044002 (2004) and private communication.
23. R.B. Wiringa, S.C. Pieper, J. Carlson, and V.R. Pandharipande, Phys. Rev. C **62**, 014001 (2000).
24. R. Lazauskas, J. Carbonell, A.C. Fonseca, A. Kievsky, M. Viviani, and S. Rosati, Phys. Rev. C **71**, 034004 (2005).
25. B.M. Fisher, C.R. Brune, H.J. Karwowski, D.S. Leonard, E.J. Ludwig, T.C. Black, A. Kievsky, M. Viviani, and S. Rosati, Phys. Rev. C **74**, 034001 (2006).
26. D.R. Entem and R. Machleidt, Phys. Rev. C **68**, 041001(R) (2003).
27. V. Efros, W. Leidemann, and G. Orlandini, Phys. Lett. **B338**, 130 (1994).
28. D. Andreasi, W. Leidemann, C. Reiss, and M. Schwamb, Eur. Phys. J. A **24**, 361 (2006).
29. S. Quaglioni, V. D. Efros, W. Leidemann, and G. Orlandini, Phys. Rev. C **72**, 064002 (2005).
30. D. Gazit, S. Bacca, N. Barnea, W. Leidemann, and G. Orlandini, Phys. Rev. Lett. **96**, 112301 (2006).
31. V.D. Efros, W. Leidemann, G. Orlandini, and E.L. Tomusiak, Phys. Rev. C **72**, 011002R (2005).
32. H. Arenhövel, W. Leidemann, and E.L. Tomusiak, Eur. Phys. J. A **23**, 147 (2005).
33. F. M. Lev, E. Pace, and G. Salmè, Phys. Rev. C **62**, 064004 (2000).
34. A. Kievsky, E. Pace, and G. Salmè, to be published.
35. M. Gari and W. Krümpelmann, Z. Phys. **322**, 689 (1985).
36. I. Sick, Prog. Part. Nucl. Phys. **47**, 245 (2001).

THREE-NUCLEON CONTINUUM STATES WITHIN THE HYPERSPHERICAL ADIABATIC METHOD

P. BARLETTA

Department of Physics and Astronomy, University College London, Gower Street, London WC1E 6BT, United Kingdom and Laboratory of Molecular Spectroscopy, Institute of Chemistry, Eötvös University, H-1518 Budapest 112, P.O. Box 32, Hungary

A. KIEVSKY

INFN, Sezione di Pisa, Largo B. Pontecorvo 3, 56127 Pisa, Italy

The Hyperspherical Adiabatic Method are used to describe the scattering states of a three-body system. We exploit the well know fact that the Adiabatic Functions describe asymptotically the scattering states of the system, both below and above the breakup threshold. In order to obtain an accurate description of the Adiabatic basis at large ρ we follow closely the procedure indicated by Nielsen and co-workers. The scattering wave function is expanded in terms of the Adiabatic basis and the S-matrix is obtained variationally by means of the Khon variational principle. We show two simple numerical applications of the method for the study of a three-nucleon system with a pure central interaction and a realistic potential.

1. Introduction

The Hyperspherical Adiabatic (HA) method is based on the parametrization of the internal degrees of freedom with hyperspherical coordinates. The method consists in expanding the wave function on a basis made of hyperangular optimized functions (the adiabatic basis set) times hyperradial functions. The hyperangular basis elements are taken as the convectors of the Hamiltonian operator for a fixed hyperradius. Once those eigenvectors have been calculated, the hyperradial functions are obtained as the solutions of a system of (infinite) coupled one-dimensional differential equations. The advantages of such an approach are that the HA basis should drive a quick convergence for the expansion, due to its optimization, the payback is represented by the necessity and the difficulty in calculating accurately the first and second derivatives of the adiabatic basis set with respect to the hyperradius. Those terms are crucial to the method as they represent the coupling terms between the various hyperradial differential equations. In some applications of the HA method it was shown that the strong coupling between elements of the adiabatic basis makes the hyperradial problem particularly hard to solve.

The properties of the adiabatic basis functions have been object of several studies and are well-known [1,2]. In particular, in the asymptotic limit of large hyperradius those functions are known to converge towards the scattering states of the three-body system, both below and above breakup. This characteristics makes the adiabatic expansion an optimum choice to describe the three-body continuum states. In this paper we investigate the possibility of such an approach. In fact in the literature there are several studies of the bound spectrum of a three-body system by means of the HA method, and very few dealing with the continuum. In the present contribution we will use the Khon variational principle to calculate variational bounds for the scattering observables [3]. We will show that a reduced number of adiabatic basis elements are necessary to get convergence. However, the extreme situation in which only one term of the expansion is retained, as proposed in Ref. [2], is not sufficient to obtain the desired accuracy.

This contribution is organized as follows: in the next Section the HA method is briefly outlined. In Section 3 the Khon variational method is introduced, and in Section 4 a simple numerical example is discussed, illustrating how calculations of scattering observables can be performed. The final Section is devoted to the conclusions.

2. Hyperspherical Adiabatic method

In this Section we will give a short presentation of the hyperspherical adiabatic basis set. A more complete description of the basis and of the adiabatic method can be found in Ref. [4], and references therein. Let us consider a system of three identical nucleons of mass m, in a state of total angular momentum J, J_z. The HA basis elements $\{\Phi_\nu\}$ are defined as the eigenfunctions of the hyperangular part of the system's Hamiltonian written in hyperspherical coordinates:

$$\mathcal{H}_\Omega \Phi_\nu = \left[\frac{\hbar^2}{2m\rho^2} G^2 + V\right] \Phi_\nu(\rho;\Omega) = U_\nu(\rho)\Phi_\nu(\rho;\Omega), \tag{1}$$

where $U_\nu(\rho)$ is the eigenvalue associated with Φ_ν, G^2 is the grand-angular operator, V is the potential operator, which is assumed to be a function of the internal coordinates only, and ρ and Ω are the hyperradius and the set of hyperangles, respectively. The hyperspherical coordinates are defined in terms of the set of Jacobi coordinates $\{\mathbf{x}_i, \mathbf{y}_i\}$:

$$x_i = \rho \cos\phi_i, \quad y_i = \rho \sin\phi_i, \tag{2}$$

with

$$\mathbf{x}_i = \frac{1}{\sqrt{2}}(\mathbf{r}_j - \mathbf{r}_k), \mathbf{y}_i = \frac{1}{\sqrt{6}}(\mathbf{r}_j + \mathbf{r}_k - 2\mathbf{r}_i), \tag{3}$$

where $\{\mathbf{r}_i\}$ are the coordinates of the three particles and $i, j, k = 1, 2, 3$ cyclic. The index i refers to the possibility of building three different set of coordinates by changing the reference pair. The set Ω_i of hyperangles consists of the hyperangle ϕ_i and the four angles $\hat{\mathbf{x}}_i, \hat{\mathbf{y}}_i$.

3. Scattering observable calculations

For a system energy above the two-body binding energy, the associated wavefunction can be written in general as

$$\Psi = \Psi_c + \Psi_a, \tag{4}$$

where the first term describes the system when the three particles are close to each other, whereas the second term is the solution of the Schrödinger equation in the asymptotic region. The core part of the wavefunction, which one expects to be not much dissimilar from the wavefunctions of the bound trimer states having the same quantum numbers, can be expanded onto a square integrable basis set. In particular, we can chose the adiabatic basis Φ_ν in combination with Laguerre polynomials $\{P_m\}$ times an exponential:

$$\Psi_c = \sum_{m\nu} A_{m\nu} P_m(\beta\rho) \Phi_\nu(\rho; \Omega) \exp\left[-\beta\rho/2\right], \tag{5}$$

This basis is complete in the limit of infinite adiabatic functions and infinite number of polynomials, and it can be effectively used to describe bound states of the system. In practice, however, the number of terms included has to be kept finite, let us say N, with $N = \nu_{max} \times m_{max}$, where ν_{max} is the number of adiabatic basis functions and m_{max} the number of Laguerre polynomials. The parameter β is a non-linear variation parameter which can be tuned to optimize the expansion [5].

The term Ψ_a describes the asymptotic configuration in which a nucleon approaches the deuteron. It can be written as a linear combination of the following antisymmetric states:

$$\Omega^R_{LSJ} = A \sum_i \mathcal{R}^R_L(y_i) \{[\Psi_d(\mathbf{x}_i) \otimes \frac{1}{2}]_S \otimes Y_L(\hat{\mathbf{y}}_i)\}_{JJ_z} \tag{6}$$

and

$$\Omega^I_{LSJ} = A \sum_i \mathcal{R}^I_L(y_i)\{[\Psi_d(\mathbf{x}_i) \otimes \frac{1}{2}]_S \otimes Y_L(\hat{\mathbf{y}}_i)\}_{JJ_z} \tag{7}$$

where $\Psi_d(\mathbf{x}_i)$ is the deuteron wavefunction and the functions $\mathcal{R}^R_L$, $\mathcal{R}^I_L$ are the regular and irregular solutions of the two-body Schrödinger equation. A is a normalization factor chosen so that the Wronskian results

$$W(\Omega^R_{LSJ}, \Omega^I_{LSJ}) = \frac{1}{2} \tag{8}$$

The irregular function $\mathcal{R}^I_L$ has been regularized with the factor $[1 - \exp(-\gamma y)]^{L+1}$. The final result should be independent of the value chosen for γ but a wrong choice can slow down the convergence significantly.

A general choice for the scattering states is given by

$$\Omega^0_{LSJ} = u_{00}\Omega^R_{LSJ} + u_{01}\Omega^I_{LSJ}, \quad \Omega^1_{LSJ} = u_{10}\Omega^R_{LSJ} + u_{11}\Omega^I_{LSJ}, \tag{9}$$

where the different options for the matrix u are described in Ref. [3]. Then a general scattering wavefunction is thus written as

$$\Psi_{LSJ} = \Psi_c + \Omega^0_{LSJ} + \sum_{L'S'} \mathcal{L}^{SS'}_{LL'}\Omega^1, \tag{10}$$

and the meaning of the scattering matrix $\mathcal{L}$ depends on the choice of u. For example, possible choices often used are the reactance matrix K, its inverse K^{-1}, the S−matrix and the T−matrix. Once convergence is achieved for the particular choice of $\mathcal{L}$, the reactance matrix can be obtained from the relation

$$K = (u_{01} + u_{11}\mathcal{L})(u_{00} + u_{01}\mathcal{L})^{-1}. \tag{11}$$

The Khon variational principle can be written as

$$[\mathcal{L}^{SS'}_{LL'}] = \mathcal{L}^{SS'}_{LL'} - \frac{2}{\det u}\langle\Psi_{LSJ}|\mathcal{H} - E|\Psi_{L'S'J}\rangle. \tag{12}$$

If we define an array of unknowns $(\{A_{m\nu}\}, \mathcal{L})$ of dimension $N + M$, where N is the size of the expansion for the core part of the wavefunction and M is the dimension of the scattering matrix, the numerical implementation of the variational principle leads to a first order approximation to $\mathcal{L}$ obtained through the solution of a linear system of equations of size $N + M$. The linear system can be pictorially written as:

$$\begin{pmatrix} \{H_{m'\nu',m\nu}\} & \{H_{m'\nu',\Omega^1}\} \\ \{H_{\Omega^1,m\nu}\} & H_{\Omega^1,\Omega^1} \end{pmatrix} \begin{pmatrix} \{A_{m\nu}\} \\ \mathcal{L} \end{pmatrix} = \begin{pmatrix} \{H_{m'\nu,\Omega^0}\} \\ -\frac{1}{2}H_{\Omega^1,\Omega^0} \end{pmatrix}, \tag{13}$$

where $H_{x',x}$ stands for the matrix element

$$H_{x',x} = \langle x'|\mathcal{H} - E|x\rangle. \tag{14}$$

The second order estimate for $\mathcal{L}$ is then given by

$$\mathcal{L}^{2nd} = \mathcal{L}^{1st} - \frac{2}{\det u}\delta_{\mathcal{L}} = \mathcal{L}^{1st} - \frac{2}{\det u}\langle\Psi^{1st}|\mathcal{H} - E|\Psi^{1st}\rangle, \tag{15}$$

where Ψ^{1st} is the wavefunction built with the coefficients obtained from the solution of the linear system.

4. Results

In order to illustrate with a numerical example the method presented in the previous Section, we have chosen to study a system of three-nucleons acting with a central potential, expressed as a sum of three pairwise interactions.

The pairwise potential used is

$$V(r) = V_0 \exp[-(r/d)^2], \tag{16}$$

where r is the nucleon-nucleon distance, and $V_0 = -51.5$ MeV, and $d = 1.6$ fm. This potential supports only one bound state for the dimer, with zero angular momentum, of energy $E_1^{2b} = -0.397742816$ MeV. The zero-energy scattering length is $a_s = 11.356$ fm. For the nucleon mass we have used a value such that $\hbar^2/m = 41.4696$ MeV fm^2.

Let us now discuss in more detail the linear system (13). The top left part contains matrix elements between the functions of the HA basis times the hyperradial basis onto which the Ψ_c is expanded. The eigenvalues of this N^2 sub-matrix provides upper bounds to the energy levels of the bound three-body states, as a consequence of the Rayleigh-Ritz variational principle. In Table 1 the convergence of those variational estimates, for the $J = 1/2^+$ state, is shown for the two bound states supported by the unrealistic interaction used in terms of the HA channels. For each HA channel 80 Laguerre polynomials have been included in order to make the expansion of Eq. (5) independent of nonlinear parameter β. As it can be seen from Table 1, the lowest adiabatic channel provides more than 99% of the energy of the two bound states. One could conclude that limiting the HA expansion to the lowest adiabatic basis element could provide at the same time a noticeable simplification of the numerics and a very good approximation to the exact value. However, in the present case we have discussed a potential much simplified. When we consider a realistic potential as the Argonne AV14 the situation slightly changes, as it is reported in Table 2 for a restricted case in which the ground state wave function has been expanded in a three angular–spin–isospin channels. The exact result for the ground state energy in this case is -6.928 MeV. In order to reach an accuracy of 1 keV with the HA expansion is necessary to include 20 adiabatic channels.

In the case of a central potential the scattering matrix has dimension one and the solution of the complete $(N+1)\times(N+1)$ linear system provides a first order variational bound for the scattering observable $\mathcal{L}$. In order to evaluate the second order expression for $\mathcal{L}$, one needs to compute the matrix elements between the basis functions and the asymptotic solutions, which include the calculation of the of the hyperangular integrals

$$\langle\Phi_\nu|\mathcal{H}-E|\Omega^0\rangle_\Omega, \quad \langle\Phi_\nu|\mathcal{H}-E|\Omega^1\rangle_\Omega, \tag{17}$$

between the asymptotic functions and the HA basis elements. In the case of continuum states, in a previous work [4] we have presented results in which the HA basis has been truncated to its first term. Moreover the asymptotic functions were approximated by the asymptotic form of the first adiabatic eigenvector times an hyperradial ingoing or outgoing function. This approach, proposed by Fabre de la Ripelle, is justified by the fact that the HA expansion is normally very fast to converge and much of the physics is given by the lowest adiabatic potential. However the asymptotic form is correct only at very large distances and the rate of convergence is slow. The complete approach requires the evaluation of the matrix elements of Eq. (17). In Table 1 we present the convergence pattern for several observables obtained by solving the system (13) (with $\gamma = 0.25$ fm^{-1}), in terms of the number of coupled differential equations included. Besides the energies of the two bound states, we report values for the doublet scattering length a_d and the phase shifts δ for the state $J = 1/2^+$ at four different energies below three-body breakup. For this simple potential, it is possible to see that the value obtained with just one equation is normally again a very good approximation to the converged result. Figure 1 shows the phase-shift δ for a wide energy range. The continuum line corresponds to phase-shifts obtained with the procedure outlined in the previous contribution [4], in which the asymptotic state was constructed as the deuteron wave function times and hyperradial ingoing or outgoing function. The circles corresponds to the fully converged results, some of which are also reported in Table 1.

5. Conclusions

In this contribution we have applied the Khon variational principle in conjunction with an expansion on the HA basis to obtain variational bounds for scattering observables at energies below the three-body breakup. We

Table 1. Convergence of the energies of the two bound states (in MeV), E_0 and E_1, the doublet scattering length a_d (in fm), and the $1/2^+$ phase-shifts (in degrees) at four different energies below the breakup threshold in terms of the number of HA states ν_{max}.

ν_{max}	E_n		a_d	δ			
	$n=0$	$n=1$		-0.3	-0.2	-0.1	-10^{-4}
1	-9.7559	-0.4783	–	101.775	82.22	70.5	62.1
2	-9.7767	-0.4811	-34.	103.353	84.094	72.72	64.73
3	-9.7794	-0.4815	-31.9	103.625	84.412	73.048	65.064
4	-9.7795	-0.4816	-31.6	103.690	84.480	73.120	65.143
8	-9.7796	-0.4817	-29.8	103.772	84.571	73.218	65.247
12	-9.7796	-0.4817	-29.343	103.781	84.579	73.227	65.255
16	-9.7796	-0.4817	-29.269	103.783	84.581	73.228	65.257
20	-9.7796	-0.4817	-29.250	103.783	84.581	73.228	65.257

Table 2. Convergence of the three-nucleon bound state, in MeV, using a three--channel wavefunction and the AV14 potential, as a function of the number of HA states ν_{max} and the Laguerre polynomials m_{max}.

$m_{max}\backslash\nu_{max}$	1	5	10	15	20	25	30
8	-6.663	-6.813	-6.839	-6.842	-6.843	-6.843	-6.843
16	-6.730	-6.887	-6.914	-6.917	-6.918	-6.919	-6.919
24	-6.735	-6.893	-6.921	-6.924	-6.925	-6.925	-6.926
32	-6.736	-6.894	-6.923	-6.926	-6.927	-6.927	-6.927
40	-6.736	-6.894	-6.923	-6.926	-6.927	-6.928	-6.928

have also studied the convergence pattern for the bound state of three nucleons in the case of a simple central potential or a realistic one. In this last case the truncation of the HA basis can be done safely by retaining at most 20 adiabatic channels.

In the case of scattering states, we observed that the asymptotic wave function, describing a two body bound state and an incoming nucleon, cannot be replaced by the asymptotic form of the lowest adiabatic eigenvector times a hyperradial ingoing or outgoing function. It is necessary to introduce the correct asymptotic configuration given in Eqs. (6,7). The advantage of this procedure, if compared to the hyperspherical harmonic (HH) approach, is given by the much smaller size of the problem. Both approaches (HA and HH) lead to the solution of a system of coupled differential equations, whose size is $m_{max} \times \nu_{max}$ in the case of the HA method, and $m_{max} \times k_{max}$ for the HH, where k_{max} is the total number of HH states considered, and in general $\nu_{max} \ll k_{max}$.

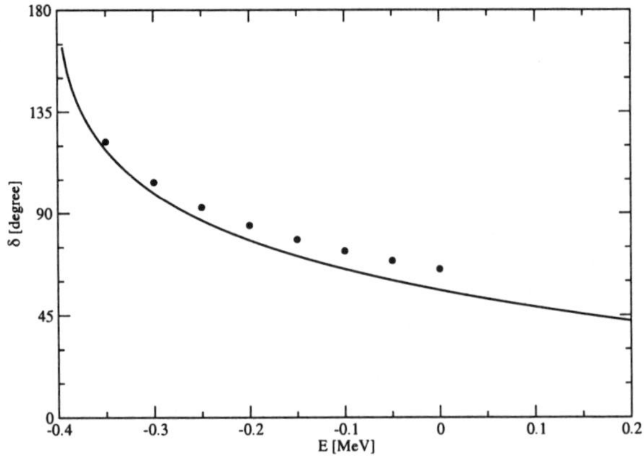

Fig. 1. Phase-shift δ for different energies of the three-body system, calulated with the approximated method of Ref. [4] (continuum line), and with the approach outlined in this contribution (dots).

References

1. E. Nielsen, D. V. Fedorov, and A. S. Jensen, J. Phys. B: Mol. Opt. Phys. **31**, 4085 (1998).
2. M. Fabre de la Ripelle, Few-Body Syst. **14**, 1 (1993).
3. A. Kievsky, Nucl. Phys. **A624**, 125 (1997).
4. P. Barletta and A. Kievsky, " *Proceedings of the 10th Conference on Problems in Theoretical Nuclear Physics*", October 2004, Cortona (Italy), Ed. S. Boffi, A. Covello, M. Di Toro, A. Fabrocini, G. Pisent and S. Rosati, World Scientific (Singapore) 2005, p.163
5. P. Barletta and A. Kievsky, Phys. Rev. A **64**, 042514 (2001).

BAKAMJIAN-THOMAS MASS OPERATOR FOR THE FEW-NUCLEON SYSTEM: AN EFFECTIVE THEORY APPROACH

L. GIRLANDA and M. VIVIANI

INFN Sez. di Pisa
Largo B. Pontecorvo 3, 56127 Pisa, Italy
E-mail: girlanda@pi.infn.it, viviani@pi.infn.it

We present an exploratory study consisting in the formulation of a relativistic quantum mechanics to describe the few-nucleon system at low energy, starting from the quantum field theoretical chiral Lagrangian involving pions and nucleons. To this aim we construct a Bakamjian-Thomas mass operator and perform a truncation of the Fock space which respects at each stage the relativistic covariance. Such truncation is justified, at sufficiently low energy, in the framework of a systematic chiral expansion. As an illustration we discuss the low-energy phaseshifts of the nucleon-nucleon and pion-nucleon scattering.

Keywords: Relativistic dynamics; Nucleon-Nucleon interaction; Chiral Lagrangians.

1. Relativistic quantum mechanics

The quantitative understanding of the dynamics of few-nucleon systems in terms of phenomenological (or derived from chiral effective theory) potentials has reached such a high degree of accuracy that the need of including relativistic effects has been put forward. The derivation of a potential, to be considered in a Schroedinger or Lippman-Schwinger equation, starting from a relativistic (quantum field theoretical) setting is a long standing problem in nuclear physics, and we refer to Ref. [1] for a review. Alternatively, for a given set of particles, it is possible to put constraints on the mutual interactions in order to fulfill the requirements of relativity. This problem has been posed in full generality by Dirac in 1949 [2] , and developed among others by Krajcik and Foldy [3] . Both approaches lead to definite predictions for the relativistic corrections to the potential ("drift potentials") which can take the form of multinucleon forces and are currently being investigated quantitatively [4,5] . The approach we present here is rather different in that we want to describe the few-nucleon systems with a relativistic wave equation, with simple transformation properties under the Lorentz symmetry, so that the behaviour of e.g. a $N-N$ subsystem in relative motion with respect to other nucleons, as demanded by relativity, is easily accounted for. In this perspective we critically analyze the proposal of Ref. [6] consisting in a Bakamjian-Thomas construction formulated in the point-form of relativistic dynamics, which has already been applied to bound state problems in the framework of chiral quark models [7]. Compared to the above references we focus on nucleons instead of quarks and adopt a systematic ChPT-inspired point of view (see Ref. [8] for a recent review), instead of relying on specific models.

2. Bakamjian-Thomas mass operator

In the point-form of relativistic dynamics the interactions are contained in the four-momentum P^μ whereas the generators of Lorentz transformations are the same as in the non-interacting case. It follows that this formulation is manifestly Lorentz covariant. The requirements of Poincaré invariance can therefore be written as

$$[P^\mu, P^\nu] = 0, \quad U_\Lambda P^\mu U_\Lambda^{-1} = (\Lambda^{-1})^\mu_\nu P^\nu, \tag{1}$$

with U_Λ the unitary operator corresponding to the Lorentz trasformation Λ, which is "kinematical", i.e. does not contain interactions. In this case the Bakamjian-Thomas construction takes a particularly simple form: starting from the non-interacting Poincaré generators $\left\{H_0, \vec{P}_0, \vec{J}_0, \vec{K}_0\right\}$, one introduces the auxiliary operators M_0 and V_0^μ as $M_0 = \sqrt{H_0^2 - \vec{P}_0^2}$ and $P_0^\mu = M_0 V_0^\mu$. One then adds the interactions only to the mass operator $M = M_0 + M_I$ and reconstructs the interacting four-momentum as $P^\mu = M V_0^\mu$. The dynamical equation to be solved reduces to

$$(M_0 + M_I)|\Psi\rangle = m|\Psi\rangle. \tag{2}$$

In order to satisfy Poincaré commutation relations the interacting mass operators must commute with the four-velocity and be a Lorentz scalar. Starting from a field-theoretical Lagrangian, the interacting four-momentum is not of the Bakamjian-Thomas form [9]. In the proposal of Ref. [6] this feature is enforced with the explicit inclusion, in the definition of the interacting mass operator, of a velocity-conserving δ-function,

$$\langle v', k_i', \mu_i'|M_I|v, k_i, \mu_i\rangle = \langle v', k_i', \mu_i'|\mathcal{H}(0)|v, k_i, \mu_i\rangle (2\pi)^3 \delta^3(\vec{v'} - \vec{v}) f(\Delta m). \tag{3}$$

In this equation $\mathcal{H}(x)$ is the (Lorentz scalar) density of the interaction Hamiltonian, written in terms of nucleon and pion fields, and the "velocity states" $|v, k_i, \mu_i\rangle$, introduced in Ref. [10], are eigenstates of the velocity operator, linear combinations of multiparticle momentum states with specific transformation properties under Lorentz boosts. They correspond to standard multiparticle momentum states defined in their center of mass system, boosted with a canonical boost transformation up to four-velocity v. The form factor f, depending on the mass difference, is introduced in order to compensate for the neglection of the off-diagonal terms in the four velocity, and to regulate the integrals as well. From its definition it carries dimension of $[m]^{-3}$. In the sequel we will take for f a Gaussian function centered around zero, with cutoff Λ,

$$f(\Delta m) \propto \exp\left[-\frac{(\Delta m)^2}{2\Lambda^2}\right]. \tag{4}$$

This cutoff, within the effective theory implementation of the present approach, should be understood as the short distance scale at which new degrees of freedom start to become active. One such form factor is understood for each vertex of the effective Lagrangian, and the requirement that physics at low-energy be independent of the cutoff (provided it is large enough) should fix the running of the coupling constants with Λ.

3. Low-energy power counting

The dynamical equation (2) is solved in a truncated Fock space. Due to baryon number conservation, subspaces with different numbers of nucleons will be block-diagonal and we can treat separately the 1-nucleon sector, 2-nucleon sector,... of the theory. A truncation is performed also, inside each different sector, on the number of pions. This can be justified in the framework of a low-energy power counting, as will be illustrated below, on the basis of the Goldstone's theorem, which demands that the pion couplings involve space-time derivatives. Consider for instance the leading pion-nucleon interaction, in the 2-nucleon sector,

$$\mathcal{L}_{\pi N} = -\frac{g_A}{2F_\pi}\bar{\psi}\gamma^\mu\gamma_5\partial_\mu\pi\psi, \tag{5}$$

where ψ is the isodoublet nucleon field and $\vec{\pi}$ the isovector pion field. The interacting mass operator connects states which differ in the number of pions by ± 1. Physical states are linear combinations of free-particle velocity states which solve the eigenvalue problem

$$\begin{pmatrix} D_2 & K & 0 & \\ K & D_3 & K & \\ 0 & K & D_4 & \\ & & & \ddots \end{pmatrix} \begin{pmatrix} \phi_2 \\ \phi_3 \\ \phi_4 \\ \vdots \end{pmatrix} = \sqrt{s} \begin{pmatrix} \phi_2 \\ \phi_3 \\ \phi_4 \\ \vdots \end{pmatrix}, \tag{6}$$

where the subscripts in the component of the Fock space denote the corresponding number of particles (starting from 2 nucleons) and the diagonal entries D_n represent the (bare) relativistic energies of the n-particle states. The formal solution of the above equation corresponds to

$$\left[\sqrt{s} - D_2 - K \frac{1}{\sqrt{s} - D_3 - K \frac{1}{\sqrt{s} - D_4 - \ldots} K} K\right] \phi_2 = 0. \tag{7}$$

Counting the pion mass and the typical three-momenta as small parameters, much smaller than the typical hadronic scale Λ at which new physics arises, $M_\pi, |\vec{p}| << \Lambda$, it is possible to show that the successive terms of this series are more and more suppressed at low energy. The series can then be Taylor-expanded

$$(\sqrt{s} - D_2) - K \frac{1}{\sqrt{s} - D_3} K - K \left[\frac{1}{\sqrt{s} - D_3}\right]^2 K \frac{1}{\sqrt{s} - D_4} K K \ldots, \tag{8}$$

and shown to correspond to the hierarchy

$$O(p^{-2}) + O(p^0) + O(p^2) + \ldots. \tag{9}$$

The leading order wave equation in the 2-nucleon sector, denoting $\phi_2(\mathbf{k})_{\mu\mu',ii'} = \langle \mathbf{k}, \mu, i; -\mathbf{k}, \mu', i' | \phi_2 \rangle$, with μ (i) spin (isospin) indices and $\omega_\mathbf{k} = \sqrt{m_N^2 + \mathbf{k}^2}$, reads

$$2\omega_\mathbf{k} \phi_2(\mathbf{k})_{\mu\mu',ii'} + \int \omega_\mathbf{q} \frac{d^3\mathbf{q}}{(2\pi)^3} V(\mathbf{k}, \mathbf{q})^{\rho\rho',jj'}_{\mu\mu',ii'} \phi_2(\mathbf{q})_{\rho\rho',jj'} = \sqrt{s} \phi_2(\mathbf{k})_{\mu\mu',ii'} \tag{10}$$

with

$$V(\mathbf{k}, \mathbf{q}) = \left[B(\mathbf{k}, \mathbf{q}) + (2\pi)^3 \delta^3(\mathbf{k} - \mathbf{q}) A(\mathbf{k})\right] - (\mathbf{q} \leftrightarrow -\mathbf{q}, \rho \leftrightarrow \rho', j \leftrightarrow j'), \tag{11}$$

and

$$\begin{aligned} B(\mathbf{k}, \mathbf{q}) &= \frac{g_A^2}{4F_\pi^2} \frac{(\omega_\mathbf{k} + \omega_\mathbf{q} + \omega^\pi_{\mathbf{q}-\mathbf{k}})^3}{\omega^\pi_{\mathbf{q}-\mathbf{k}}} \frac{f^*(\omega_\mathbf{q} + \omega^\pi_{\mathbf{q}-\mathbf{k}} - \omega_\mathbf{k}) f(\omega_\mathbf{k} + \omega^\pi_{\mathbf{q}-\mathbf{k}} - \omega_\mathbf{q})}{\sqrt{s} - \omega_\mathbf{q} - \omega_\mathbf{k} - \omega^\pi_{\mathbf{q}-\mathbf{k}}} \times \\ &\quad \times \bar{u}(\mathbf{k}) \gamma^\mu \gamma_5 u(\mathbf{q}) \bar{u}(-\mathbf{k}) \gamma^\nu \gamma_5 u(-\mathbf{q}) p^\pi_\mu p^\pi_\nu \vec{\tau}_{ij} \cdot \vec{\tau}_{i'j'} \\ A(\mathbf{k}) &= \frac{3g_A^2}{4F_\pi^2} \int \frac{d^3\mathbf{q}}{(2\pi)^3} \frac{(\omega_\mathbf{q} + \omega_\mathbf{k} + \omega^\pi_{\mathbf{q}-\mathbf{k}})^3}{\omega_\mathbf{k} \omega_\mathbf{q} \omega^\pi_{\mathbf{q}-\mathbf{k}}} \omega^2_\mathbf{k} \frac{|f(\omega_\mathbf{q} + \omega^\pi_{\mathbf{q}-\mathbf{k}} - \omega_\mathbf{k})|^2}{\sqrt{s} - \omega_\mathbf{q} - \omega_\mathbf{k} - \omega^\pi_{\mathbf{q}-\mathbf{k}}} \times \\ &\quad \times \bar{u}(\mathbf{k}) \gamma^\mu \gamma_5 u(\mathbf{q}) \bar{u}(\mathbf{q}) \gamma^\nu \gamma_5 u(\mathbf{k}) p^\pi_\mu p^\pi_\nu. \end{aligned} \tag{12}$$

The momentum-conserving component of the potential corresponds to disconnected diagrams in which a pion is emitted and absorbed by the same nucleon. These diagrams can be resummed and renormalize the energies of the asymptotic states. The same diagram describes the renormalization of the nucleon mass in the 1-nucleon sector, which corresponds to the eigenvalue problem (with superscript ^{r} denoting renormalized quantities)

$$\begin{pmatrix} m_N & K \\ K & D_2 \end{pmatrix} \begin{pmatrix} \phi_1 \\ \phi_2 \end{pmatrix} = m_N^r \begin{pmatrix} \phi_1 \\ \phi_2 \end{pmatrix} \tag{13}$$

and reads explicitly

$$\frac{3g_A^2}{4F_\pi^2} \int \frac{d^3\mathbf{k}}{(2\pi)^3} \mathbf{k}^2 \frac{(\omega_\mathbf{k}+\omega_\mathbf{k}^\pi)^3}{4\omega_\mathbf{k}\omega_\mathbf{k}^\pi} \frac{[\omega_\mathbf{k}^\pi+\omega_\mathbf{k}+m_N]^2}{\omega_\mathbf{k}+m_N} \frac{|f(\omega_\mathbf{k}+\omega_\mathbf{k}^\pi-m_N)|^2}{m_N-\omega_\mathbf{k}-\omega_\mathbf{k}^\pi} = \frac{m_N^r-m_N}{m_N^3} \tag{14}$$

If the theory were to respect the cluster decomposition principle, these two renormalizations should be the same. In order to realize this condition, at least in the leading order of the low-energy expansion, one has to tune the form factor f to be different in the 1-nucleon and in the 2-nucleon sectors, in this case $f^{(2)}(\Delta m) = 1/8 f^{(1)}(\Delta m) + ...$, where the dots stand for higher orders. This is a first constraint for the (otherwise arbitrary) form factor. It remains to be seen whether such constraint can be respected order by order in the low-energy expansion.

4. Leading order results

We have performed a study of the πN and NN systems to the respective leading orders, which amount to the diagrams shown in Fig. 1. The

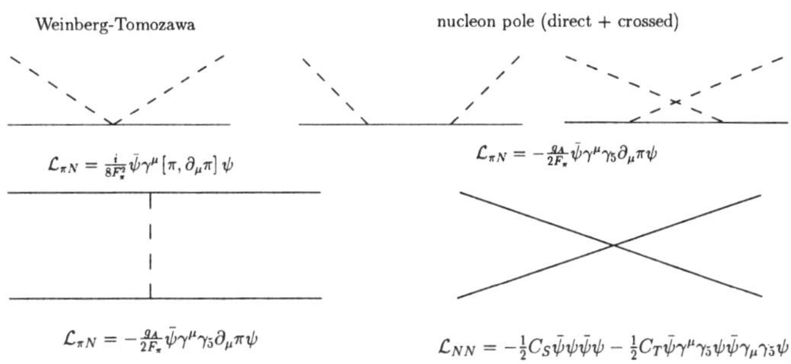

Fig. 1. Diagrams corresponding to the πN and NN scattering at the leading order of our scheme.

phaseshifts are obtained by solving the Lippman-Schwinger equation

$$T(\mathbf{q}, \mathbf{k}) = V(\mathbf{q}, \mathbf{k}) + \int \omega_{\mathbf{p}} \frac{d^3\mathbf{p}}{(2\pi)^3} \frac{V_{\sqrt{s}}(\mathbf{q}, \mathbf{p}) T(\mathbf{p}, \mathbf{k})}{\sqrt{s} - 2\omega_{\mathbf{p}} + i\epsilon}, \qquad (15)$$

in which the momentum integration is carried up to our cutoff Λ.

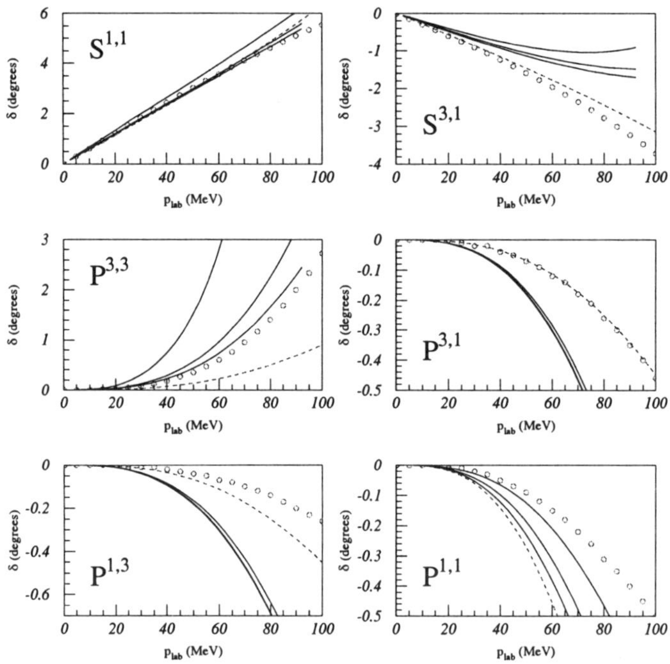

Fig. 2. S and P waves of πN scattering as function of laboratory momentum for three values of the cutoff $\Lambda = 300 - 500$ MeV (full lines). Also shown are the experimental analyisis [11] (circles) and the non relativistic limit (dashed).

We fix g_A from the peripheral NN phases of Ref. [12], the contact coupling C_S from the ${}^3S_1$ phase and C_T from the ${}^1S_0$ phase, the Weinberg-Tomozawa vertex from the $I = 3/2$ πN S-wave, for different values of the cutoff Λ =300 MeV–500 MeV. In Figs. 2-4 we compare with experimental analyses [11,12]. Strong dependence in the cutoff signals the importance of higher order corrections, as is the case e.g. for the πN scattering in the channel of the Δ resonance. We also compare with the non-relativistic limit

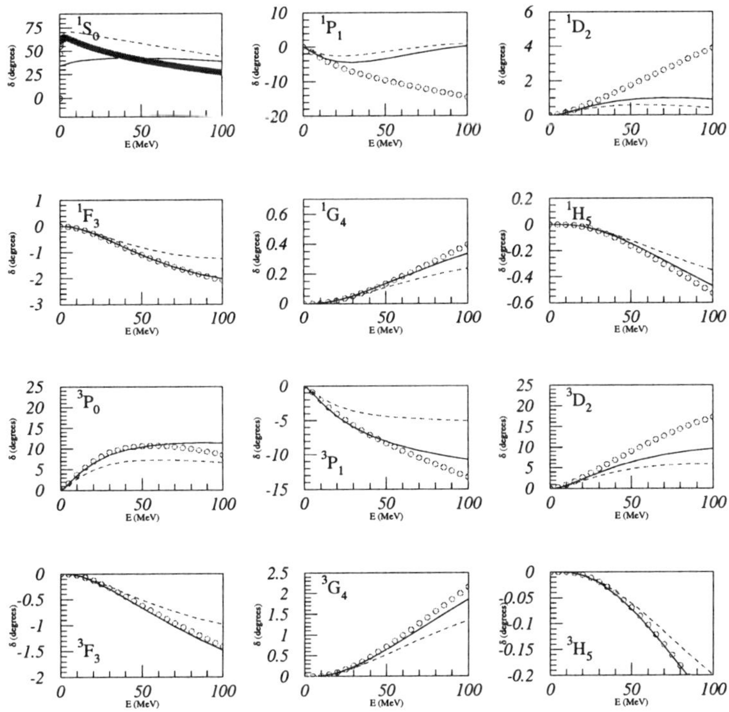

Fig. 3. NN phaseshifts in the non-mixed cases as function of laboratory kinetic energy for $\Lambda = 300$ MeV (full lines). Also shown are the experimental analyisis [12] (circles) and the non relativistic limit (dashed).

(realized in our framework as $m_N \to \infty$), thus quantifying the size of the relativistic corrections included in our scheme.

References

1. R. J. N. Phillips, Rep. Prog. Phys. **22**, 562 (1959).
2. P. A. M. Dirac, Rev. Mod. Phys. **21**, 392 (1949).
3. L. L. Foldy, Phys. Rev. **122**, 275 (1961).
   R. A. Krajcik and L. L. Foldy, Phys. Rev. D **10**, 1777 (1974).
4. R. Higa and M. R. Robilotta, Phys. Rev. C **68**, 024004 (2003) [arXiv:nucl-th/0304025].
   R. Higa, M. R. Robilotta, and C. A. da Rocha, Phys. Rev. C **69**, 034009 (2004) [arXiv:nucl-th/0310011].

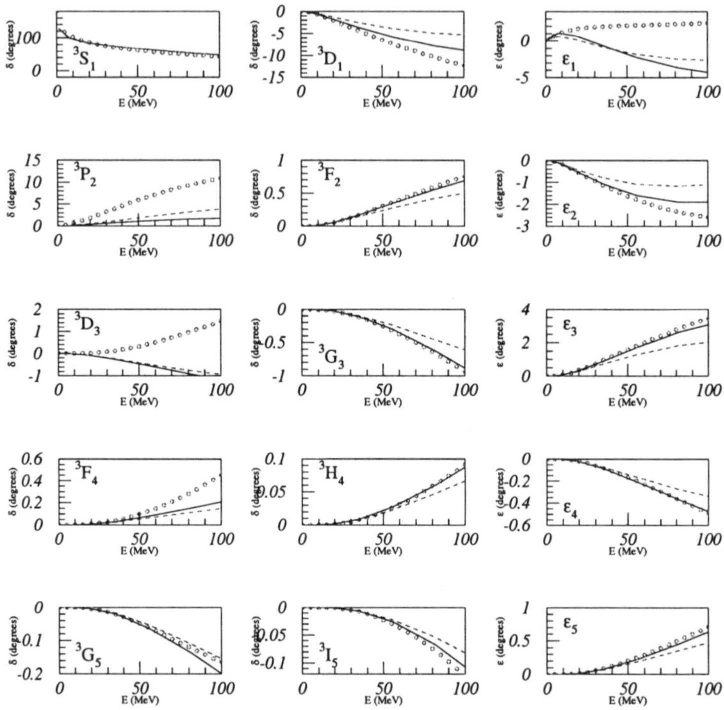

Fig. 4. *NN* phaseshifts in the mixed cases. Legend is the same as for Fig. 3.

5. J. L. Forest, V. R. Pandharipande, J. Carlson, and R. Schiavilla, Phys. Rev. C **52**, 576 (1995) [arXiv:nucl-th/9410013].
6. W. H. Klink, Nucl. Phys. **A716**, 123 (2003) [arXiv:nucl-th/0012031].
7. A. Krassnigg, W. Schweiger, and W. H. Klink, Phys. Rev. C **67**, 064003 (2003) [arXiv:nucl-th/0303063].
8. E. Epelbaum, Prog. Part. Nucl. Phys. **57**, 654 (2006) [arXiv:nucl-th/0509032].
9. S. Fubini, A. J. Hanson, and R. Jackiw, Phys. Rev. D **7**, 1732 (1973).
10. W. H. Klink, Phys. Rev. C **58**, 3617 (1998).
11. SAID on-line program, R. A. Arndt, W. J. Briscoe, R. L. Workman, and I. I. Strakovsky, see the website http://gwdac.phys.gwu.edu/
12. V. G. J. Stoks, R. A. M. Klomp, M. C. M. Rentmeester, and J. J. de Swart, Phys. Rev. C **48**, 792 (1993); see the website http://nn-online.org

ELECTROMAGNETIC REACTIONS ON FEW-NUCLEON SYSTEMS

M. SCHWAMB

Dipartimento di Fisica, Università degli Studi di Trento,
Via Sommarive 14, I-38100 Povo (Trento), Italy
Istituto Nazionale di Fisica Nucleare, Gruppo Collegato di Trento,
Via Sommarive 14, I-38100 Povo (Trento), Italy
European Center for Theoretical Studies in Nuclear Physics and Related Areas (ECT),*
Villa Tambosi, Strada delle Tabarelle 286, I-38100 Villazzano (Trento), Italy
E-mail: schwamb@ect.it

An overview over the field of few-nucleon systems is presented. Special emphasis is laid on the construction of a unified approach to hadronic and electromagnetic reactions on few-nucleon systems, necessary for studying the borderline between quark-gluon and effective descriptions. We briefly outline the Lorentz integral transform (LIT) method which offers a unique possibility to tackle this ambitious task.

Keywords: Few-Body Systems; Electromagnetic processes; Integral transformations

1. Introduction

Despite the large efforts in the last decades, our present understanding of hadronic systems is still far from being satisfactory. Due to the non-abelian gauge structure of QCD, preventing simple solutions in the non-perturbative domain, one uses in "conventional" nuclear physics not the fundamental quarks and gluons of QCD, but nucleons, isobars and mesons as relevant degrees of freedom. These so-called "effective" approaches are presently still the most promising ones for reaching a quantitative understanding of hadronic physics at low and intermediate energies below about 1 GeV excitation energy. On the other side, it is clear that the effective description will break down at some sufficiently high energy/momentum transfer. Moreover, it is presently not clear whether a clear cut borderline exists or whether even at relatively small energies quark and gluon degrees of freedom manifest themselves in specific reactions and observables.

It is obvious that for a detailed study of such fundamental questions a profound understanding of few-nucleon systems is inevitably necessary because the corresponding theoretical treatment is naturally the most cleanest one. Apart from that, the study of few-nucleon systems like the deuteron or ^{3}He is very important with respect to the question whether they may serve as effective neutron targets so that a better understanding of few-nucleon systems may also lead to a better understanding of elementary particle properties. On the other hand, the study of more complex nuclei (mass number A $\geq$ 4) will allow to build a bridge between few- and many-body physics and provide tests for typical many-body approximations.

2. The two-nucleon sector

Despite the small amount of participating nucleons, *microscopic* studies of few-nucleon systems are far from being trivial. This becomes obvious even in the two-nucleon sector in which beyond pion-threshold quite a large number of different reactions is possible, e.g. elastic NN-scattering, Compton-scattering, electromagnetic deuteron breakup, elastic electron-deuteron scattering, electromagnetic as well as hadronic pion production. These different reactions are linked by unitarity via the optical theorems for hadronic and electromagnetic reactions

$$\mathrm{Im}\, T(NN \to NN, \theta = 0) \sim \sigma_{tot}(NN \to NN, \pi d, \pi NN, ...) \,, \tag{1}$$
$$\mathrm{Im}\, T(\pi d \to \pi d, \theta = 0) \sim \sigma_{tot}(\pi d \to NN, \pi d, \pi NN, ...) \,, \tag{2}$$
$$\mathrm{Im}\, T(\gamma d \to \gamma d, \theta = 0) \sim \sigma_{tot}(\gamma d \to NN, \pi d, \pi NN, ...) \,, \tag{3}$$

where the left sides are understood to be evaluated in forward direction ($\theta = 0$). This means, for example, that the forward *Compton-scattering* amplitude is related to *all* possible reactions with a photon and a deuteron in the initial state. Therefore, as a consequence of unitarity, a *unified* description of *all* possible reactions is necessary.

In a first step, we may restrict ourselves to energies up to the Δ-region so that a basically nonrelativistic treatment should be sufficient and channels with at most one asymptotically free pion need solely to be studied. Despite these "simple" boundary conditions, the construction of such a unified approach is far from being trivial. For example, the requirement of a simultaneous description of breakup ($\gamma d \to NN$) and pion production reactions ($\gamma d \to \pi NN$) leads to an equitable treatment of nucleons and pions which turns out to be quite involved in practical applications. Therefore, in most existing approaches only very few of the above mentioned reactions of interest are selected and the rest is just ignored.

Within the required unified approach, a minimal Hilbert space suitable up to the Δ-region consists at least of three orthogonal subspaces $\mathcal{H}_N^{[2]}$, $\mathcal{H}_\Delta^{[2]}$ and $\mathcal{H}_X^{[2]}$, where $\mathcal{H}_N^{[2]}$ contains two nucleons, $\mathcal{H}_\Delta^{[2]}$ one nucleon and one Δ-resonance, and $\mathcal{H}_X^{[2]}$ two nucleons and one meson $X \in \{\pi, \rho, \sigma, \delta, \omega, \eta\}$. States with two Δ-isobars, higher resonances beyond the Δ and/or two or even more mesons may be omitted in a first step. In order to describe pion production/annihilation processes, πNN- and $\pi N\Delta$-vertices have to be considered as essential building blocks of the hadronic interaction V. They generate via iteration in the corresponding Lippmann-Schwinger equation effective two-body baryon-baryon potentials acting in $\mathcal{H}_N^{[2]} \oplus \mathcal{H}_\Delta^{[2]}$ which contain *retarded* one-boson exchange (OBE) mechanisms. This means that the propagation of the intermediate πNN-system in the potential is described by the exact retarded propagator $G_0^{ret}(W+i\epsilon) = (W+i\epsilon - H_0)^{-1}$, with W denoting the invariant energy of the system and H_0 the kinetic Hamilton operator for the intermediate πNN-system. Furthermore, this approach allows to incorporate the electromagnetic interaction straightforwardly via minimal substitution in the vertices, so that gauge invariance can be fulfilled – in practice at least on an approximative level.

In previous studies [1,2], it has been clearly demonstrated that such a retarded approach is indispensable for a proper description of the simplest photonuclear reaction – deuteron photodisintegration – beyond pion-threshold. On the other hand, it turns out that the widespread static approximation for the XNN-propagators in the two-body operators, i.e. the neglect of the energy transfer between the nucleon by the exchanged meson, clearly fails [3]. This important result indicates that the presently available high precision potentials, which are all energy independent, appear to be inappropriate for the description of electromagnetic reactions beyond pion-threshold.

In the most recent work [4], not only deuteron breakup or elastic NN-scattering as in [1,2,5,6] has been discussed, but also several hadronic and electromagnetic pion production/annihilation processes have been studied. Moreover, apart from other improvements not outlined here in detail, the mutual interaction $V_{\pi NN}$ of the involved particles in the propagating πNN-system has been incorporated on a much more sophisticated level compared to previous studies. In part of the reactions like elastic NN-scattering or electromagnetic breakup reactions, even a nonperturbative treatment of $V_{\pi NN}$ within a Faddeev approach has been realized. Therefore, a retarded unified approach is presently available at least in the two-nucleon system for energies up to the Δ-region. Despite some technical improvements still to be performed, it may serve as an ideal starting point for the study of more complex few-nucleon systems to be discussed in the next section.

3. More complex few-nucleon systems

One of the basic questions in the study of more complex systems is, up to which mass number A and energy/momentum transfer precise *microscopic* calculations for electromagnetic reactions can be performed. The most fundamental observable in this field is definitely the total inclusive response function $R(W)$ for the reaction $\gamma A \rightarrow X$ proportional to the total inclusive cross section. It reads for a given energy W of the reaction

$$R(W) = \int d\Psi_f \, |\langle \Psi_f | O | \Psi \rangle|^2 \, \delta(E_f - W) \tag{4}$$

with $|\Psi\rangle$ denoting the target state, $|\Psi_f\rangle$ the final states with energy E_f and O the electromagnetic transition operator.

In conventional scattering theory, an economic method to calculate $R(W)$ is to apply the optical theorem (see Fig. 1). By exploiting the completeness of the final states $|\Psi_f\rangle$ in (4) one obtains straightforwardly

$$R(W) \sim \lim_{\epsilon \rightarrow 0} \mathrm{Im}\, \tilde{T}^{\gamma A \rightarrow \gamma A}(W + i\epsilon, \theta = 0) \tag{5}$$

with the forward Compton-scattering amplitude

$$\tilde{T}^{\gamma A \rightarrow \gamma A}(W + i\epsilon, \theta = 0) = \langle \Psi | O^\dagger G(W + i\epsilon) O | \Psi \rangle \,, \tag{6}$$

in which the full resolvent

$$G(z) = \frac{1}{z - H_0 - V} \tag{7}$$

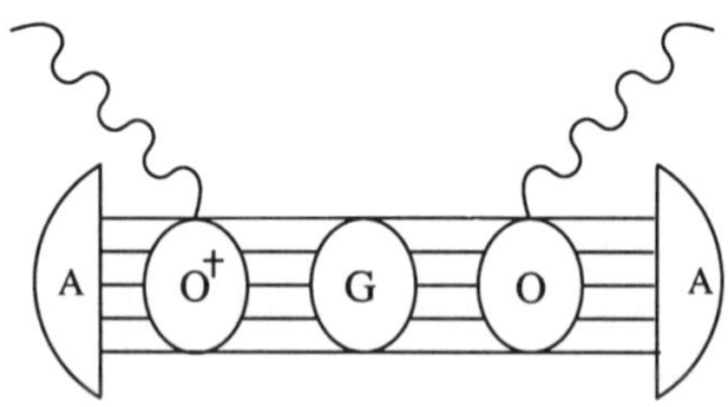

Fig. 1. Diagrammatic representation of the direct Compton-scattering amplitude for an A-nucleon system with G denoting the full resolvent (7) and O the electromagnetic transition operator.

of the system enters with H_0 denoting the kinetic part of the Hamiltonian and V the potential describing the interaction of the involved hadrons. Usually, the resolvent is expressed in terms of the so-called scattering amplitude T for the hadronic interaction

$$G(z) = G_0(z) + G_0(z)T(z)G_0(z) \tag{8}$$

with the free resolvent

$$G_0(z) = \frac{1}{z - H_0} \; . \tag{9}$$

The amplitude T obeys the Lippmann-Schwinger equation

$$T(z) = V + VG_0(z)T(z) \tag{10}$$

which contains for more than two particles a disconnected kernel. It is therefore necessary to reformulate T in terms of an appropriate integral equation whose kernel is compact. For $A = 3$, this leads for example to the famous Faddeev equations [7–9]. Moreover, one needs a careful treatment of the pole structure of the amplitude $T(z)$ which is a challenging numerical task for $A \geq 3$. Due to these reasons, rigorous microscopic calculations of electromagnetic reactions on the A-body system become more and more complicated with increasing A and increasing energy/momentum transfer and finally practically impossible.

An elegant solution of this problem has been proposed about a decade ago by the Trento group together with Efros [10]. The essential idea is to perform a so-called Lorentz integral transform (LIT) of $R(W)$ according to

$$L(\sigma_R, \sigma_I) = \int dW \frac{R(W)}{|W - \sigma|^2} \tag{11}$$

with a complex parameter $\sigma = \sigma_R + i\sigma_I$. After some algebra, it turns out that $L(\sigma_R, \sigma_I)$ has the same formal structure as the optical theorem for Compton-scattering (see equation (5)), i.e.

$$L(\sigma_R, \sigma_I) \sim \mathrm{Im}\, \tilde{T}^{\gamma A \to \gamma A}(\sigma_R + i\sigma_I, \theta = 0) \; . \tag{12}$$

The essential difference between (5) and (12) lies in the argument $W + i\epsilon$ versus $\sigma_R + i\sigma_I$. In the optical theorem, the quantity ϵ has to be treated as infinitesimal small yielding in consequence the abovementioned complicated pole structure as well as the requirement of solving an appropriate integral equation for the description of the hadronic interaction in the intermediate state of Fig. 1 after photon absorption and before photon emission. On the other hand, its counterpart in the LIT, σ_I, is finite and at our disposal. It can, at least in principle, be chosen arbitrarily. This has far reaching consequences, because the pole structure in (11) vanishes for a fixed $\sigma_I \neq 0$. This yields enormous numerical simplifications, because one needs only *bound state techniques.* In consequence, no integral equation needs to be tackled. Very efficient techniques, like the Effective Interaction method in the Hyperspherical Harmonics formalism [11,12] or the Lanczos algorithm [13], have been applied in the meantime to solve the remaining bound state problem. In order to obtain the desired inclusive cross section σ_{tot}, one has to perform finally a numerical inversion of the LIT. Recently, a variety of different reliable inversion methods has been presented [14] so that this problem appears to be very well under control. An important cross check for the inversion is that the resulting cross section should be independent from the parameter σ_I so that the LIT method is completely parameter free.

Due to these features, it is not very surprising that the LIT has been applied with considerable success to microscopic calculations of quite a few electroweak reactions on various nuclei ranging form $A = 2 - 7$ like inclusive electron scattering (see e.g. [15–17]) and total photoabsorption cross sections (see e.g. [18–20]). In the meantime, it has also been extended to exclusive reactions [21–24] as well as weak processes [25]. A recent highlight is reported in [26] where the first microscopic treatment of the four-body problem with realistic two- and three-body forces beyond three-body breakup threshold is discussed.

This list of applications shows that the LIT approach constitutes an important progress opening up the possibility to carry out ab initio microscopic calculations not only for reactions on the classical few-body systems like the deuteron or ^{3}He, but also for reactions on more complex nuclei. However, apart from very few simplifying exceptions [27,28], the presently available LIT calculations consider only nucleonic degrees of freedom so that they can be applied only in the low-energy region. In future work, the Trento group plans to extend the LIT approach up to the Δ-region. The final aim in the long-term run is the construction of a unified approach in this energy regime for all possible reactions not only in the two-body sector as in [4], but also for more complicated systems. In order to treat properly the occurring nonlocalities in the Δ-propagation, the calculations will be performed in momentum and not in configuration space as in the bulk of existing LIT studies. A suitable starting point is naturally the three-nucleon system where presently available Faddeev calculations are restricted, similar as the LIT approach, to energies below pion-threshold, see [29] and the references therein. Concerning possible applications, we intend to study a variety of different interesting topics like (i) the inclusive cross section for electron scattering and photoabsorption on ^{3}He in the Δ-region and (ii) the investigation of the Gerasimov-Drell-Hearn Sum rule (GDH) [30,31] on ^{3}He which requires a unified approach for polarized inclusive cross sections both below as well as above pion-threshold. This topic is moreover of specific importance with respect to the important question whether it will allow to draw conclusions with respect to the GDH sum rule on the neutron. Moreover, also semi-inclusive reactions can be tackled. Examples are pion electroproduction on polarized ^{3}He, which is an attractive tool to study nucleon properties in pion production [32], or the study of the electric neutron form factor G_{En} in $^3\vec{\mathrm{He}}(\vec{e}, e'n)$ at large momentum transfer beyond pion-threshold. These planned studies will of course be of specific importance for various experimental activities (MAMI, TJNAF) in the field of few-nucleon systems. Apart from these studies in the electromagnetic sector, the Trento group plans to extend the LIT approach also to closely related *hadronic* reactions like inclusive proton deuteron scattering.

4. Summary and outlook

The study of few-nucleon systems is a very active field both from the experimental as well as theoretical point of view. It is of particular importance for testing present theoretical frameworks in terms of effective degrees of freedom. In this context, the Lorentz integral transform method allows nowadays for the first time purely microscopic calculations of electromagnetic reactions on few-nucleon systems beyond the deuteron or ^{3}H/^{3}He in which the final state interaction can be taken into account nonperturbatively. In future work, it is planed to extend this method to higher energies beyond pion-threshold by considering explicitly nonnucleonic degrees of freedom. Hopefully, this may lead in the long-term run to a much more profound understanding of nuclear dynamics in general.

Acknowledgement

I would like to thank W. Leidemann for his careful reading of the manuscript and for various stimulating discussions.

References

1. M. Schwamb and H. Arenhövel, Nucl. Phys. **A690**, 682 (2001).
2. M. Schwamb and H. Arenhövel, Nucl. Phys. **A696**, 556 (2001).
3. P. Wilhelm and H. Arenhövel, Phys. Lett. **B318**, 410 (1993).

4. M. Schwamb, Habilitation Thesis (Johannes Gutenberg-Universität Mainz, 2006).
5. M. Schwamb and H. Arenhövel, Nucl. Phys. **A690**, 647 (2001).
6. M. Schwamb and H. Arenhövel, Phys. Lett. **B588**, 49 (2004).
7. L.D. Faddeev, *Mathematical Aspects of the Three-Body Problem in Quantum Scattering Theory* (Israel Program for Scientific Translations, Jerusalem, 1965).
8. E.W. Schmid and H. Ziegelmann, *The Quantum Mechanical Three-Body Problem* (Pergamon Press, Oxford and Friedrich Vieweg & Sohn, Braunschweig, 1974).
9. W. Glöckle, *The Quantum Mechanical Few-Body Problem* (Springer Verlag, Berlin 1983).
10. V.D. Efros, W. Leidemann, and G. Orlandini, Phys. Lett. **B338**, 130 (1994).
11. N. Barnea, W. Leidemann, and G. Orlandini, Phys. Rev. C **61**, 054001 (2000).
12. N. Barnea, W. Leidemann, and G. Orlandini, Nucl. Phys. **A693**, 565 (2001).
13. M. A. Marchisio *et al.*, Few-Body Syst. **33**, 259 (2003).
14. D. Andreasi *et al.*, Eur. Phys. J. A **24** 361 (2005).
15. V.D. Efros, W. Leidemann, and G. Orlandini, Phys. Rev. Lett. **78**, 432 (1997).
16. V.D. Efros *et al.*, Phys. Rev. C **69**, 044001 (2004).
17. V.D. Efros *et al.*, Phys. Rev. C **72**, 011002 (2005).
18. V.D. Efros, W. Leidemann, and G. Orlandini, Phys. Rev. Lett. **78**, 4015 (1997).
19. S. Bacca *et al.*, Phys. Rev. Lett. **89**, 052502 (2002).
20. S. Bacca *et al.*, Phys. Lett. **B603**, 159 (2004).
21. A. La Piana and W. Leidemann, Nucl. Phys. **A677**, 423 (2000).
22. S. Quaglioni *et al.*, Phys. Rev. C **69**, 044002 (2004).
23. S. Quaglioni *et al.*, Phys. Rev. C **72**, 064002 (2005).
24. D. Andreasi *et al.*, Eur. Phys. J. A **27**, 47 (2006).
25. D. Gazit and N. Barnea, Phys. Rev. C **70**, 048801 (2004).
26. D. Gazit *et al.*, Phys. Rev. Lett. **96**, 112301 (2006); see also the contribution by S. Bacca to these proceedings.
27. Ch. Reiss *et al.*, Eur. Phys. J. A **17**, 589 (2003).
28. Ch. Reiss, H. Arenhövel, and M. Schwamb, Eur. Phys. J. A **25**, 171 (2005).
29. J. Golak *et al.*, Phys. Rep. **415**, 89 (2005).
30. S.B. Gerasimov, Yad. Fiz. **2**, 598 (1965) (Sov. J. Nucl. Phys. **2**, 430 (1966)).
31. S.D. Drell and A.C. Hearn, Phys. Rev. Lett. **16**, 908 (1966).
32. R.-W. Schulze and P.U. Sauer, Phys. Rev. C **55**, 1051 (1997).

ISOSPIN MIXING IN THE ^{4}He BOUND STATE AND THE NUCLEON STRANGE FORM FACTOR *

M. VIVIANI, L. GIRLANDA, and A. KIEVSKY

INFN, Sezione di Pisa, 56127 Pisa, Italy

L.E. MARCUCCI and S. ROSATI

Department of Physics, University of Pisa, 56127 Pisa, Italy
INFN, Sezione di Pisa, 56127 Pisa, Italy

R. SCHIAVILLA

Department of Physics, Old Dominion University, Norfolk VA 23529, USA
and Jefferson Lab, Newport News, VA 23606, USA

The contribution of isospin admixtures in the ground state of the ^{4}He nucleus is studied using wave functions derived from the most modern nuclear interactions, including isospin symmetry breaking terms. The present calculations show that this contribution is larger than previous estimates had indicated. Its effect on parity violating elastic scattering of polarized electrons from ^{4}He is investigated. In particular, a simple analysis of the recently measured left-right asymmetry at low Q^2 shows that the contribution of these isospin admixtures is of comparable magnitude to that associated with strangeness components in the nucleon electric form factor.

Keywords: Isospin admixtures in complex nuclei, parity violating electron scattering

1. Introduction

Understanding the structure of the nucleon is one of the great challenges of modern hadron physics. In particular, an important issue is to determine, at a quantitative level, the role of the $q\bar{q}$ sea, expecially of the strange quark. Parity-violating (PV) elastic scattering of polarized electrons from nucleons and nuclei offers the opportunity to investigate this issue experimentally. The PV asymmetry is due to the interference between the γ- and Z-induced

*In memory of our colleague and friend Adelchi Fabrocini.

amplitudes. The Z couples to the neutral current (NC) of the nucleon, which is sensitive to its flavor content in a different way than the electromagnetic current. Therefore, from measurements of the PV asymmetry it is possible to disentangle the strange electric and magnetic form factors $G_E^{(s)}$ and $G_M^{(s)}$ of the nucleon, the analogs of the usual Sachs form factors G_E and G_M.

Indeed, new results of strangeness in the nucleon have been reported recently from the HAPPEX [1] and G0 [2] Collaborations at Jlab, which complemented earlier work at MIT-Bates [3] and Mainz [4]. The experiments scatter longitudinally polarized electrons from an unpolarized target and measure the asymmetry $A_{LR} = (\sigma_R - \sigma_L)/(\sigma_R + \sigma_L)$, where σ_R (σ_L) is the cross section for right-handed (left-handed) electrons. The asymmetry is tiny, of the order of a few parts per million.

Most of these experiments study PV electron scattering from protons. They are sensitive to linear combinations of $G_E^{(s)}$ and $G_M^{(s)}$ (and also to the axial form factor G_A^Z) [5]. It would be desiderable to have enough measurements to determine the contribution of the individual form factors. From this perspective, the study of PV electron scattering on ^{4}He is very interesting. This is because ^{4}He is a $J^\pi = 0^+$ state and therefore no contributions from magnetic and axial-vector currents are allowed [5]. However, there are nuclear effects which could mask the strange quark content of the nucleon.

In a nuclear target, these effects have different origins. There are contributions from meson exchange currents (MEC). This problem was studied in Ref. [6] using a realistic ^{4}He wave function and a realistic model of MEC. The authors of that paper concluded that for $Q^2 < 0.15$ GeV^2, where $Q^2 = q^2 - \omega^2$ denotes the square of the 4-momentum transfer (ω and q are the energy and the magnitude of the three-momentum transfer to the target), the MEC contributions are negligible. If, in addition to MEC, radiative corrections and a tiny relativistic contribution (associated with the spin-orbit term in the single-nucleon strange charge operator) are ignored, the PV asymmetry for electron scattering on a ^{4}He target is simply given by [5]

$$A_{LR} = \frac{G_\mu Q^2}{4\pi\alpha\sqrt{2}} \left[4\sin^2\theta_W + \frac{G_E^{(s)}}{(G_E^p + G_E^n)/2} - 2R_{T=1} \right] , \qquad (1)$$

where G_μ is the Fermi constant measured in muon decay, α the fine structure constant, and θ_W the weak mixing angle. The term $R_{T=1}$ is due to the presence of $T = 1$ components in the nuclear wave function. Since ^{4}He is to a very good approximation an isoscalar target, $R_{T=1}$ is usually neglected. The aim of this paper is to estimate the order of magnitude of $R_{T=1}$. The additional contribution arising from isospin admixtures in the *nucleon* [7] is not shown explicitly in Eq. (1) and will be discussed elsewhere. In this paper we will consider the proton and neutron as an isospin doublet.

An approximate calculation of this quantity was performed more than 10 years ago [8], by taking into account only the isospin admixture induced by the Coulomb interaction and, furthermore, by generating the relevant $T = 1$ component in the ^{4}He wave function in first order perturbation theory. The calculated value for $R_{T=1}$ was found to be rather small, and it produced a less than 1% correction in Eq. (1) [8] at low Q^2, using the value for $G_E^{(s)}(Q^2)$ known at the time.

Three new facts have prompted us to re-analyze this issue. First, new refined measurements of A_{LR} have shown that $G_E^{(s)}(Q^2 \approx 0.1\text{GeV}^2)$ is considerably smaller than previously assumed [9].

Second, there now exist several accurate models of nucleon-nucleon (NN) interactions [10–14] which include explicit charge symmetry breaking (CSB) terms, in addition to the Coulomb repulsion between protons. These CSB terms have been determined by fitting both the *pp* and *np* scattering data sets, and therefore are tightly constrained by experiment. In order to study ^{4}He, three-nucleon (NNN) interactions have also to be included in the Hamiltonian. These are still not well known, and the most commonly used models [15–17] do not contain isospin-symmetry violating terms. However, the NNN interaction gives much smaller effects relative to the NN: for example, its expectation value in ^{4}He is around 5% than that of the NN interaction. Therefore, one would expect that NNN CSB terms to be very tiny indeed.

Third, several accurate methods have been developed to compute ^{4}He wave functions starting from a given realistic model of NN and NNN interactions [18]. In these calculations, $T > 0$ isospin components are generated non-perturbatively. The $T = 1$ percentage in the ^{4}He wave function is typically found to be of the order of 10^{-3} %, systematically larger than estimated in Ref. [8] (see below).

For these reasons, we have reconsidered the problem of determining the magnitude of the term $R_{T=1}$. We have also studied the "model dependence" of this quantity, namely its dependence on the nuclear Hamiltonian used to generate the ^{4}He wave function.

In this paper, we use the Hyperspherical Harmonic (HH) expansion method to compute the ^{4}He wave function. The accuracy of this method is briefly discussed in Sec. 2, where some selected results obtained for ^{4}He observables are reported and compared with corresponding ones given in

Table 1. The α–particle binding energies B (MeV), the expectation values of the kinetic energy operator $\langle T\rangle$ (MeV), the mean square radii $\sqrt{\langle r^2\rangle}$ (fm), and the P, D, $T=1$ and $T=2$ probabilities (%) for the four Hamiltonian models considered in this paper.

Interaction	Method	B	$\langle T\rangle$	$\sqrt{\langle r^2\rangle}$	P_P	P_D	$P_{T=1}$	$P_{T=2}$
AV18	HH	24.210	97.84	1.512	0.347	13.74	0.0028	0.0052
	FY [22]	24.25	97.80	-	0.35	13.78	0.003	0.005
	FY [25]	24.223	97.77	1.516	-	-	-	-
	FY [23]	24.24	-	-	-	-	-	-
	EIHH [28]	24.268	-	1.518	-	-	-	-
AV18/UIX	HH	28.47	113.30	1.428	0.73	16.03	0.0025	0.0050
	FY [22]	28.50	113.21	-	0.75	16.03	-	-
	GFMC [26]	28.34(4)	110.7(7)	1.44	-	-	-	-
CDBonn/UIXb	HH	28.30	85.82	1.423	0.53	12.89	0.0020	0.0099
N3LO	HH	25.38	69.24	1.516	0.172	9.289	0.0035	0.0024
	FY [29]	25.37	69.20	-	0.172	9.293	0.0033	0.0024
	FY [23]	25.38	-	-	-	-	-	-
	NCSM [24]	25.36(4)	-	-	-	-	-	-

literature. The results for $R_{T=1}$ are presented and discussed in the last section.

2. The Hyperspherical Harmonic Method

The HH method has been presented in detail in Refs. [19–21]. Here, we shall limit ourselves to discuss the accuracy of the method. To this end, the fully converged values for the ^{4}He binding energies and other ground-state properties are reported in Table 1. The model Hamiltonians considered are: 1) the Argonne V18 NN interaction [11] (AV18 model); 2) the Argonne V18 NN + Urbana IX NNN interactions [15] (AV18/UIX model); 3) the CD Bonn [12] NN + Urbana IXb NNN interactions (CDBonn/UIXb model); and 4) the chiral N3LO [13] NN interaction (N3LO model). The Urbana UIXb NNN interaction is a slightly modified version of the Urbana UIX potential to be used together with the CD Bonn potential (in that model, the parameter U_0 in the central repulsive term has been multiplied by the factor 0.812). The CD Bonn and N3LO NN potentials are non-local and are given in momentum space. These results are compared with those obtained with other methods, including the Faddeev-Yakubovsky (FY) technique [22,23,25], No Core Shell Model (NCSM) approach [24], Green's Function Monte Carlo (GFMC) method [26], and Effective Interaction HH (EIHH) technique [27,28]. There is good numerical agreement, at the level of 0.1 %, among the results obtained with these different techniques for all quantities considered in the Table.

We also report the calculated percentages of $T>0$ components in the

^{4}He ground state wave function. In this case too, the results obtained with the HH method are in excellent agreement with those obtained using the FY technique. We find that, depending on the Hamiltonian model adopted, the values of the $T = 1$ percentage vary little, while those of $T = 2$ percentage display significantly larger variation. The origin of these $T > 0$ components was discussed in detail in Ref. [20]. We are confident to have full control of the $T > 0$ components in the ground state wave function.

3. Results and Conclusions

In ^{4}He the PV asymmetry reads (ignoring radiative corrections)

$$A_{LR} = \frac{G_\mu Q^2}{4\pi\alpha\sqrt{2}} \left[4\sin^2\theta_W + \frac{F^{(s)}(Q^2)}{F^{(T=0)}(Q^2)} - 2\frac{F^{(T=1)}(Q^2)}{F^{(T=0)}(Q^2)} \right] , \tag{2}$$

where

$$F^{(a)}(Q^2) = \frac{1}{\sqrt{4\pi}} \langle \Psi | \rho_\mu^{(a)}(q_\mu) | \Psi \rangle , \qquad (a) = (T=0), (T=1), (s), \tag{3}$$

are the various "charge form factors" of ^{4}He. In the equation above, Ψ is the ^{4}He ground state wave function and $\rho^{(a)}$ are charge operators. Their one-body components are given by

$$\rho^{(a)}(q_\mu) = \sum_{k=1}^{A} e^{i\boldsymbol{q}\cdot\boldsymbol{r}_k} \left[\frac{G_E^{(a)}(Q^2)}{\sqrt{1+\tau}} \right] , \quad (a) = (T=0), (s) , \tag{4}$$

$$\rho^{(T=1)}(q_\mu) = \sum_{k=1}^{A} e^{i\boldsymbol{q}\cdot\boldsymbol{r}_k} \tau_z(k) \left[\frac{G_E^{(T=1)}(Q^2)}{\sqrt{1+\tau}} \right] , \tag{5}$$

where $\tau = Q^2/4m_N^2$ and spin-orbit relativistic corrections [6] have been ignored. The isoscalar $G_E^{(T=0)}$ and isovector $G_E^{(T=1)}$ combinations of the electric nucleon form factors are defined as

$$G_E^{(T=0)}(Q^2) = \frac{1}{2} \left[G_E^p(Q^2) + G_E^n(Q^2) \right] , \tag{6}$$

$$G_E^{(T=1)}(Q^2) = \frac{1}{2} \left[G_E^p(Q^2) - G_E^n(Q^2) \right] . \tag{7}$$

In this limit (*i.e.*, one-body charge operators), it follows that

$$\frac{F^{(s)}(Q^2)}{F^{(T=0)}(Q^2)} = \frac{G_E^{(s)}(Q^2)}{G_E^{(T=0)}(Q^2)} , \tag{8}$$

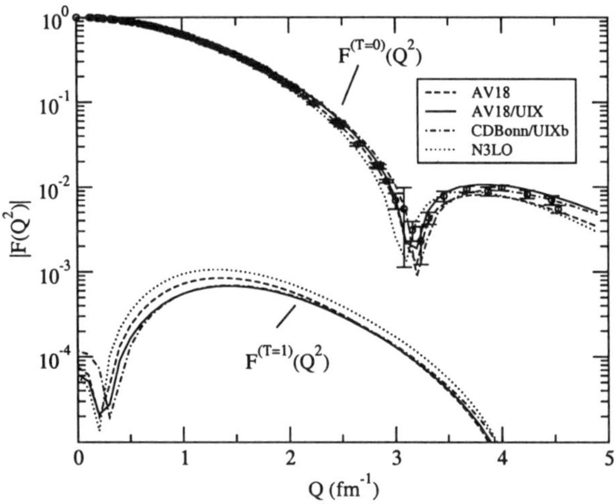

Fig. 1. The $F^{(T=0)}(Q^2)$ and $F^{(T=1)}(Q^2)$ form factors plotted as functions of Q for the Hamitonian models considered in this paper. The $F^{(T=0)}(Q^2)$ is compared with the experimental ^{4}He charge form factor [30,31]. Note that the calculated form factors reported here include also MEC contributions to the charge density operators. However, their effect is significant only for $Q > 2$ fm^{-1}.

namely nuclear structure effects cancel out, and by comparison with Eq. (1), we identify

$$R_{T=1} \equiv \frac{F^{(T=1)}(Q^2)}{F^{(T=0)}(Q^2)} \,. \tag{9}$$

We have performed the calculation of the form factors $F^{(T=0)}$ and $F^{(T=1)}$ using Monte Carlo techniques [6] and the ^{4}He wave functions discussed in the previous section. The calculated values of $F^{(T=0)}$ and $F^{(T=1)}$ are reported in Fig.1 for four Hamiltonian models (the AV18, AV18/UIX, CD-Bonn/UIXb and N3LO models). The theoretical predictions for $F^{(T=0)}$, *i.e.* the ^{4}He charge form factor, are compared with the experimental data of Refs. [30,31]. As can be seen, the form factor $F^{(T=1)}$ is rather small and the various Hamiltonian models give predictions quite close to each other, the remaining differences reflect the different percentages of the $T = 1$ component in the ^{4}He wave function.

Let us now consider the effect of the present form factors predictions on PV electron scattering from ^{4}He. The most accurate measurements have been recently reported at $Q^2 = 0.077$ GeV^2, corresponding to $Q \approx 1.41$ fm^{-1}. The experimental value for A_{LR} is [9]

$$A_{LR} = +6.40 \pm 0.23(\text{stat}) \pm 0.12(\text{syst}) \; ppm \,, \tag{10}$$

Table 2. Calculated values of $F^{(T=0)}$, $F^{(T=1)}$, and $R_{T=1}$ for $Q^2 = 0.077$ GeV^2 for Hamiltonian models considered in this paper. The values in parentheses have been obtained by including MEC in the charge operators.

Interaction	$F^{(T=0)}$	$F^{(T=1)}$	$R_{T=1}$
AV18	0.835	−0.00174	-0.00208 (-0.00211)
AV18/UIX	0.890	−0.00140	-0.00157 (-0.00160)
CDBonn/UIXb	0.897	−0.00141	-0.00157 (-0.00158)
N3LO	0.800	−0.00218	-0.00275 (-0.00274)

which, after inserting the values for G_μ, α, and θ_W, gives (see Eq. (1))

$$\frac{G_E^{(s)}(Q^2)}{G_E^{(T=0)}(Q^2)} - 2\,R_{T=1} \equiv \Gamma = 0.0048 \pm 0.04\ , \qquad Q^2 = 0.077\ \text{GeV}^2\ . \quad (11)$$

This result is consistent with zero. We report the calculated values of $R_{T=1}$ at $Q^2 = 0.077$ GeV^2 in Table 2.

The $R_{T=1}$ term is found to be of the order of -0.002. Note that the value previously estimated in Ref. [8] was $|R_{T=1}(Q^2 = 0.077\ \text{GeV}^2)| \approx 0.0003$, an order of magnitude less than the present result. The value obtained with N3LO is slightly larger than for other potentials, reflecting the larger percentage of $T = 1$ admixtures predicted by this potential. The inclusion of a NNN interaction tends to decrease the value of $R_{T=1}$ and the effect of MEC is confirmed to be, at this value of Q^2, negligible.

Finally, the computed $R_{T=1}$ is of the same order of magnitude as the value of Γ in Eq. (11). Using in Eq. (11) the value $R_{T=1} \approx -0.00157$ obtained with the Hamiltonians including NNN forces, one would obtain $G_E^{(s)}(Q^2 = 0.077\ \text{GeV}^2) = 0.001 \pm 0.02$, thus suggesting that the value of Γ is almost entirely due to isospin admixtures in the ^{4}He wave function. Of course, the experimental error on Γ is still too large to allow us to draw a more definite conclusion.

In summary, we have estimated the value of $R_{T=1}$ at the Q^2 value relevant for the recent HAPPEX experiment [9], and have found it to be larger than previously estimated. A recent lattice QCD estimate of $G_E^{(s)}$ reports [32] $G_E^{(s)}(0.1\ \text{GeV}^2) = +0.001 \pm 0.004 \pm 0.003$. An increase of the experimental accuracy of one order of magnitude would be necessary to be sensitive to $G_E^{(s)}$ at low values of Q^2. However, the present study shows that the effect of the $T = 1$ component needs to be taken into account. Indeed, if the present lattice QCD predictions are confirmed, the present data would suggest that the leading correction to the asymmetry is from $T = 1$ isospin admixtures in ^{4}He. As shown in Ref. [20], this component does not only originate from the Coulomb repulsion between protons, but arises also from CSB terms in the nuclear interaction. The study of CSB is a very interesting problem, since it is linked to fundamental properties of the u and d quarks, namely their mass and charge differences [33].

References

1. K.A. Aniol *et al.* [HAPPEX Coll.], Phys. Rev. Lett. **96**, 022003 (2006).
2. D.S. Amstrong *et al.* [G0 Coll.], Phys. Rev. Lett. **95**, 092001 (2005).
3. D.T. Spayde *et al.* [SAMPLE Coll.], Phys. Lett. **B583**, 79 (2004).
4. F.E. Maas *et al.* [A4 Coll.], Phys. Rev. Lett. **93**, 022002 (2004).
5. M.J. Musolf *et al.*, Phys. Rep. **239**, 1 (1994).
6. M.J. Musolf, R. Schiavilla, and T.W. Donnelly, Phys. Rev. C **50**, 2173 (1994).
7. See, for example, B. Kubis and R. Lewis, Phys. Rev. C **74**, 015204 (2006) and references therein.
8. S. Ramavataran, E. Hadjimichael, and T. W. Donnelly, Phys. Rev. C **50**, 1175 (1994).
9. A. Acha *et al.* [HAPPEX Coll.], `nucl-ex/0609002`.
10. V. G. J. Stoks *et al.*, Phys. Rev. C **49**, 2950 (1994).
11. R. B. Wiringa, V. G. J. Stoks, and R. Schiavilla, Phys. Rev. C **51**, 38 (1995).
12. R. Machleidt, Phys. Rev. C **63**, 024001 (2001).
13. D.R. Entem and R. Machleidt, Phys. Rev. C **68**, 041001 (2003).
14. E. Epelbaum, W. Gloeckle, and Ulf-G. Meissner, Nucl. Phys. **A747**, 362 (2005).
15. B. S. Pudliner *et al.*, Phys. Rev. C **56**, 1720 (1997).
16. S. A. Coon *et al.*, Nucl. Phys. **A317**, 242 (1979).
17. H.T. Coelho, T.K. Das and M.R. Robilotta, Phys. Rev. C **28**, 1812 (1983).
18. See, for example, H. Kamada *et al.*, Phys. Rev. C **64**, 044001 (2001).
19. M. Fabre de la Ripelle, Ann. Phys. (N.Y.) **147**, 281 (1983).
20. M. Viviani, A. Kievsky, and S. Rosati, Phys. Rev. C **71**, 024006 (2005).
21. M. Viviani *et al.*, Few-Body Syst. **39**, 159 (2006).
22. A. Nogga *et al.*, Phys. Rev. C **65**, 054003 (2002).
23. A. Deltuva and A.C. Fonseca, `nucl-th/0611029`.
24. P. Navrátil and E. Caurier, Phys. Rev. C **69**, 014311 (2004).
25. R. Lazauskas and J. Carbonell, Phys. Rev. C **70**, 044002 (2004).
26. R. B. Wiringa *et al.*, Phys. Rev. C **62**, 014001 (2000).
27. N. Barnea, W. Leidemann, and G. Orlandini, Phys. Rev. C **61**, 054001 (2000).
28. D. Gazit *et al.*, `nucl-th`/0512038.
29. A. Nogga, private communications.
30. R.F. Frosch *et al.*, Phys. Rev. **160**, 874 (1968).
31. R.G. Arnold *et al.*, Phys. Rev. Lett. **40**, 1429 (1978).
32. D.B. Leinweber *et al.*, Phys. Rev. Lett. **97**, 022001 (2006).
33. G. Miller, B.M.K. Nefkens, and I. Slaus, Phys. Rept. **194**, 1 (1990).

Nuclear Matter and Nuclear Dynamics

NUCLEAR MATTER AND NUCLEAR DYNAMICS

M. BALDO

INFN Sezione di Catania, Corso Italia 57, 95129 Catania, Italy

1. Introduction

One of central issue in the theory of astrophysical compact objects and heavy ion reactions at intermediate and relativistic energy is the Nuclear Equation of State (EOS). On one hand, the large and expanding set of experimental and observational data are expected to constraint the behavior of the nuclear EOS, especially at density above saturation. On the other hand, theoretical predictions of the EOS are essential for modeling heavy ion collisions, at intermediate and relativistic energy, and the structure of neutron stars, supernova explosions, binary collisions of compact stellar objects and their interactions with black holes. An enormous range of scales in size, characteristic time, energy and temperature is involved in all these phenomena and processes. In the astrophysical context the dynamics is slow enough and the size scale large enough to assure the local equilibrium of nuclear matter, i.e. hydrodynamics can be applied, and therefore the very concept of EOS is extremely useful. On the other hand, in nuclear collisions the time scale is the typical one for nuclear processes and the size of the system is only one order of magnitude larger than the interaction range or possibly of the particle mean free path. The physical conditions in the two contexts are therefore quite different. Despite that, by a careful analysis of experimental data on heavy ion collisions and astrophysical observations it is possible to connect the two realms of phenomena which involve nuclear processes at fundamental level, and the EOS provides the crucial concept to establish this link.

From the theoretical point of view the microscopic theory of nuclear matter has a long history and impressive progress has been made along the years. In this general talk I will first review the many-body theory of nuclear matter and compare the predictions of different approaches, Sec. 2 and

3. In the following Sections possible hints from astrophysical observations and heavy ion reactions on the nuclear EOS will be critically reviewed, with emphasis on the connections that can be established between the two fields.

2. Many-body theory of EOS

The many-body theory of nuclear matter, where only nucleonic degrees of freedom are considered, has developed since several decades along different lines and methods. We summarize the most recent results on this field and compare the different methods at formal level, as well as the final EOS calculated within each one of the considered many-body schemes. The outcome of this analysis should help in restricting the uncertainty of the theoretical predictions for the nuclear EOS.

Within the non–relativistic approach the main microscopic methods are the Bethe-Brueckner-Goldstone (BBG) approach and the variational method (VM).

The BBG expansion is a general many-body method particularly suited for nuclear systems. It has been extensively applied to homogeneous nuclear matter since many years and it has been presented in several review articles and textbooks. For a pedagogical review see ref. [1], where a short historical introduction can be found. The basic quantity of the scheme is the in–medium G–matrix, which can be interpreted as an effective nucleon–nucleon (NN) interaction and which replaces the original bare NN interaction in the diagrammatic expansion of the ground state energy. The diagrams are classified according to the number of hole lines which they contain and are expected to describe correlations of increasing order. The diagrams up to the three hole–line diagrams are depicted in Fig. 1, where a wavy line indicates a Brueckner G–matrix. The box labeled by $T^{(3)}$ indicates the in-medium three–body scattering matrix. The two–body G–matrix and the three–body scattering matrix satisfy each one an integral equation which sum up the two– and three–particle ladder diagram series

$$\langle k_1 k_2 | G(\omega) | k_3 k_4 \rangle = \langle k_1 k_2 | v | k_3 k_4 \rangle +$$

$$+ \sum_{k_3' k_4'} \langle k_1 k_2 | v | k_3' k_4' \rangle \frac{\left(1-\Theta_F(k_3')\right)\left(1-\Theta_F(k_4')\right)}{\omega - e_{k_3'} + e_{k_4'}} \langle k_3' k_4' | G(\omega) | k_3 k_4 \rangle \quad , \qquad (1)$$

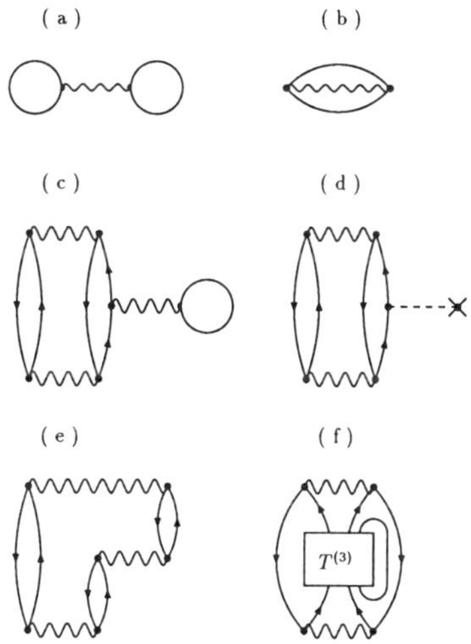

Fig. 1. Diagrams up to the three hole–line diagrams in the BBG expansion.

$$
\begin{aligned}
T^{(3)} &= G + G\,X\,\tfrac{Q_3}{e}\,T^{(3)} \\
\langle k_1 k_2 k_3 | T^{(3)} | k_1' k_2' k_3' \rangle &= \langle k_1 k_2 | G | k_1' k_2' \rangle \delta_K(k_3 - k_3') + \\
&\quad + \langle k_1 k_2 k_3 | G_{12}\,X\,\tfrac{Q_3}{e}\,T^{(3)} | k_1' k_2' k_3' \rangle \quad , \qquad (2)
\end{aligned}
$$

where e_k is the single particle spectrum, the factor $Q = (1 - \Theta_F(k))(1 - \Theta_F(k'))$ projects the two intermediate momenta out of the Fermi sphere and the factor Q_3 is the analogous projection factor for three particles. Notice that in the integral equation for $T^{(3)}$ the G–matrix appears as a kernel instead of the bare interaction v. The factor X is typical of the Fadeev equation for three–body problem. Its inclusion avoids double counting. Equation (2) is usually called the Bethe–Fadeev equation, since it is the generalization of the Fadeev equation to the in–medium case.

To illustrate the relevance of the Pauli operator Q in Fig. 2 is reported a comparison between the free T–matrix and the in–medium G–matrix for pure neutron matter. One can see the drastic renormalization of the scattering matrix even at a this relatively low density [2].

The BBG expansion has been extensively applied to nuclear matter, both symmetric and asymmetric, and it has been shown [3–5] that with the inclusion of the three hole–line contribution convergence of the expansion is likely to be achieved up to several times saturation density. In particular

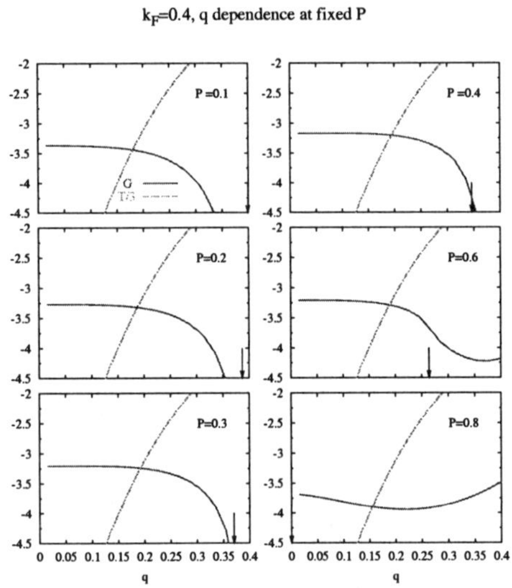

Fig. 2. Comparison of the free T–matrix (T) and the in–medium G–matrix (G) at $k_F = 0.4\ \text{fm}^{-1}$ for different total momentum P and different relative momenta q.

in the so called continuous choice for the single particle potential the contributions of the tree hole–line diagrams turn out to be small and negligible in first approximation.

The BBG expansion can be obtained also within the Coupled-Cluster Method (CCM) [6,7], a general many-body theory which has beeen extensively applied in a wide variety of different physical systems, both bosonic and fermionic ones. The connection between the CCM and the BBG expansions has been clarified by D. Day [7]. In the CCM method one starts from a particular ansatz on the form of the exact ground state wave function Ψ in terms of the unperturbed ground state Φ, $|\Psi\rangle = e^{\hat{S}}|\Phi\rangle$, where the hermitean operator $\hat{S}$ is expanded in terms of n-particle and n-hole unperturbed states

$$\hat{S} = \sum_n \sum_{k_1,k_2...k_n,k'_1,k'_2...k'_n} \frac{1}{n!^2} \langle k'_1, ...k'_n|S_n|k_1, ...k_n\rangle a^\dagger(k'_1)...a^\dagger(k'_n)a(k_n)...a(k_1) \tag{3}$$

where all the k ' s are hole momenta, i.e. inside the Fermi sphere, and all the k' ' s are particle momenta, i.e. outside the Fermi sphere. For translational invariant systems the term S_1 vanishes due to momentum conservation. The exponential form is chosen in order to include, as much as possible, only "linked" terms in the expansion of $\hat{S}$, in the spirit of the linked-cluster theorem. This form also implies the normalization $\langle\Phi|\Psi\rangle = 1$. The functions

S_n are expected to describe the n-body correlations in the ground state. As an illustration, let us consider only S_2 for simplicity and let us assume that it can be considered local in coordinate space, $S_2(r_i - r_j) = \chi_{ij}$, where the labels ij include spin-isospin variables. Then the correlated ground state can be written

$$\Psi(r_1, r_2, ...) = \Pi_{i<j} f_{ij} \Phi(r_1, r_2, ...) \qquad (4)$$

where the product runs over all possible distinct pairs of particles and $f_{ij} = \exp(2\chi_{ij})$. In general, however, the functions S_n are highly non-local in coordinate space and the expression for the ground state wave function cannot be written in such a simple form, and in fact the function S_2 turns out to be the "defect function" of the Brueckner method. Despite that, Eq. (3) suggests a connection of the BBG expansion with the variational method (VM). Indeed, in the VM one *assumes* that the ground state wave function has the form of Eq. (3), since the correlation factor are assumed to take into account the short range repulsion produced by the repulsive core of the NN bare interaction. The correlation function $f(r_{ij})$ is here determined by the variational principle, i.e. by imposing that the mean value of the hamiltonian gets a minimum (or in general stationary) $\frac{\delta}{\delta f} \frac{\langle\Psi|H|\Psi\rangle}{\langle\Psi|\Psi\rangle} = 0$. In principle this is a functional equation for the correlation function f, which however can be written explicitly in a closed form only if additional suitable approximations are introduced. A practical and much used method is to assume a parametrized form for f and to minimize the energy with respect to the set of parameters which constrain its form. Since the wave function is expected to decrease strongly whenever two particles are at distance smaller than the repulsive core radius of the NN interaction, the function $f(r_{ij})$ is assumed to converge to 1 at large distance and to go rapidly to zero as $r_{ij} \rightarrow 0$, with a shape similar to the one for the defect function in the BBG expansion. For nuclear matter it is necessary to introduce a channel dependent correlation factor, which is equivalent to assume that f is actually a two-body operator $\hat{F}_{ij}$. One then assumes that $\hat{F}$ can be expanded in the same spin-isospin, spin-orbit and tensor operators which appear in the NN interaction. Momentum dependent operators, like spin-orbit, are usually treated separately. The product in Eq. (4) must be then symmetrize since the different terms do not commute anymore. For each correlation channel there will be then a different correlation factor. The most flexible assumption on the F's is to impose that they go to 1 at a given "healing" distance d with zero derivative. The healing distances, which eventually can be defined for each spin-isospin and tensor channels, is then taken as a variational parameters, while the functions for $r < d$ are determined directly from the variational procedure. In principle, the condition of energy minimum (or extremal) should produce a set of Euler-Lagrange equations which determine the correlation factors. In practice, a viable explicit form can be used only for the two-body cluster terms.

If the two-body NN interaction is local and central, its mean value is directly related to the pair distribution function $g(\mathbf{r})$

$$< V >= \frac{1}{2}\rho \int d^3 r v(r) g(\mathbf{r}) \tag{5}$$

where

$$g(\mathbf{r_1} - \mathbf{r_2}) = \frac{\int \Pi_{i>2} d^3 r_i |\Psi(r_1, r_2 ....)|^2}{\int \Pi_i d^3 r_i |\Psi(r_1, r_2 ....)|^2} \tag{6}$$

The main job in the variational method is to relate the pair distribution function to the correlation factors F. In general this cannot be done exactly, and one has to rely on some suitable expansion. For the central part of the correlations, the physical quantity which describes the main perturbation with respect to the free Fermi gas is the function $1 - F(r)^2 = h(r)$, which is a measure of the strength of the short range part of te correlation. One can then expand the square of the correlated wave function in the components with a given number of h-factors, and correspondingly the energy mean value can be expanded in the different terms, each one with a given number of h-functions. For the non-cental part of the correlation factors F_{nc}, the smallness factors are F_{nc}^2 and the product $F_{nc} \cdot h$, since they are expected to be small and vanish at large distance. The different terms can be represented graphically by diagrams to help their classifications and identify their possible cancelations. It turns out that the mean value, at least in the thermodynamic limit, is the summation of the so-called "irreducible" diagrams, in strong similarity with the linked-cluster theorem of the BBG expansion. Indeed, the reducible diagrams are canceled exactly by the expansion of the denominator in Eq. (6). The problem of calculating $g(r)$ from the asnsatz of Eq. (3) has also a strong similarity with the statistical mechanics of a classical gas at finite temperature, where different methods to sum up infinite series of diagrams in the so-called "virial expansion" have been developed, noticeably the Hypernetted Chain (HNC) summation method. These methods can be almost literarily translated to the case of boson systems. With some modifications due to the different statistics, they can be extended to fermion systems (FHNC), provided in this case the correlations are taken to be only central ("Jastrow type" correlations) in Eq. (4), i.e. the correlation factors are assumed to be only dependent on the coordinates. Unfortunately, in nuclear matter, as already mentioned, correlations are of complicated structure due to the NN interaction, and the HNC method can be applied only within approximate schemes, like the Single Operator Chain (SOC) summation method, called also Variational Summation Method (VSM). In VMS only chains with a given correlation operator are considered. In general, the correlation functions are calculated at the two-body cluster level, where one gets the Euler-Lagrange coupled equations for all operator channels, that can be solved exactly for the set of correlation functions $F^p(r_{ij})$ at fixed values of the "healing distances". The index p labels the different two-body operators, spin-spin, spin-isospin, tensor, and so on. The VMS method is then applied, keeping the same set of correlation functions, to calculate the total energy. The procedure is repeated for different values of the healing distances and the energy minimum is found within this parameter space. The minimization gives of course automatically the ground state wave function. The VSM allows to include a definite class of higher order "clusters" beyond the two-body ones. However, particle clusters in the variational method are physically quite different from the ones in the CCM method as well as from the BBG ones, where particle "clusters" are defined in terms of diagrams with a given number of hole-lines, according to the hole-line expansion. Generally speaking, the summation of clusters performed by chain summations are expected to include long range correlations, while the variational procedure leading to the Euler-Lagrange equations should include mainly short range correlations. Indeed, in the low density limit the Euler-Lagrange equation reduces to the Schroedinger equation for two particles in free space.

In summary the main formal difference between the BBG expansion and the VM can be identified as follows

1. The correlation factor in the VM are local and momentum independent, with possible corrections introduced by means of gradient terms, while in the BBG they turn out to be strongly non local and momentum dependent.

2. In the BBG method an essential ingredient is the auxiliary single particle potential, which allows to speed up the rate of convergence. In the VM no such potential is introduced within the minimization procedure. It can be of course extracted after the wave function has been calculated. This indicates also that the meaning of "cluster" is not the same in the two methods.

3. The VM chain summations include physically long range correlations, short range correlations being included in the irreducible n–vertex diagrams

[8]. Long range correlations in the VM are, in principle, summed up to all order, while in the BBG expansion they are included ("ring diagrams") at the same order as the other diagrams.

Let us now compare the calculated EOS within the two methods, starting from the pure neutron matter case. In Fig. 3 are reported the two EOS in a wide density range for the NN interaction Argonne v_{18}. The overall agreement looks satisfactory and give confidence on the results obtained from the microscopic many–body theory. In ref. [9] a comparison was made of the two methods for two simplified interactions, the v_6 and the v_8', for which Green's function Monte Carlo (GFMC) calculations are available, at least up to the density $n = 0.24$ fm^{-3}. Excellent agreement between BBG and VM, on one hand, and with the Monte Carlo results, on the other, is found. Since GFMC is, in principle, an exact numerical method, this agreement gives further confidence on the accuracy of the employed many-body theories. A similar agreement is found for symmetric nuclear

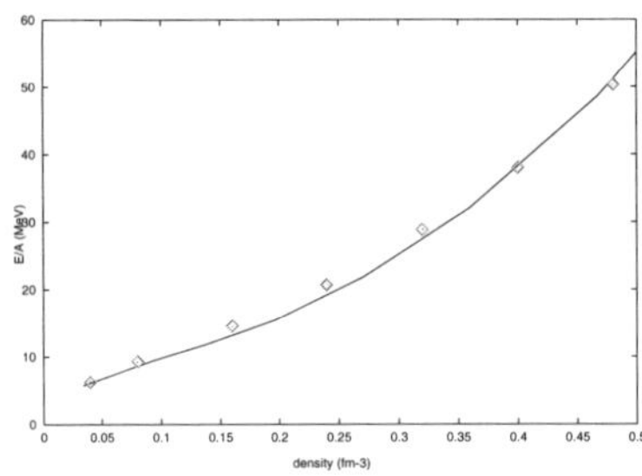

Fig. 3. Pure neutron matter EOS in the BBG (full line) and variational method (open diamonds).

matter. However, as it is well known, no NN interactions has been found that is able to reproduce the empirical saturation point (SP) of symmetric nuclear matter, at least within the non–relativistic many–body approach. Therefore, modifications of the hamiltonian has to be invoked, in particular the addition of three–body forces (TBF) or relativistic effects. First we will discuss TBF and later we will consider relativistic effects. The type of agreement between BBG and VM that has been found with two–body forces only is still present if three-body forces (TBF) are introduced to the purpose of getting a SP in agreement with the empirical findings. This can be seen in Fig. 4, where calculations with Argonne v_{18} interaction and the Urbana model for three-body forces is used. A wider density range is consider and pure neutron matter EOS is also reported. This TBF contains an attractive and repulsive part, whose structure is suggested by elementary processes which involve meson exchanges and three nucleons but cannot be separated into two distinct nucleon-nucleon interaction processes. This TBF are phenomenological in character since the two parameters, namely the strength of the attractive and the repulsive terms, cannot be fixed from first principle but they are adjusted to reproduce accurately experimental data. Since the VM produces an EOS which is superluminal at high density, in ref. [10] a modification of this EOS was proposed, which is also reported in Fig. 4 for comparison. Few observations are in order. It turns out that the parameters of the three-body force which have been fitted to data on few nucleon systems (triton, ^3He and ^4He) have to be modified if the SP has to be reproduced within the phenomenological uncertainty. Generally speaking the repulsive part has to be reduced substantially. It could be argued that this change of parameters poses a serious problem, since if the TBF model makes sense one must keep the same parameters at all density, otherwise higher order many-body forces should be invoked. Actually this is a false problem. In fact the discrepancy on the SP cannot be reduced more then few hundred MeV (typically 200-300 KeV), due to the intrinsic uncertainty in its position. This discrepancy remain essentially the same for few-body systems if one use the same TBF fitted in nuclear matter, and actually the contribution of TBF in few-body system is quite small. Therefore TBF which allows to describe both few-body system and the nuclear matter SP do exist, if the accuracy is kept at the level of 200-300 KeV in energy per particle, and indeed they are not unique. To try a very precise overall fit at the level of 10 KeV or better, as it is now possible in the field of few-body system, appears definitely to be too challenging. There are surely higher order terms (four-body forces, retardation effects in the NN interaction, other relativistic effects, etc.) which could contribute at this level of precision. In fact the contribution of TBF to the energy per particle around saturation is in all cases about 1-2 MeV, while for few-body systems it is one order of magnitude smaller. Therefore a TBF tuned to fit the binding energy of few-body system with an accuracy of 1-10 KeV cannot be extended to fit nuclear matter SP, since this would correspond to a quite unbalanced fitting procedure. In other words, 200-300 KeV for the energy per particle is the limit of accuracy of the TBF model applied to the nuclear EOS in the considered wide range of density.

Secondly, it has to be noticed that once the SP is reproduced by adjusting the TBF, it turns out that the parameters are not the same in the two

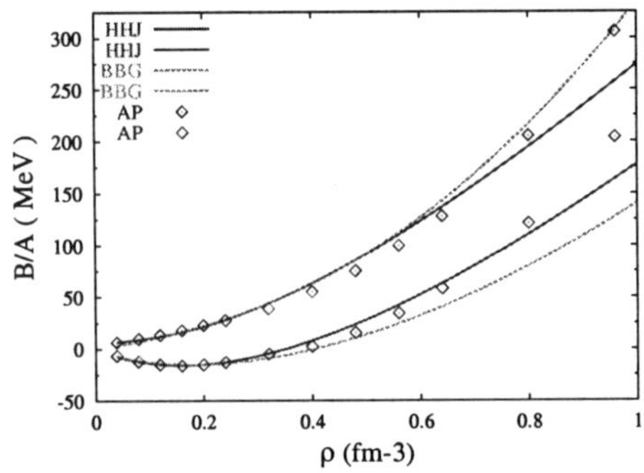

Fig. 4. Pure neutron matter (upper curves) and symmetric nuclear matter (lower curves) EOS, calculated within the BBG (dotted line), the VM (diamonds) schemes for the v_{18} NN interaction plus TBF. The curves labeled HHJ correspond to the EOS proposed in ref. [10].

methods, i.e. the TBF are not the same. Finally the method to incorporate TBF is simplified in the BBG method, namely the TBF is reduced to a density dependent two-body force by a suitable average over the position and spin-isospin quantum numbers of the third particle. The results presented in Fig. 4 are Brueckner calculations with TBF included following this procedure. The results and the agreement between the two methods seem to indicate that once the SP is reproduced correctly, the full EOS is determined to a large extent up to density as high as 0.6 fm^{-3}. However the conclusion is restricted to the particular model for the two-body and three-body forces. The possible dependence on the considered forces will be considered in the next section.

3. Dependence on the Two and three-body forces

The most common approach to nuclear interaction is the scheme where the NN interaction is determined, at least in an effective way, by the exchange of different mesons, and practically all modern NN interactions take inspiration for their structure by this assumption, in an explicit or implicit way. A really large set of two-body interactions has been developed along the years, but nowadays it is mandatory to restrict the possible choice to the most modern ones, since they are the only ones which fit the widest and most accurate experimental data, on one hand, and are more accurate in the fitting, on the other. One can then restrict the set of possible NN two-body interactions to the ones which fit the latest data (few thousands of data points) with an accuracy which gives a χ^2 close to 1. With these requirements, the number of NN interactions reduces quite a bit, and only few ones can be considered acceptable. The one which is constructed more explicitly from meson exchange processes is the recently developed CD Bonn potential by Machleidt, which is the latest one of the Bonn potential series. In principle this interaction is the one with the best χ^2 value. However the experimental data are not always consistent to the needed degree of accuracy and some selection must be done. In the same work one can find a detailed analysis of the different data set together with the method of selection which has been followed for the most accurate NN interactions and the values of the corresponding χ^2, if one includes the data up to the year 2000. Among the most accurate NN interactions one has to include the Argonne v_{18}, already discussed. It is constructed by a set of two-body operators which arise naturally in meson exchange processes, but the form factors are partly phenomenological (except, of course, the one-pion exchange). This interaction has been recently modified in the $^1S_0$ and $^3S-D_1$ channels with the inclusion of a purely phenomenological short range non-local force, which substitutes the original potentials below 1 fm. This allows to reproduce the binding energy of three and four nucleon systems very accurately without the inclusion of any TBF, at variance with the original Argonne v_{18}. Also the radii of ^3H and ^3He is accurately reproduced, while the radius of ^4He is slightly underestimated. This potential is phase-equivalent to the original interaction, but the off-shell behavior is modified. Finally one can mention the latest potentials of the Nijmegen series. They have also the ideal value $\chi^2 \approx 1$. However the fit was performed separately in each partial wave and the corresponding two-body operator structure cannot have the simple form as expected from meson exchange processes. A larger set of interactions, which includes the old NN potentials, can be found in ref. [11], where the resulting symmetric nuclear matter EOS' s at BHF level are compared.

It has to be noticed that the three selected NN interaction,v_{18}, CD Bonn and IS, give an increasingly better reproduction of the three-body binding energy and radii, and at the same time their non-locality is increasing in the same order. They are all phase equivalent to a good accuracy, so that the differences appearing in the nuclear matter EOS can be solely due to their different off-shell behavior. It is well known that phase equivalent potentials can give quite different saturation point and overall EOS, but here the comparison is restricted to a very definite class of realistic accurate NN interactions, with an operatorial structure which is quite similar and suggested by meson exchange processes. In Fig. 5 we compare [12] the three corresponding EOS at the BHF level and with the inclusion of three hole-line diagrams (no TBF). First of all one can notice that also for CD

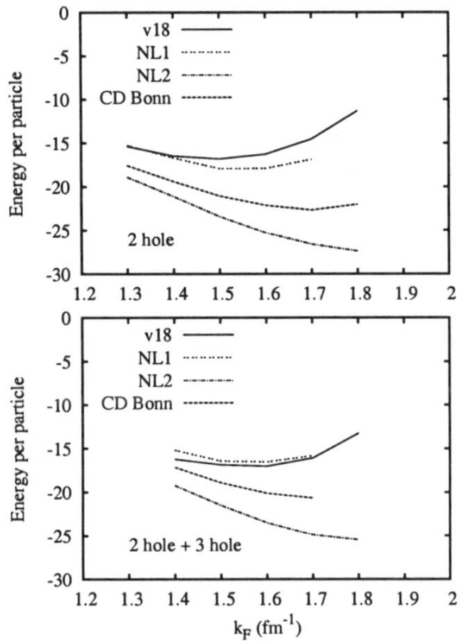

Fig. 5. Symmetric nuclear matter EOS for the NN interactions Argonne v_{18} (full line), CD Bonn (dotted line)and the non–local interaction IS1 (dash–dotted and double–dotted lines). In the upper panel only two hole–line diagrams are included, in the lower one the three hole–line diagrams are added.

Bonn and IS interactions the three hole-line contributions are still relatively small (especially if one compares the interaction energies at two and three hole levels). Convergence of the hole expansion appears a general feature. Furthermore, at increasing non-locality the EOS become softer and the SP tends to run away from the empirical SP and unreasonably large values for the binding energy and the SP density are obtained. The results indicate that the problem of reproducing with two-body realistic and accurate interactions the empirical SP cannot be solved by introducing non-locality, which modify the off-shell properties of the two-body potentials. Furthermore, the requirement of a very accurate fitting of the binding energy and radii of three (and eventually four) nucleon systems makes the reproduction of nuclear matter SP more challenging.

The necessity of introducing TBF around saturation is in any case definitely confirmed. According to the results shown in Fig. 5 it is apparent that TBF cannot be unique, but they depend on the two-body forces employed, since each one of the two-body force gives a different discrepancy for the SP and therefore needs a different correction. This is not surprising, since both two-body force and TBF should originate within the same physical framework and therefore they are intimately related. In particular, in the nucleon-meson coupling models the TBF should be generated by processes which involve the coupling constants which are already present in the two-body forces. The problem of consistency between two-body interaction and TBF has been taken systematically in ref. [13] and further developed in more recent works [14]. The TBF calculated along these lines do not contain in principle any additional parameters. It is not surprising then that the corresponding EOS has a SP which is appreciably more shifted away from the empirical one if compared with the EOS with the same two-body force (v_{18}) but with the phenomenological TBF [11]. In any case the TBF can be a good starting point for further improvements. The effect of these TBF in few-body systems is not known, but it is expected to be not very large, since at lower density also the contribution of these TBF in nuclear matter becomes quite small. The most significant difference with the phenomenological TBF is the stiffness of the EOS at high density, which turns out to be much higher, as can be seen also in the case of pure neutron matter. The energy and pressure rise steeply above saturation, and this can create some problems, as discussed later. The TBF suitable for the CD Bonn and the IS1 potential is still an open question.

As already mentioned, one of the deficiency of the considered interactions is the use of the non-relativistic limit. The relativistic framework is of course the framework where the nuclear EOS should be ultimately be based. The best relativistic treatment developed so far is the Dirac-Brueckner (DBHF) approach. Excellent review papers on the method can be found in the literature and in textbook [15]. The main ingredient of the method is the use of the Dirac equation for the single particle states. Despite the T–matrix and the corresponding in–medium G–matrix are actually three dimensional reductions of the original full relativistic Bethe–Salpeter equations for the two body scattering, the introduction of the Dirac spinors has a substantial saturating effect on the calculated nuclear matter EOS. It has been shown [16] that this is due to the "rotation" of the Dirac spinors, with respect to the free ones, produced by the relativistic mean field. This is equivalent to the virtual excitation of a nucleon antinucleon pairs in the intermediate states of the two–body scattering processes ("Z diagram"), which corresponds to a well defined three–body force. Indeed, if this single TBF is added to a non–relativistic BHF calculation, the resulting EOS turns out to be quite similar to the corresponding DBHF, as it has been systematically checked in ref. [11]. Generally speaking, DBHF gives an improved saturation point and a stiff EOS at high density, even if the compressibility does not increase dramatically since the energy per particle has an almost linear increase with density well above saturation.

4. Hints on the EOS from astrophysics and heavy ion collisions

Experimental data and observational phenomenology cannot give detailed indications on nuclear matter EOS but only possible hints on some of its characteristics. We will discuss some of them, without the pretension to be exhaustive.

4.1. *Compressibility above saturation*

In heavy ion collisions at intermediate energy nuclear matter is expected to be compressed and to reach densities few times larger than the saturation value. Several observable quantities have been devised that should be sensitive to the stiffness of the nuclear EOS. In particular the measure of different types of "flow" is considered particularly useful and this line has been followed in many experiments. A more ambitious, and probably more questionable, analysis was performed in ref. [17], where the authors consider both the in–plane transverse flow and the elliptic flow measured in different experiment on $Au + Au$ collision at energies between 0.2 and 10 GeV/A. According to relativistic BUU simulations it is claimed that density up to 7 times saturation density is reached during the collisions (at the highest energy), and from the data an estimate of the pressure is extracted. Together with an evaluation of the uncertainty, the analysis results in the determination of a region in the pressure–density plane where the nuclear EOS should be located. In this way it appears easy to test a given EOS, and it became popular to confront the various microscopic and phenomenological EOS with this region, which is assumed to be the allowed one (essentially for symmetric nuclear matter). If one believes to the validity of this analysis, it turns out that the test is quite stringent, despite the fact that in ref. [17] it is also shown that the value of the compressibility at saturation is not at all well determined. In ref. [18] some microscopic and phenomenological EOS are indeed checked with this method and most of them do not pass the test. In particular it is noticeable that the DBHF

Equation of State is only marginally compatible with the test and fails at the highest density. As shown in ref. [19], the EOS from BHF calculations with v_{18} interaction and phenomenological TBF, previously discussed, is fully compatible with the test. It turns out that the variational EOS of ref. [20] is also compatible with the test.

It has to be noticed that in general the flow values, and other observable quantities, in general do not depend only on the nuclear EOS, as embodied in the single particle potential, but also on the in–medium nucleon–nucleon collision cross section and the effective mass (i.e. momentum dependence of the single particle potential). The extraction of meaningful information from the experiments requires a careful analysis and interpretation of the data. Other quantities which are related to the EOS are the rate of particle productions, in particular K^+ and K^- and their ratio. In fact the strange particle production is intimately related to the density reached during the collision and therefore to the nuclear matter compressibility. However in order to reach meaningful conclusions, it is necessary to have a reasonably good description of the behavior of kaons in the nuclear medium at high density, which is not an easy task theoretically.

On the astrophysical side, as it is well known each EOS for asymmetric matter gives rise to a definite relationship between the mass and radius of neutron stars. This is because ordinary NS are bound by gravity and the solution of the Tolmann–Opennheimer–Volkoff (TOV) equation, based only on general relativity and the adopted EOS, provides the full density profile of the star. Unfortunately it is quite difficult to get from observation both the mass and the radius of a single NS, and up to now the accuracy, especially of the radius value, is not good enough to discriminate among different EOS. The quantity which has been given mostly the attention is then the maximum mass of NS. The mass vs. radius plot has indeed a maximum value at the smaller radius, beyond which the star configuration is unstable towards collapse to a black hole. This maximum is characteristic of each EOS, and the observation of a NS with a mass larger than the predicted one would rule out the corresponding EOS. If one assumes that only nucleonic degrees of freedom are present inside the NS, than one can adopt the EOS discussed above. It turns out that BHF EOS give maximum mass close to two solar masses, while DBHF and variational one a value slightly larger, around 2.2 – 2.3 solar masses. However, as already discussed, the variational EOS of ref. [20] becomes superluminal at high density, and actually the corrected one of ref. [10] gives a maximum mass very close to 2 solar masses. Also the BHF EOS with the TBF calculated in ref. [21]

gives a maximum mass larger than 2, but unfortunately this EOS become also superluminal already at relatively low density. From the astrophysical observations (mainly binary systems) the masses of NS was found, up to few years ago, mostly concentrated around 1.5 solar masses, the most precise one being 1.44 [22]. These values look compatible with the theoretical predictions. However, the situation for the maximum mass is by far much more complex. First of all it is likely that other degrees of freedom, besides the nucleonic ones, can appear inside a NS, in particular hyperonic matter. The BBG scheme has been extended to matter containing hyperons in several papers. If the most recent hyperon–nucleon and hyperon–hyperon interactions are used the maximum mass of NS drops to values below the observational limit, as shown in Fig. 6, taken form ref. [23], where references to previous works can be found. There are two possibilities to overcome this

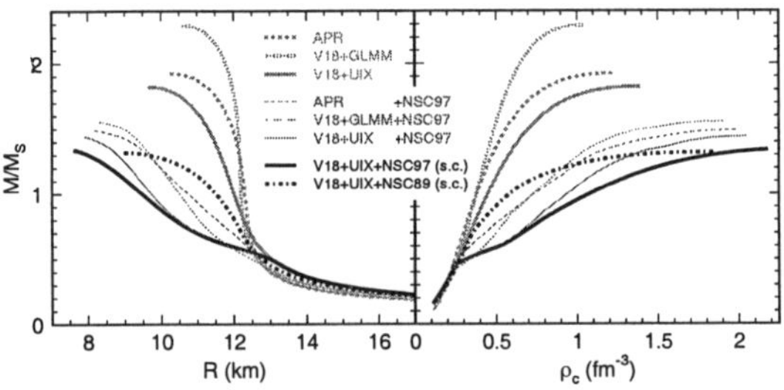

Fig. 6. Neutron star mass as a function of the radius for nuclear matter EOS including hyperons with different forces. The upper curves correspond to masses calculated with only nucleonic degrees of freedom and different TBF models. The left panel reports the same mass values as a function of the central density.

failure in reproducing the observational constraint. The hyperon–nucleon and hyperon–hyperon interactions are poorly known from laboratory experiments, which presently are able to provide only few data points to be fitted. Different interactions, still compatible with phenomenology, could be able to produce a stiffer EOS at high density and consequently larger mass values. Another possibility is the appearance od other degrees of freedom, in particular the transition to quark matter could occur in the core of NS. This possibility has been studied extensively by many authors and indeed it has been found that the onset of the deconfined phase is able to increase the maximum mass to values compatible with the observational limit and ranging from 1.5 to 1.8 solar masses. These results have been obtained within simple quark matter model, like the MIT bag model, the Color Dielectric

Model and the Nambu–Jona Lasinio model, with the possible inclusion of color superconductivity [24]. If perturbative–like corrections to the simple MIT model are introduced, masses up to about 2 can be obtained [25]. In any case, all that shows clearly the great relevance of the nowadays standard observations on NS to our knowledge of the high density nuclear EOS. The astrophysical observations are able to rule out definite EOS or put constraints on them.

As a final remark on this subject one has to mention the recent claims of the observation of NS with mass definitely larger than 2 [26]. If confirmed, these observations would put serious constraints on the nuclear EOS and would point to an additional repulsion which should be present at high density, i.e. a larger stiffness of the quark matter EOS. It has to be stressed that the nuclear EOS appropriate to NS cannot be directly applied to heavy ion collisions. The NS matter is in beta equilibrium and the strange content is determined by chemical equilibrium, which cannot be established during the collision time of heavy ions. In fact the hyperon multiplicity in heavy ion collisions is much smaller than one and no strange matter can be actually formed. Furthermore, the asymmetry of NS matter is much larger than the values reachable in laboratory experiments. Of course a good microscopic theory must be able to connect the two different physical situations within the same many–body scheme, which is one of the main challenge of nuclear physics. Similar considerations apply to the quark onset, where the connection between NS and heavy ion collisions, in the physical regime planned at the future facility of GSI, can be studied with numerical simulations and simple quark matter models [27].

4.2. *Symmetry energy*

In recent years great attention has been given to the symmetry energy of the nuclear EOS, in relation with the planned radioactive beam facilities and the interest on exotic nuclei. Also the structure and composition of NS is affected by the density dependence of the symmetry energy. Furthermore, the structure of NS crust involves highly asymmetric nuclear matter and exotic nuclei. The nuclei which are present near dripping and the "blobs" of nuclear matter in the inner crust are characterized by values of the asymmetry that cannot be reached in laboratory and are a challenge for nuclear structure.

At sub–saturation density the symmetry energy of nuclear matter seems to be under control from the theoretical point of view since the different microscopic calculations agree among each other and the results look only marginally dependent on the adopted nuclear interaction. As an example we report in Fig. 7 the symmetry energy calculated within the BBG scheme, taken from ref. [28], for different NN interaction (TBF has a negligible effect) in comparison with some of the most used phenomenological Skyrme forces. Variational or DBHF calculations give very similar results. The approximate agreement of the phenomenological calculations with the microscopic ones does not hold for other Skyrme forces, as shown e.g. in ref. [29], and a wide spread of values is actually found below saturation. It

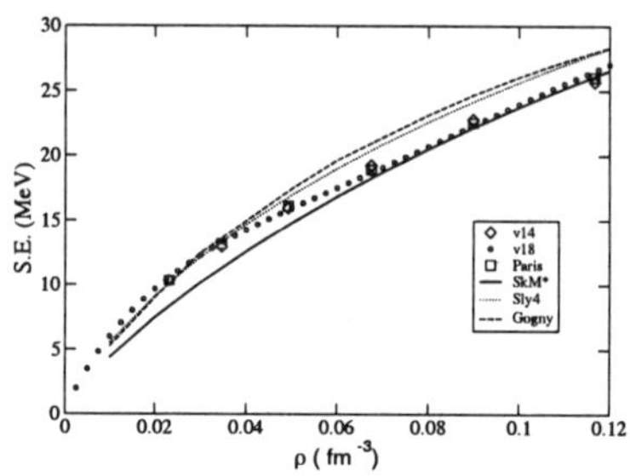

Fig. 7. Symmetry energy of nuclear matter calculated within the BBG scheme and different NN interaction and in comparison with some calculations with Skyrme forces.

appears of great interest to test the behavior of the symmetry energy C_{sym} below saturation density, which, according to the microscopic calculations, can be approximately described by $C_{sym} = 31.3 \cdot (\rho/\rho_0)^{0.6}$ where ρ_0 is the saturation density. Notice that the exponent is close to the one for a free Fermi gas (of course the absolute values are quite different, approximately a factor 2 smaller). Few methods in heavy ion collision experiments have been devised to extract from the data the value of C_{sym} in this density range. In particular it has been found that the experimental isotopic distribution in multifragmentation processes can be described by a so called "isoscaling" low of the type $R_{21} = Y_1(N,Z)/Y_2(N,Z) = C\exp(\alpha N + \beta Z)$; $\alpha = 4C_{sym}/T(Z_1^2/A_1^2 - Z_2^2/A_2^2)$, where the yields Y_1 and Y_2 of the fragment (N,Z) is measured in two different reactions with two different emitting sources (Z_1, A_1) and (Z_2, A_2) respectively. The accuracy of such a phenomenological fit can be seen in Fig. 8, taken from ref. [30], where data from National Laboratory of Legnaro and South National Laboratory (Catania) of INFN were analyzed. In the same work further detail and extensive references on the subject can be found. This form of distribution is suggested by a grand–canonical description of the multifragmentation process. However it can be explained also by simplified model of multifragmentation where no thermal equilibration is assumed [31]. In the latter work it is also shown that the isotopic yield is indeed dependent on the isostiffness of the nuclear EOS and it is suggested that a detailed analysis of the isotopic distribution of a given Z–value could give indications on the density dependence of C_{sym}. Values of the symmetry energy extracted in several experiments, in particular in ref. [30], appear compatible with the theoretical predictions. Furthermore these values decrease with increasing temperature T. This is in line with the intuitive picture where a higher excitation corresponds to a lower density of the fragmenting system. Unfortunately the uncertainty is still too large to really test quantitatively the expected trend, but the method is surely promising. The low density symmetry energy is crucial for

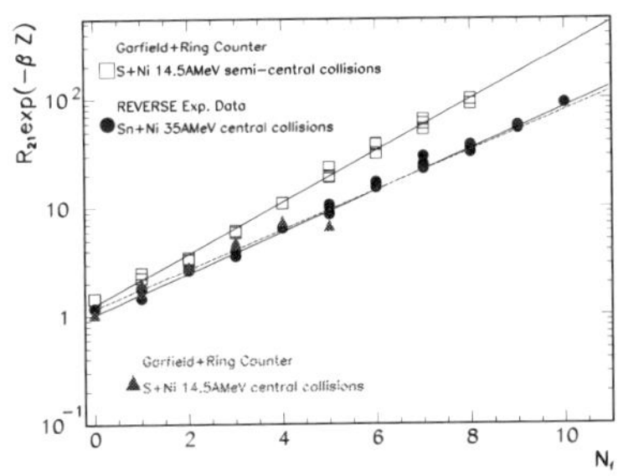

Fig. 8. Isoscaling fit of isotopic distributions.

the structure of NS crust, but the connection with astrophysical observations is quite indirect and much work has still to be done in this field, both at the theoretical level and in the interpretation of data.

The symmetry energy at density above saturation can also be studied with heavy ion reactions in central or semi–central collisions where nuclear matter can be compressed. Particle emissions and productions are among the processes which have been widely used to this purpose. Generally speaking the signal coming from these studies are weak because several competing effects are very often present at the same time and they largely cancel out among each others. In ref. [32] the ratio between K^+ and K^0 rate and between π^- and π^+ has been studied through simulations of $Au + Au$ central collisions at energy in the range 0.8 – 1.8 GeV. These ratios seem to be dependent on the strength of the isovoctor part of the single particle potentials, but the dependence is not so strong due to the compensation between symmetry potential effect and threshold effect. In any case it has to be stressed that the behavior of K mesons, or even π mesons, in nuclear matter is a complex many–body problem, which complicates the interpretation of the experimental data.

To this respect one has to notice that it was suggested in ref. [33] that in NS it could be possible the onset of a kaon condensate. This can happen due to the steep increase inside the star of the electron chemical potential which can finally equal the in–medium mass of K^- mass. Since this condensate produces a substantial softening of the EOS, the NS maximum mass turns out to be limited to about 1.5 solar masses. The possibility of kaon condensation was re–examined recently in ref. [34]. In any case, this value looks in contradiction with the latest observational data and shows once again the great value of the astrophysical studies on NS for our knowledge of dense nuclear matter.

Another process in NS which is sensitive to symmetry energy is cooling. The main mechanism of cooling is the direct Urca (DU) process $n \to p + e^- + \overline{\nu}_e$, $p + e^- \to n + \nu_e$, where neutrinos and antineutrinos escape from the star, cooling the object with a time scale of the order of million years. Since the chemical potentials of neutrons and protons are quite different because of the large asymmetry of the NS matter, the conservation of energy and momenta forbid these reactions when the percentage of protons is below 14 % (when muons are also included). The percentage of protons is directly determined by the symmetry energy, and therefore the density at which the threshold for DU occurs is directly determined by the density dependence of the symmetry energy. At density above saturation this threshold can be different for different EOS. In some case, as for the EOS of Akmal and Pandharipande [20] is practically absent up to almost the central density of NS, even for the largest masses. In this case other processes, like the indirect Urca process, are the dominant cooling mechanism, which are however much less efficient than the DU process. Models of NS which do not include DU process are only marginally compatible with cooling data on NS [35]. If the DU threshold is at too "low" density, the cooling process can be too fast, even with the inclusion of nuclear matter pairing [18], which hinders DU process. This is what occurs for the EOS derived from the DBHF method, whose symmetry energy rises steeply with density. The EOS from BHF, with the inclusion of TBF, gives a threshold density for DU process intermediate between these two cases and seems to be compatible with cooling data. For EOS with a similar behavior of the symmetry energy the scenario of NS cooling involves a slow cooling for low masses and a fast one for the higher masses. Of course a more detailed description of NS cooling requires the many–body treatment of the different processes which can contribute [36]. Finally, the possible onset of quark matter could again change the whole cooling scenario, which has then to be reconsidered, but the above general considerations can be still applied.

4.3. *Finite temperature EOS*

One of the characteristic of symmetric nuclear matter is the occurrence of a liquid gas phase transition. This can be seen both in EOS derived from phenomenological Skyrme forces or relativistic mean field and in microscopic calculations. All the many–body methods mentioned above can be extended to finite temperature, and the phase transition is indeed present. The characteristic Van der Waals shape of a fluid that undergoes a liquid gas phase transition is indeed in all cases apparent. However the critical temperatures for BBG and variational methods are similar, around 18-19 MeV, while for the DBHF the critical temperature is smaller by about a factor 2.

Below the critical temperature the Maxwell construction gives the densities of the liquid and of the gas that coexist, and the set of these points define the coexistent line. However, nucleonic matter is a two component system and for unequal proton and neutron fractions one has to consider also the asymmetry as a new thermodynamical parameter. This gives rise to the distillation phenomenon, i.e. the asymmetry of the gas is different from the one of the liquid which is in equilibrium with it. If the multifragmentation process can be associated with the phase transition, one can try to identify the gas and liquid components and check the presence of the two different asymmetries. In this way one could get some indirect evidence of the transition [37]. Unfortunately the signal is weak, since the asymmetry values of nuclei available in laboratory are not large enough. Maybe with the future exotic beam facilities it will be possible to get a stronger indication.

Evidence of the phase transition come from other quantities extracted from the data, like the heat capacity, which is guessed to be negative if a transition is present. A plot of the temperature of the nuclear system as a function of excitation energy ("caloric curve") has given some indication of a plateau [38], which was considered evidence of the transition. However this is correct only if the transition occurs at constant pressure, which is not expected to happen in heavy ion collisions. In any case, some evidence of a "limiting temperature" has been accumulated [38], or at least of a change of regime at a specific excitation energy. If one assumes that a well defined emitting source is formed during the collision, which is close to equilibrium, one can interpret this excitation energy as the maximum excitation energy this composite nucleus can sustain. The corresponding temperature has been also indicated as limiting temperature. According to this interpretation the limiting temperature is the one at which a nucleus becomes mechanical unstable and should mark the onset of the multifragmentation regime. This temperature can be estimated on the basis of the nuclear EOS and some simplified model like the liquid drop model. In Fig. 9, taken form ref. [39] is reported the limiting temperature for different EOS as a function of the mass number, along the stability valley, in comparison with the values extracted from experiments [38].

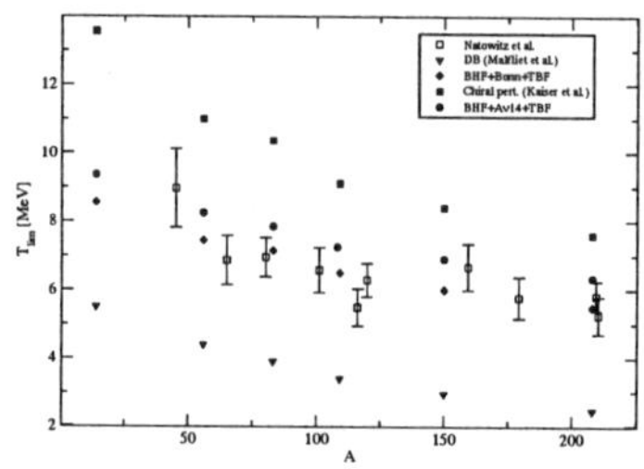

Fig. 9. Limiting temperature as a function of the mass number.

One can see that the comparison with phenomenology offers another stringent test for the microscopic EOS. The DBHF seems to give a too low limiting temperature, in line with the low value of T_c, while the EOS derived form chiral perturbation theory in ref. [40] gives a too high limiting temperature, as expected from the corresponding high value of $T_c = 25$ MeV.

The finite temperature EOS is of great relevance for the physics of proto–neutron stars, which will be discussed in another general talk at the Conference [41].

References

1. *Nuclear Methods and the Nuclear Equation of State*, Edited by M. Baldo, World Scientific, Singapore, International Review of Nuclear Physics Vol. 9, 1999.
2. M. Baldo and C. Maieron, to be published.
3. H.Q. Song *et al.*, Phys. Rev. Lett. **81**, 1584 (1998).
4. M. Baldo *et al.*, Phys. Lett. **B473**, 1 (2000).
5. M. Baldo *et al.*, Phys. Rev. C **65**, 017303 (2001).

6. H. Kümmel, K.H. Lührmann and J.G. Zabolitzky, Phys. Rep. **36**, 1 (1978).
7. B.D. Day, *Brueckner–Bethe Calculations of Nuclear Matter*, Proceedings of the School E. Fermi, Varenna 1981, Course LXXIX, ed. A. Molinari, (Editrice Compositori, Bologna, 1983), p. 1.
8. J. Morales, V.R. Pandharipande, and D.G. Ravenhall, Phys. Rev. C **66**, 054308 (2002).
9. M. Baldo and C. Maieron, Phys. Rev. C **69**, 014301 (2004).
10. H. Heiselberg and M. Horten-Jensen, Astrophys. J. **525**, L45 (1999).
11. Z.H. Li *et al.*, Phys. Rev. C **74**, 047304 (2006).
12. M. Baldo and C. Maieron, Phys. Rev. C **72**, 034005 (2005).
13. P. Grangé *et al.*, Phys. Rev. C **40**, 1040 (1989).
14. W. Zuo *et al.*, Nucl. Phys. **A706**, 418 (2002); Eur. Phys. J. **A14**, 469 (2002).
15. See e.g. R. Brockmann and R. Machleidt, in ref. [1].
16. G.E. Brown *et al.*, Comm. Nucl. Part. Phys. **17**, 39 (1987).
17. P. Daneliewicz, R. Lacey, and W.G. Lynch, Science **298**, 1592 (2002).
18. T. Klähn *et al.*, nucl-th/0609067.
19. Kh. Gad, Nucl. Phys. **A747**, 655 (2005).
20. A. Akmal, V.R. Pandharipande, and D.G. Ravenhall, Phys. Rev. C **58** 1804 (1998).
21. X.R. Zhou *et al.*, Phys. Rev. C **69**, 018801 (2004), and references therein.
22. R.H. Hulse and J.H. Taylor, Astrophys. J. **195**, L51 (1975).
23. H.-J. Schulze *et al.*, Phys. Rev. C **73**, 058801 (2006).
24. A. Drago, A. Lavagno, and G. Pagliara, Nucl. Phys. **B138**, 522 (2005).
25. M. Alford *et al.*, ApJ **629**, 969 (2005).
26. David J. Nice *et al.*, ApJ **634**, 1242 (2005) ; F. Ozel, astro-ph/0605106.
27. M. Di Toro *et al.*, Nucl. Phys. **A775**, 102 (2006).
28. M. Baldo *et al.*, Nucl. Phys. **A736**, 241 (2004).
29. Lie-Wen Chen, Che Ming Ko, and Bao-An Li, nucl-th/0610057.
30. E. Geraci *et al.*, NUCL-EX Collaboration, Proceedings of RNB7 Conference, Cortina d' Ampezzo, July 2006, Eur. Phys. J. **A**, in press.
31. M. Colonna and F. Matera, Phys. Rev. C **71**, 064605 (2005).
32. G. Ferini *et al.*, Phys. Rev. Lett. **97**, 202301 (2006).
33. G.Q. Li, C.H. Lee, and G.E. Brown, Phys. Rev. Lett. **79**, 5214 (1997).
34. A. Li *et al.*, Phys. Rev. C **74**, 055801 (2006).
35. D.G. Yakovlev and C.J. Pethick, Ann. Rev. Astron. Astrophys. **42**, 169 (2004).
36. D. Blaschke, H. Grigorian, and D.N. Voskresensky, Astron. Astrophys. **424**, 979 (2004).
37. S.S. Avancini *et al.*, Phys. Rev. C **74**, 024317 (2006), and reference therein.
38. J.B. Natowitz *et al.*, Phys. Rev. Lett. **89**, 212701 (2002).
39. M. Baldo, L.S. Ferreira, and O. Nicotra, Phys. Rev. C **69**, 034321 (2004).
40. S. Fritsch, N. Kaiser, and W. Weise, Phys. Lett. **B545**, 73 (2002).
41. E. Vigezzi, "Nuclear Astrophysics", Proceedings of this Conference.

QUANTUM MONTE CARLO CALCULATIONS OF SYMMETRIC NUCLEAR MATTER

S. GANDOLFI and F. PEDERIVA

Dipartimento di Fisica and INFN, University of Trento, via Sommarive 14, I-38050 Povo, Trento Italy

S. FANTONI

S.I.S.S.A., International School of Advanced Studies and INFN via Beirut 2/4, 34014 Trieste, Italy

K. E. SCHMIDT

Department of Physics, Arizona State University, Tempe, AZ, USA

We present an accurate numerical study of the equation of state of nuclear matter based on realistic nucleon–nucleon interactions by means of Auxiliary Field Diffusion Monte Carlo (AFDMC) calculations. The AFDMC method samples the spin and isospin degrees of freedom allowing for quantum simulations of large nucleonic systems and represents an important step forward towards a quantitative understanding of problems in nuclear structure and astrophysics.

1. Introduction

The knowledge of the properties of symmetric and asymmetric nuclear matter, is one of the main ingredients of the equations predicting the structure, the dynamics and the evolution of stars, in particular during their last stages, when they become ultra–dense neutron stars. The large variety of available results, partly due to the use of many different approximations, and partly to the limited knowledge of the nucleon–nucleon interaction, lead to different interpretations and hypothesis. Depending on the model used, the density of nuclear matter in the inner shells can reach up to 9 times the core density of stable nuclei, $\rho_0 = 0.16$ fm^{-3} [1], or be limited to 4-5 times ρ_0.

One important step towards the understanding of these astrophysical problems is the study of the symmetric nuclear matter, related to the various model NN interactions available. While considerable advances have been made [2,3], it is still impossible to understand how different results depend on the model interaction used rather than on the particular approximation employed for computing energies.

Therefore, it is not possible to draw conclusions when comparing with astronomical observations. However, the recent success in predicting the properties of light nuclei gives us some confidence that the non–relativistic description of nuclear matter based on effective potentials fitted to reproduce NN data and the binding energy of light nuclei can be reliable enough. The main feature of such nucleon–nucleon interactions, besides the short range repulsion, is the explicit dependence on the relative quantum state of the nucleons which can be described using spin and isospin, angular momentum, spin–orbit, and tensor operators [4].

Properties of light nuclei ($A \leq 6$) can be efficiently computed with high accuracy using modern few–body techniques [5–7] or with the ab initio no-core nuclear shell model (with $A \leq 12$ [8]). Quantum Monte Carlo techniques based on recasting the Schroedinger equation into a diffusion equation (Diffusion Monte Carlo or Green's Function Monte Carlo), allowed for performing calculations up to $A \leq 12$ [9,10]. However, the computational resources needed for such simulations are very large, because of the summation over all the possible states necessary to evaluate the terms of the Hamiltonian with a quadratic dependence on spin and isospin. The number of such terms, and the CPU time needed to calculate them, grows exponentially with the number of nucleons; ^{12}C [9] or 14 neutrons [11] is the limit for the currently available computational resources.

Since the spatial degrees of freedom are already sampled, one would like to replace the sum over the spin–isospin states with an efficient sampling method. The simulation of symmetric nuclear matter requires a minimum of 28 nucleons in a box replicated in space (7 nucleons for each spin–isospin state) to obtain a wave function with closed shells of momenta, and this is out of the reach of the standard Quantum Monte Carlo methods.

Recently [12] we shown that the spin–isospin is efficiently sampled by using the Auxiliary Field Diffusion Monte Carlo method [13], which is based on the use of auxiliary variables to linearize the quadratic spin–isospin operators of the nuclear matter Hamiltonian, making them treatable in a diffusion Monte Carlo scheme. Up to now it has been applied to simulate pure neutron matter (up to 114 neutrons) [14,15], and neutron drops [16,17] interacting with realistic two– plus three–body interactions.

Here we extend calculations to include isospin degrees of freedom and to deal with the strong tensor–isospin force, responsible of the nuclear binding.

The method can readily handle an asymmetry in the number of neutron and protons or the deformation of heavy nuclei.

2. Method

Auxiliary Field Diffusion Monte Carlo (AFDMC) [13] is an extension of the standard Diffusion Monte Carlo method in which the ground state of an Hamiltonian H is obtained by solving the imaginary time dependent Schroedinger equation. The solution is obtained by evolving a population of configurations of the system ("walkers") $X = \{R, \sigma, \tau\}$, where $R = \{\vec{r}_1, \ldots, \vec{r}_N,\}$, $\sigma = \{\vec{\sigma}_1, \ldots, \vec{\sigma}_N\}$, and $\tau = \{\vec{\tau}_1, \ldots, \vec{\tau}_N\}$, with $F(X,t) = \Psi_T(X)\Psi(X,t)$, according to

$$F(X,t) = \int dX' \frac{\Psi_T(X)}{\Psi_T(X')} G_0(X,X',t)F(X',0) \tag{1}$$

The function Ψ_T is a "trial" wave function, usually determined by means of variational calculations, and G_0 is an approximation to the Green's function of the imaginary time Schroedinger equation:

$$G_0(X,X',t) = (4\pi Dt)^{-3A/2} e^{-(R-R')^2/4Dt} e^{-t(V(X)-E_0)}, \tag{2}$$

where $D = \hbar^2/2m$, E_0 is an estimate of the ground state energy of the system, and $V(X)$ is the nucleon–nucleon interaction. For a long enough imaginary time the distribution of the walkers converges to the product $\Psi_T(X)\Psi_0(X)$ where Ψ_0 is the wave function of the ground state of H. This fact allows the computation of matrix elements $\langle \Psi_T | \hat{O} | \Psi_0 \rangle$ of observables $\hat{O}$ of interest in a Monte Carlo way. When $\hat{O} \equiv \hat{H}$ the value obtained is the exact ground state energy of the system.

The presence of an interaction $V(X)$ including operators like $(3\vec{\sigma}_i \cdot \hat{r}_{ij} \vec{\sigma}_j \cdot \hat{r}_{ij} - \vec{\sigma}_i \cdot \vec{\sigma}_j)$ and $\vec{\tau}_i \cdot \vec{\tau}_j$ is the origin of the computational cost in the standard approaches. AFDMC uses the Hubbard–Stratonovich method to transform the operators $\hat{S}$ which are quadratic in the spin and isospin into linear operators:

$$e^{-(1/2)t\lambda \hat{S}^2} = \frac{1}{\sqrt{2\pi}} \int_{-\infty}^{+\infty} dy e^{-y^2/2} e^{y\sqrt{-\lambda t}\hat{S}} \tag{3}$$

Then $\hat{S}$ are operators which are linear combinations of the spin and isospin operators for each nucleon, and λ depend on the interaction. The transformed Green's Function is applied to the spin-isospin part of the wave function, and its effect consists of a rotation of the spin and isospin degrees of freedom (written as four-component spinors in the proton-neutron up-down basis) by a quantity that depends on the auxiliary variable y along with multiplication of the state by an overall factor. The sum over spin and isospin is replaced by sampling a set of rotations of the variables. This procedure reduces the dependence of the computational time on the number of nucleons necessary for performing a simulation step from exponential to cubic. It is therefore possible to perform on a regular workstation or on a modest PC cluster calculations that would require Tflop supercomputers with the standard methods. This method, like other diffusion Monte Carlo methods, suffers from the so–called "sign problem" when it is applied to fermions, and when complex wave functions need to be used. In our calculations we apply the fixed-phase approximation to overcome this problem [18]. While this method has already been successfully applied to pure neutron matter [15], it has not been previously used for mixed proton and neutron systems. It should be noted that it does not guarantee an upperbound to the mixed energy used here. As a test for the correctness and the efficiency of our approach we reproduced within 0.3 MeV total energy the existing results for the binding energy of ^{4}He with potentials of the AV6 class [19]. We have also been able to compute binding energies for ^{16}O and for ^{40}Ca [20] with this method.

3. Interaction

A crucial point in dealing with nuclear matter is the choice of the interaction among nucleons. As already mentioned, several modern two–body potentials are available nowadays, all fitting the NN data with $\chi^2 \sim 1$. We use the potentials of the Argonne class with n operators (AVn) [21]. While the full version contains $n = 18$ operators, most of the physics is reasonably well described by the first 6 operators made up of 4 central spin–isospin dependent components and two tensor ones, which include the long range one–pion–exchange force. The most important missing terms are the spin–orbit components. In nuclei and neutron drops the spin–orbit contribution to the energy amounts to a few tenths of MeV/nucleon. A correction of the order 1MeV/nucleon can be attributed at low densities to the remaining terms included in AV18. Specifically, we have used the interaction Argonne v_8' [19] truncated by dropping the spin–orbit terms, and including only the first six operators, which we denote as AV6'.

It is well known that two–body NN interaction underbind light nuclei, and one needs to add a specific effective three–body potential to reproduce their low energy properties. Semi–phenomenological three–nucleon interactions following the lowest order three–nucleon diagrams with one and two intermediate Delta resonance states provide a very satisfactory description of the ground state energy and the low level spectra of light nuclei up to ^{12}C [9]. We have disregarded such three–body forces in our simulations. In nuclear matter they are essential to reproduce the experimental saturation density, and, in general, they contribute about 10% of the total binding energy. A full comparison with the available experimental data goes beyond the scopes of the present paper. Here we are interested in showing the efficiency of the AFDMC methods in dealing with nuclear matter models which include realistic tensor interactions like in the AV6' potential.

4. Results

The results of the calculations with A=28 include box corrections that have computed by adding to the two body sums contribution of nucleons in the first shell of periodic cells. Such procedure is effective. In order to assess the magnitude of finite size effects we performed calculations with 76 and 108 nucleons at densities $\rho = 0.08$ fm^{-3} and $\rho = 0.48$ fm^{-3}. The results [12] coincide with the ones obtained with 28 nucleons within 3 percent.

We computed the EOS of symmetric nuclear matter in the range of densities $0.5 \leq (\rho/\rho_0) \leq 3$, and compared it with previous available results obtained with the same potential using Fermi Hypernetted Chain in the Single Operator Chain approximation (FHNC/SOC) and the Brueckner-Hartree-Fock (BHF) in the two–hole line approximation [22]. AFDMC calculations were performed with 28 nucleons, filling the shell of plane waves with momentum of modulus 1 and providing a wave function yielding an isotropic density.

The results are summarized in Fig. 1 and reported in Ref. [12]. The comparison of the various EOS suggests the following comments: FHNC/SOC leads to an overbinding at high density. A similar indication was found by Moroni *et al.* [23] after a DMC calculation of the EOS of normal liquid ^{3}He at zero temperature, with a guiding function including triplet and backflow correlations. The comparison with the equivalent FHNC/SOC calculations of Refs. [24,25] have shown similar discrepancies. There are two main intrinsic approximations in variational FHNC/SOC calculations, which violate the variational principle. The first one consists in neglecting a whole class of cluster diagrams, the so called *elementary diagrams*, which cannot be summed up by means of FHNC integral equations. We have calculated the lowest order diagram of this class, namely the one having only one correlation bond and four exchange bonds. The results obtained show a substantial effect from this diagram and bring the FHNC/SOC estimates very close to AFDMC results, as shown in Fig. 1. The second approximation

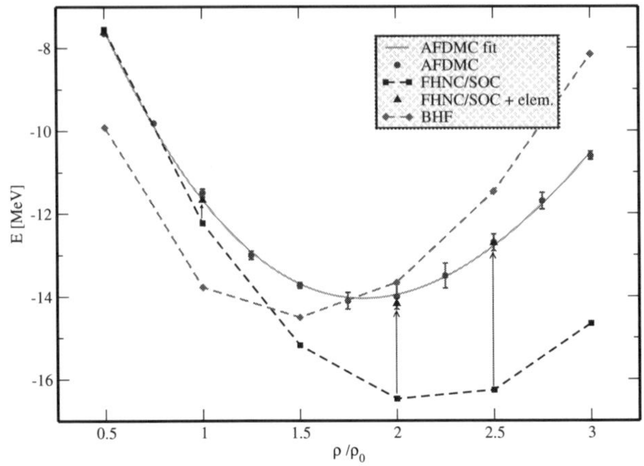

Fig. 1. (color online). Equation of state of symmetric nuclear matter calculated with different methods. Red circles represent AFDMC results with statistical error bars and the green line is the fitted functional form described in the text. Dashed lines correspond to calculations performed with other methods [22] (blue line with squares: FHNC/SOC; magenta with diamonds: BHF). Blue triangles represent the FHNC/SOC energies corrected by including the low order of elementary diagrams as described in the text. Blue arrows show the corresponding energy shift, which increases at higher densities.

is related to the non-commutativity of the correlation operators entering the variational wave function. The only class of cluster diagrams contributing to such non-commuting terms, which can be realistically calculated, is that characterized by single operator chains. It is believed that such an approximation is reliable in nuclear matter, but there is no clear proof of this.

BHF calculations of Ref. [22] predict an EOS with a shallower binding than the AFDMC one. It has been shown for symmetric nuclear matter, using the AV18 and AV14 potentials, that contributions from three hole–line diagrams add a repulsive contribution up to $\sim$ 3MeV at densities below ρ_0 [26], and decrease the energy at high densities [27]. Such corrections, if computed with the AV6' potential, would probably preserve the same general behavior, and bring the BHF EOS closer to the AFDMC one. Therefore, our calculations show that the two hole–line approximation used in Ref. [22] is too poor, particularly at high density.

The AFDMC equation of state was fitted with the following functional form:

$$\frac{E}{A} = \frac{E_0}{A} + \alpha(x - \bar{x})^2 + \beta(x - \bar{x})^3, \tag{4}$$

where $x = \rho/\rho_0$ and the various coefficients are given by E_0/A = -14.04(4) MeV, α = 3.09(6) MeV, β = -0.44(8) MeV, and $\bar{x}$ = 1.83(1). The resulting compressibility $K = 9\bar{x}^2 \left(\partial^2 (E/A) / \partial x^2\right)_{\bar{x}}$ at saturation density $\bar{x}$ is $\sim$ 190 MeV. The fit of the EOS allows for computing the pressure vs. density for symmetric nuclear matter.

5. Conclusions

The availability of an efficient and relatively fast projection algorithm for the computation of energies and other observables of dense hadronic matter enables the possibility of a more quantitative understanding of the properties of neutron stars and supernovae, as well as that of medium–heavy nuclei. Computations on such systems are at present out of reach of the standard GFMC methods and available supercomputers. Therefore, the extension of AFDMC algorithm to deal with nuclear matter is a significant step forward. Some technical improvements on the calculations presented here, such as the addition to the AV6' of spin–orbit terms and three–body interactions are already underway. The treatment of asymmetric nuclear matter, particularly important for the determination of the properties of neutron stars, is also straightforward, and will be the subject of future exploration.

We acknowledge helpful conversations with M.H. Kalos, P. Faccioli, W. Leidemann, E. Lipparini, and G. Orlandini. This work was in part supported by NSF grant PHY-0456609. Calculations were performed on the HPC facility "BEN" at ECT* in Trento under a grant for Supercomputing Projects.

References

1. J. Piekarewicz, Phys. Rev. C **69**, 041301 (2004).
2. J. Morales, Jr., V. R. Pandharipande, and D. G. Ravenhall, Phys. Rev. C **66**, 054308 (2002).
3. A. Akmal, V. R. Pandharipande, and D. G. Ravenhall, Phys. Rev. C **58**, 1804 (1998).
4. V. R. Pandharipande and R. B. Wiringa, Rev. Mod. Phys. **51**, 821 (1979).
5. H. Kamada, *et al.*, Phys. Rev. C **64**, 044001 (2001).
6. D. Gazit, S. Bacca, N. Barnea, W. Leidemann, and G. Orlandini, Phys. Rev. Lett. **96**, 112301 (2006).
7. J. Carlson and R. Schiavilla, Rev. Mod. Phys. **70**, 743 (1998).
8. P. Navrátil, J. P. Vary, and B. R. Barrett, Phys. Rev. Lett. **84**, 5728 (2000).
9. S. C. Pieper, Nucl. Phys. **A751**, 516 (2005).
10. S. C. Pieper, K. Varga, and R. B. Wiringa, Phys. Rev. C **66**, 044310 (2002).

11. J. Carlson, J. Morales, Jr., V. R. Pandharipande, and D. G. Ravenhall, Phys. Rev. C **68**, 025802 (2003).
12. S. Gandolfi, F. Pederiva, S. Fantoni, and K. E. Schmidt, nucl-th/0607022.
13. K. E. Schmidt and S. Fantoni, Phys. Lett. **B446**, 99 (1999).
14. S. Fantoni, A. Sarsa, and K. E. Schmidt, Phys. Rev. Lett. **87**, 181101 (2001).
15. A. Sarsa, S. Fantoni, K. E. Schmidt, and F. Pederiva, Phys. Rev. C **68**, 024308 (2003).
16. S. Gandolfi, F. Pederiva, S. Fantoni, and K. E. Schmidt, Phys. Rev. C **73**, 044304 (2006).
17. F. Pederiva, A. Sarsa, K. E. Schmidt, and S. Fantoni, Nucl. Phys. **A742**, 255 (2004).
18. S. Zhang and H. Krakauer, Phys. Rev. Lett. **90**, 136401 (2003).
19. R. B. Wiringa and S. C. Pieper, Phys. Rev. Lett. **89**, 182501 (2002).
20. S. Gandolfi, F. Pederiva, S. Fantoni, and K. E. Schmidt, in preparation.
21. R. B. Wiringa, V. G. J. Stoks, and R. Schiavilla, Phys. Rev. C **51**, 38 (1995).
22. I. Bombaci, A. Fabrocini, A. Polls, and I. Vidaña, Phys. Lett. **B609**, 232 (2005).
23. S. Moroni, S. Fantoni, and G. Senatore, Phys. Rev. B **52**, 13547 (1995).
24. E. Manousakis, S. Fantoni, V. R. Pandharipande, and Q. N. Usmani, Phys. Rev. B **28**, 3770 (1983).
25. M. Viviani, E. Buendia, S. Fantoni, and S. Rosati, Phys. Rev. B **38**, 4523 (1988).
26. H. Q. Song, M. Baldo, G. Giansiracusa, and U. Lombardo, Phys. Rev. Lett. **81**, 1584 (1998).
27. M. Baldo, A. Fiasconaro, H. Q. Song, G. Giansiracusa, and U. Lombardo, Phys. Rev. C **65**, 017303 (2001).

Nuclear Structure

NUCLEAR STRUCTURE

F. ANDREOZZI

Dipartimento di Scienze Fisiche, Università di Napoli Federico II, and Istituto Nazionale di Fisica Nucleare, Complesso Universitario di Monte S. Angelo, Via Cintia - I-80126 Napoli, Italy
E-mail: andreozzi@na.infn.it

The research activity performed during the last two years by the Italian physicists in the field of theoretical nuclear structure is reviewed. Though covering the whole range of nuclear structure, the different approaches appear to complement each other forming a coherent research project.

1. Introduction

Starting from its 7th edition, the Conference on Problems in Theoretical Nuclear Physics, held in Cortona every other year, includes in its program overview talks on the different branches of research in theoretical nuclear physics performed in Italy. The present report, the fifth of a series [1–4], is just meant to outline the studies performed in the last two years on nuclear structure.

The report covers necessarily a large variety of approaches and models corresponding to the many degrees of freedom which come into play in the nuclear medium. The more so, now that the spectacular performances of the new accelerators and detectors have created a new frontier of research represented by the nuclei far from the β-stability line and have refined and enriched enormously our knowledge in the realm of the more traditional nuclear spectroscopy.

This rich and growing harvest of new experimental data represents a challenge for the theory community. It is, in fact, necessary to check if these new data fit naturally in the existing well established models, including shell model, at most with few adjustments, or call for completely new approaches.

On the other hand, the equally dramatic advances in computer power have provided the theoretical community with powerful tools for fully exploiting and extending the existing models and have stimulated the search for new powerful methods of investigation. Both routes have been pursued by the members of the Italian nuclear physics community.

A significant part of the research activities reviewed in the following sections has been made in the framework of the PRIN (Research Project of National Interest) *Teoria della Struttura dei Nuclei e della Materia Nucleare*, supported by the Italian Ministero dell'Istruzione, dell'Universitá e della Ricerca (MIUR). For this reason, rather than pointing out the achievement of the single authors, I will show that, beyond their specificity and originality, the different approaches complement with each other merging into a unitary project which contributes considerably to clarify the different new facets of nuclear structure. This unitary view will emerge from following a route which starts from the two-nucleon system and interaction and goes through the many-body complexity of light and heavy nuclei.

2. Realistic Nucleon-Nucleon potentials in nuclear structure

A renewed interest toward ab initio calculations in nuclei starting from the free nucleon-nucleon (NN) potential has been stimulated by the proliferation, during the last years, of NN potentials, all capable of reproducing with high precision $\left(\chi^2/datum \approx 1\right)$ the NN scattering data and the deuteron properties [5,6]. These potentials, though equally accurate in reproducing NN data, are very different from each other. As an example, the Argonne V18 potential is non-relativistic and local in all its spin, isospin and angular momentum components. On the other hand, the CD-Bonn is a one-boson exchange non local potential, based on relativistic meson field theory [7]. The application of the concepts of the effective field theory to the low-energy QCD is at the origin of so called chiral potentials. These arise from the expansion of the most general chiral Lagrangian, consistent with QCD [8], into an infinite series of terms with increasing number of derivatives and/or nucleon fields. Weinberg has shown [9] that, once applied to NN scattering, the series can be ordered in powers of $(Q/\Lambda_\chi)^n$, $(n \geq 0)$, where Q denotes a momentum or pion mass and $\Lambda_\chi \approx 1GeV$ is a cutoff. For a given n the number of terms is finite, so that the scattering amplitude is calculable to any desired accuracy. Following these ideas, Epelbaum *et al.* [10] have constructed the first next-to-next-to-leading order $(NNLO, n = 3)$ potential in momentum space. More recently, Entem and Machleidt [11] have generated the N^3LO $(n = 4)$ potential, which reproduces the NN low-energy scattering data and the deuteron properties with an accuracy comparable to the other high precision potentials. These chiral potentials not only have a sounder theoretical basis but also offer the advantage of generating at once two- and many-body forces, without a hoc prescriptions.

The availability of so many NN potentials, of comparable quality in reproducing the phase shifts, renders compelling ab initio nuclear structure calculations. These are, in fact, a unique testing ground for the off-shell behavior of each potential. On the other hand, they represent a reliable test only if a high accuracy is reached, a task far from trivial.

The first obstacle to be faced is represented by the strong repulsive core contained in each realistic NN potential. This long standing problem has eventually found two alternative and, in many ways, complementary solutions, one within perturbation theory, the other of variational nature.

2.1. *Perturbative approach:* $V_{\mathrm{low-k}}$

In the perturbative approach the nuclear eigenvalue problem for the bare Hamiltonian within the infinte dimensional Hilbert space is turned into an equivalent one formulated within a finite space where the original Hamiltonian is replaced by an effective one. This is given by an infinite series in the bare NN potential V_{NN}, composed of connected diagrams only, i.e. the linked-cluster expansion. Because of its strongly repulsive core, this series is badly divergent. The solution to this puzzle is due to Brueckner and, in a more formal approach, to Bethe and Goldstone. They have shown that the diverging series in the bare V_{NN} can be replaced by a smooth one in the so-called Brueckner reaction matrix G. This is obtained by summing an infinite sequence of ladder diagrams originating from each nucleon-nucleon vertex in V_{NN} and representing all possible multiple scattering processes of a pair of nucleons in the nuclear medium. The smooth series in G has been shown by Bloch-Horowitz and Brandow to represent the shell model effective interaction and can be evaluated by the well known Q-box plus folded diagrams technique [12].

The linked cluster expansion theory has many virtues. It provides the theoretical foundations of shell model and an elegant and flexible tool for solving the eigenvalue problem in nuclear matter and actual nuclei. Its actual implementation, however, especially for ab initio calculations in finite nuclei, offers a series of technical difficulties. One concerns, for instance, the actual derivation of the G matrix for finite nuclei. Another one has to do with the generation of the nuclear mean field potential which requires the Brueckner-Hartree-Fock method, much more involved than Hartree-Fock. For these reasons, an extensive consistent use of Brueckner theory in nuclei is not conceivable.

A shorter route for overcoming the problems connected with the strong short range repulsion of the bare NN interaction has been proposed recently [13]. The basic idea is to define a cut-off Λ for the relative momentum of the two-nucleon system and, then, partition the full momentum space in a low- ($k \leq \Lambda$) and a high-momentum ($k > \Lambda$) subspaces. Under the request of complete decoupling between the two sub-spaces, it is possible through a similarity transformation [14] to derive from the bare V_{NN} smooth effective Hamiltonians and operators for each subspace, which exactly reproduce two-body observables for $k \leq \Lambda$ and $k > \Lambda$, respectively. It has therefore been proposed [13] to keep only the effective potential obtained for the low-energy subspace by the ansatz

$$\begin{aligned} V(k', k) &= V_{low-k}(k', k) \quad && k', k \leq \Lambda \\ &= 0 && k', k > \Lambda \end{aligned} \tag{1}$$

This approximate potential is required to reproduce the deuteron binding energy, low-energy phase shifts, and low-momentum half-on-shell T matrix, namely

$$T(p', p, p^2) = T_{\mathrm{low-k}}(p', p, p^2) \quad (p, p' \leq \Lambda), \tag{2}$$

where

$$T(k', k, k^2) = V_{NN}(k', k) + \mathcal{P} \int_0^\infty q^2 dq V_{NN}(k', q) \frac{1}{k^2 - q^2} T(q, k, k^2) \tag{3}$$

and

$$\begin{aligned} &T^\Lambda_{\mathrm{low-k}}(p', p, p^2) \\ &= V_{\mathrm{low-k}}(p', p) + \mathcal{P} \int_0^\Lambda q^2 dq V_{\mathrm{low-k}}(p', q) \frac{1}{p^2 - q^2} T_{\mathrm{low-k}}(q, p, p^2). \end{aligned} \tag{4}$$

It has also been shown in Ref. [14] that the effective potential $V_{\mathrm{low-k}}$ can be evaluated by means of non-perturbative techniques dictated by the similaty transformation. The effective potential so computed incorporates high-momentum renormalizations to all orders in V_{NN}. It follows that all $V_{\mathrm{low-k}}$ potentials, even if deriving from strongly different bare V_{NN}, are very similar to each other, at least in their diagonal matrix elements. Because of its smooth behavior, $V_{\mathrm{low-k}}$ can be used directly to derive a self-consistent Hartree-Fock potential, thereby avoiding the complexities and ambiguities of Brueckner-Hartree-Fock. This has been done recently [15] . Various phase-shift equivalent NN potentials have been employed to construct the $V_{\mathrm{low-k}}$ in ^{4}He, ^{16}O, and ^{40}Ca. The $V_{\mathrm{low-k}}$ is then used to generate a Hartree-Fock mean field. Corrections up to third-order in the Goldstone expansion are included. The binding energies and rms charge radii have been computed. The results appear to be only slightly dependent on the different bare potentials, thereby suggesting that also off-shell properties of these potentials are similar. The connection between the cutoff momentum Λ and the size of the harmonic oscillator space used in the calculation has also been investigated, suggesting the validity of the proposed criterion for choosing the two-nucleon harmonic oscillator state space in correspondence of a given cutoff momentum Λ.

The smoothness of $V_{\text{low-k}}$ has been exploited not only for performing HF calculations in the ground-state, but also in shell-model at least for light p-shell nuclei [16]. A $V_{\text{low-k}}$ potential defined up to $\Lambda = 2.1 fm$ is derived starting from the CD-Bonn potential and then used to construct a self-consistent HF basis. This basis is then employed to derive both the single-particle energies and the residual two-body interaction for the shell-model Hamiltonian. Binding energies, spectra and electromagnetic properties of light p-shell nuclei, from ^{6}Li to ^{8}B are calculated and compared to the experimental ones, obtaining a quite satisfactory agreement.

Obviously, $V_{\text{low-k}}$ can also be used, in the same way as the Brueckner G matrix, to construct the shell-model effective interaction through the Q-box plus folded diagrams technique [12]. This has been done both for the study [17] of nuclei in proximity of the unstable double-closed shell nucleus ^{132}Sn and for investigating the spectroscopy [18] of complex nuclei like ^{120}Sn . Of particular relevance is the very good agreement obtained for the neutron-proton multiplets $\pi g_{7/2}\ \nu f_{7/2}$ and $\pi d_{5/2}\ \nu f_5/2$ in ^{134}Sb. This, together with the results obtained for the ^{134}Sn and ^{135}Sb nuclei, allows to conclude that there is no need to invoke modifications in the shell structure in order to explain the avalaible data in nuclei near the neutron drip line.

2.2. *The Correlated Basis Functions approach*

In its present formulation, the $V_{\text{low-k}}$ approach, while effective in low-energy spectroscopy, is not adequate at high energy when short range correlations are to be treated explicitly. This can be done in Brueckner theory by the use of effective operators which fully account for the short range correlations. In practice, this can be a very hard task. A technique, specifically suited to the study of ground state short range correlations, is given by the the Correlated Basis Functions (CBF) theory, developed by *Pisa* and *Lecce* groups.

The CBF is a variational approach whose aim is to determine the minimum of the energy functional

$$E\left[\Psi\right] = \frac{\langle\Psi|H|\Psi\rangle}{\langle\Psi|\Psi\rangle} \tag{5}$$

in the space of Jastrow-like trial wave functions of the form

$$\Psi(1,2,\cdots,A) = S\left[\prod_{i<j} F_{ij}\right] \quad \Phi(1,2,\cdots,A), \tag{6}$$

where a symmetrized product of two-body correlation operators

$$F_{ij} = \sum_{p=1}^{8} f^p(r_{ij}) O^p_{ij} \ , \tag{7}$$

$$O^p_{ij} = [1, \sigma_\mathbf{i} \cdot \sigma_\mathbf{j}, \mathbf{S_{ij}}, (\mathbf{L}\cdot\mathbf{S})_\mathbf{ij}] \otimes [1, \tau_\mathbf{i}\cdot\tau_\mathbf{j}] \tag{8}$$

acts on a Slater determinant of single-particle wave functions. The evaluation of the multidimension integrals entering into the energy functional is done by performing a cluster expansion of the expectation values of the kinetic energy operator and of the two- and three-body potentials appearing in the Hamiltonian H and, then, applying the Fermi Hypernetted Chain [19] (FHCN) resummation techniques. This idea, originally developed for infinite systems, has been applied [20] to finite nuclei at various levels of sophistication. In the most recent contribution, [21] the formalism has been extended, within the Single Operator Chain approximation, to treat nuclei with different proton and neutron wave functions in the jj coupling scheme. In particular, it has been used to calculate ground-state energies and density distributions of various nuclei, from ^{12}C to ^{208}Pb. The results, also presented in a contribution to this volume [22] , show that this kind of microscopic variational calculations for finite nuclei have reached an accuracy comparable to nuclear matter calculations.

3. Collective excitations

3.1. *Random Phase Approximation*

A typical aspect of nuclear physics is the occurrence of collective excitations. These can be treated either phenomenologically, through geometric or algebraic approaches, or microscopically. The simplest and most popular microscopic approach to collective motion in nuclei is the Random Phase Approximation (RPA). In this framework a set of phonon operators $Q^\dagger_\nu$, expressed as a linear superposition of one particle - one hole creation and annihilation operators, is introduced. Their action on the ground state $|0\rangle$, defined as the vacuum of the Q_ν operators, creates the collective states $|\nu\rangle$. The excitation energies are solutions of a system of equations which can be derived by using the equation-of-motion method. Crucial to the derivation of the RPA equations is the Quasi-Boson Approximation (QBA), i.e. the expectation value in the ground state $|0\rangle$ of any operators is replaced with the corresponding value in the uncorrelated reference state. Strictly connected to the QBA is the RPA prediction of an harmonic spectrum with regularly spaced multiphonon states.

In spite of its elegance and effectiveness in treating nuclear collective excitations, one has to go beyond the RPA harmonic approximation to interpret the recently discovered properties of collective excitations.

The RPA, however, offers a sufficiently simple, if well grounded, framework to elucidate aspects of nuclear structure still lacking of a systematic exploration. An example are the collective isovector giant resonances seen in charge-exchange nuclear rections like (p, n) or $\left({}^3He, t\right)$. Significant exceptions are, aside from the giant dipole resonance, the Isobaric Analog Resonances (IAR), $(\Delta J = \Delta L = \Delta S = 0)$ and the Gamow-Teller Resonances (GTR) $(\Delta L = 0 \ \Delta J = \Delta S = 1)$. A microscopic model, aimed at the description of these excitations along isotopic chains, consisting of the Quasi-particle RPA made on top of a Hartree-Fock-BCS calculation has been constructed by the *Milano* group [23]. The interaction used is a Skyrme force in the particle-hole channel and a zero-range density dependent pairing force in the particle-particle channel. Special attention is paid to the relation between the full self-consistency of the calculation and the isospin invariance. The model is tested by means of a numerical calculation of the energies of the IAR along the chain of Sn isotopes, obtaining good agreement with the experimental values in the stable isotopes. RPA wave functions generated by this model have been employed [24] in a Distorted Wave Born Approximation plus RPA analysis of the experimental cross section of the $\left(t, {}^3He\right)$ reaction on ^{48}Ca and ^{58}Ni at 43 MeV/nucleon.

A similar conceptual scheme (self-consistent HF+RPA) has been adopted by other people of the *Milano* group [25] to study giant multipole resonances in a number of exotic spherical nuclei (^{78}Ni, ^{100}Sn and ^{132}Sn). They use the D1S parametrization of the Gogny force, with results similar to the ones obtained for the stable doubly magic nuclei, the largest differences occurring in ^{78}Ni, presumably due to the large neutron-proton asymmetry of this nucleus. Still in the framework of the harmonic approximation are the discussion of sum rules of the multiple giant resonance states of Ref. [26] and the analysis of the medium polarization effects on nuclear binding energies [27] . In the first paper the energy weighted sum rule for the triple giant dipole resonance states is related to the single and double ones, a result which can be useful in the analysis of anharmonicities of collective motions. In the second one the starting point is a mean field calculation which makes use of a Skyrme-type interaction, whith a phenomenological Wigner correction term to take into account neutron-proton correlations and an added zero-range pairing interaction, which is treated in the BCS approximation. The parameters of the Skyrme interaction are fixed by a fit to a set of 121 spherical even-even nuclei. The main point is a systematic analysis of the effect of correlations on the binding energies, in particular the contribution associated with surface vibrations, treated in RPA (QRPA for superfluid nuclei). These corrections are used to refit the parameters of the interaction; in this way a new set of parameters of the Skyrme force is obtained, corresponding to a reduction of about 5% in the rms deviation of the computed from the experimental binding energies.

The $^1S_0$ pairing and the associated superfluidity is an important property of both infinite nuclear matter and finite nuclei, whose description, although discussed for many years, still remains matter of debate. Calculation of the pairing gap Δ in ^{120}Sn nucleus [28] , using the bare NN interaction Argonne v_{14} , accounts for only half of the empirical value, obtained from the odd-even mass differences. It has been shown that in the case of spherical nuclei the latter can be reproduced only if one takes into account the renormalization of the NN interaction due to the dynamical polarization of the nuclear medium, originating from the exchange of low-lying collective surface vibrations among nucleons moving in time-reversed states close to the Fermi surface. Recent papers of the *Milano* group [29–31], have evaluated various contributions to these renormalization effects and extended the calculation to the chain of *Sm* isotopes, ranging from the spherical ^{146}Sm to the well deformed ^{154}Sm, and to a number of well deformed *Gd* , *Dy* and *Er* isotopes, confirming also in these cases the appreciable (about 40 %) contribution of the induced interaction to the pairing gaps.

A QRPA approach based on a cranked Nilsson model has been adopted to study fast rotating deformed nuclei [32] as a function of rotational frequency ω. The most significant result concerns the enhancement of the orbital over the spin M1 strength at high rotational frequencies. Such an orbital motion corresponds to the scissors mode [33,34]. Indeed, the path of the M1 strength with the angular frequency parallels the one followed by the nuclear moment of inertia. Like the moment of inertia, the M1 strength increases with ω before the backbending because of the increasing axial deformation. Above the backbending critical value, when the nucleus undergoes a transition from a superfluid to an almost rigid phase, the M1 strength jumps to a plateau due to the sudden increase of the moment of inertia. This is a very important result since it may represent a further signature of the superluid to normal phase transition in deformed nuclei.

Scissors mode and moment of inertia have been shown to be closely correlated also in nuclei evolving at high rotational frequency toward a superdeformed shape [35]. It is remarkable that also the onset of deformation enhances dramatically the scissors mode.

3.2. *Beyond the mean field*

Multiple phonon excitations in nuclei appear now firmly established on experimental grounds. Double and even triple phonon excitations have been observed both in the low-energy regime [36] and in the giant resonance domain [37]. A description of double-phonon excitations as an uncorrelated sequential two-step process, although accounting qualitatively for the measured cross sections, shows discrepancies due to phonon-phonon interactions that cannot be ignored.

Evidence of multiphonon proton-neutron ($p-n$) states, i.e. states of mixed symmetry in the p-n exchange, in nearly spherical nuclei, first experimentally established [38] in ^{94}Mo, has been subsequently confirmed in several nuclei [39,40]. These experimental findings strongly call for a theoretical description going beyond the harmonic approximation.

3.2.1. *Direct extensions of the RPA*

A natural extension of the RPA for the study of two-phonon states is the so-called Second Random-Phase Approximation (SRPA), which enlarges the space of the elementary modes used to describe the collective states by adding two particle - two hole (2p-2h) excitations to the one particle - one hole (1p-1h) of the RPA. In deriving the SRPA equations, however, the quasi-boson approximation is still used, an approximation even more severe than that in the RPA. In a paper [41] of the *Catania* group, an extension of the RPA (and SRPA) is introduced, which improves on the QBA by considering a reference state containing two particle - two hole correlations. The quality of the corresponding results is studied by performing, within a three-level Lipkin model, calculations of energies, occupation numbers, strength functions and sum rules. From the comparison of these results with the exact ones, one can conclude that this approach, while not requiring an heavy computational effort, gives a good description of the ground state and of those excited states which, in the harmonic approximation, would correspond to one- and two-phonon states.

Again in a similar spirit is the model of Ref. [42] used to describe dipole states in the stable isotopes ^{208}Pb and ^{120}Sn as well as in the neutron rich unstable ^{132}Sn. The starting point is a mean field calculation performed in the Hartree-Fock-BCS plus Quasi-particle RPA (HF plus RPA for closed shell nuclei) approach, using consistently a Skyrme force in the particle-hole channel and a zero-range density dependent pairing interaction in the particle-particle one. The calculation then proceeds by considering the coupling of the 1p-1h states obtained from the solution of QRPA (or RPA) equations to states of more complex structure (2p-2h or 4 quasi-particle states), described as p-h pairs plus the low-energy collective states of multipolarity $1^-, 2+, 3-, 4+$ resulting from the RPA or QRPA calculations. In ^{208}Pb and ^{120}Sn agreement is obtained for the peak energy and width of the GDR, while in ^{132}Sn a large fragmentation of the GDR and the absence of collective states in the low-energy region is predicted.

3.2.2. *Boson expansion techniques*

A more radical approach to treat anharmonicities is given by the boson expansion methods. In these approaches bifermion creation operators $b_i^\dagger$ (pp or ph) are expressed as an infinite sum of boson operators $B_i^\dagger$,e. g.

$$b_\mu^\dagger = \sum_{ph} c_{ph}^\mu a_p^\dagger a_h \;\Rightarrow\; \sum_i x_i B_i^\dagger + \sum_{ijk} x_{ijk} B_i^\dagger B_j^\dagger B_k^\dagger + \dots \tag{9}$$
$$\left[B_i, B_j^\dagger\right] = \delta_{ij}.$$

In the operator mapping scheme [43] one requires that exact fermion commutation relations be satisfied, which fixes the coefficients x. Alternatively the Marumori [44] approach is followed, based on the correspondence between fermion and boson states and on the equality between corresponding matrix elements.

$$\mid n \rangle = b_\mu^\dagger b_\nu^\dagger b_\tau^\dagger \cdots \mid 0 \rangle \;\;\Rightarrow\;\; \mid n \,) = B_i^\dagger B_j^\dagger B_k^\dagger \cdots \mid 0 \,) \tag{10}$$
$$\langle\, n \mid O_F \mid n \,\rangle \;=\; (\, n \mid O_B \mid n \,)$$

An analysis of an exactly solvable many-level pairing model in a boson formalism has been provided by the *Catania* group [45]. They consider a system of particles distributed over a set of doubly degenerate uniformly spaced single-particle levels and interacting through a pairing interaction of constant strength. This simple, but highly non trivial model, whose exact solutions have been found by Richardson [46], has been used not only to simulate superfluid correlations in deformed nuclei but also in condensed matter physics to describe the superconducting properties of ultrasmall metallic grains. Because of its sensitivity to any violation of the Pauli principle it has been widely used for testing approximate methods. By making use of a Marumori-type mapping procedure, an hermitian image of the fermion Hamiltonian is constructed, including in the boson expansion up to four boson operators. First of all the authors investigate the properties of the boson Hamiltonian, in particular the role of the four-boson term in connection with the Pauli principle. Further, this boson formalism is applied to search for possible extensions of the RPA. Standard RPA equations have been recovered in correspondence with the first-order boson truncation, while the higher-level approximations considered are in correspondence with a truncation at second order of the boson expansion. The comparison between exact and approximate calculations shows that these new approximations considerably improve the standard RPA results, rendering of great interest an extension of these methods to more realistic cases.

3.2.3. *Algebraic models*

A widely adopted phenomenological approach is the Interacting Boson Model (IBM). In its simpler formulation the properties of low-lying nuclear levels are described in terms of monopole (s) and quadrupole (d) bosons, which can be viewed as a bosonization of correlated pairs of valence nucleons coupled to $J = 0^+$ and $J = 2^+$, respectively. The IBM Hamiltonian can be written in terms of bilinear products of the boson creation and annihilation operators, which are the 36 generators of the $U(6)$ group. By a proper choice of the parameters the eigenvalues of the Hamiltonian can be classified according to specific subgroup chains of $U(6)$. Clarification of the relations between these classifications and the nuclear shapes is still matter of interest. Contributions along these lines, also stimulated by the introduction of the concept of critical point symmetry [47] , are in Ref [48]. In these papers all the known solutions of the Bohr Hamiltonian are rewieved and new analytic solutions for a soft triaxial rotor with a minimum in the potential along the γ direction are obtained. An extension of these results to the case of a generic soft triaxial nucleus, using a number of different model potential is presented in Ref. [49], where an application to the osmium isotopes and a comparison with experimental data is also made. More directly connected to the problem of the critical point symmetries are the papers of the *Padova* group voted to the study of the phase transition around the critical point in the evolution from spherical to deformed γ-unstable shapes in odd nuclei in the framework of the Interacting Boson-Fermion Model [50] . Considering the particular case of an odd $j = 3/2$ particle coupled to an even-even boson core that undergoes a transition from spherical to γ-unstable, the energy spectrum and the electromagnetic transitions, in correspondence of the critical point, display behaviours qualitatively similar to those characterizing the phase transition in the even core. The results, in qualitative agreement with those obtained by Iachello [51] in the framework of the supersymmetric $E(5/4)$ model, allow not only the prediction of a new excited band but also to suggest the use of one-particle spectroscopic factors between neighbour even and odd nuclei as a signature of the shape phase transition.

When distinction is made between proton and neutron bosons one gets the so called IBM2. The existence of mixed symmetry $p - n$ states has been one of the major predictions of the IBM2. In its framework one can define symmetric (in the n-p exchange) and antisymmetric boson creation operators

$$Q_S^\dagger = (Q_p^\dagger + Q_n^\dagger) \quad , \quad Q_A^\dagger = (Q_p^\dagger - Q_n^\dagger), \tag{11}$$

and, as a consequence, symmetric $(Q_S^\dagger)^n|0\rangle$ and mixed symmetry n-p states $Q_A^\dagger \; (Q_S^\dagger)^{n-1}|0\rangle$. Precise signatures of this kind of states [52] exist. States with the same symmetry, differing by one boson, are connected through strong E2 transitions while mixed symmetry states undergo strong M1 decays on symmetric states with the same number of bosons. A microscopic interpretation of these and other anharmonic effects is presented in the following paragraph.

3.2.4. *The Quasi-particle Phonon Model*

The Quasi-particle Phonon Model [53] (QPM) constitutes one of the most accurate microscopic frameworks allowing the inclusion of a large number of configurations which have to be left out in any shell model description of complex nuclei.

In the Quasi-particle Phonon Model one adopts a Hamiltonian

$$H = H_0 + V^P + V^M \ , \tag{12}$$

where H_0 is an unperturbed one-body piece, V^P and V^M are the two body terms, of general separable structure, acting in the particle-particle and particle-hole channels, respectively. The Hamiltonian is first written in terms of quasi-particle creation and annihilation operators by means of a Bogoliubov canonical transformation, and then used to generate QRPA phonons. The Hamiltonian is expressed in terms of the QRPA phonons and finally diagonalized in a properly chosen multiphonon basis. This scheme has been adopted by the $Naples - Sofia$ collaboration [54] to study collectivity and neutron-proton symmetry in ^{94}Mo and, more recently, in ^{92}Zr. The calculation, performed in a very large configuration space and including up to three-phonon states in the final diagonalization, accounts for most of the ^{92}Zr data. Moreover, it confirms that F-spin, basically conserved in ^{94}Mo, is broken in ^{92}Zr to a much less extent than previous preliminary investigations suggested.

Similarly, the QPM came out to be essential in describing the collective modes in deformed nuclei. A recent example is provided by the large abundance of 0^+ excitations detected at low-energy in several rare earth and actinide nuclei through high-precision two-nucleon transfer reactions The first experiment was carried out on ^{158}Gd, where 13 excited 0^+ levels were identified up to 3.1 MeV [55]. The number of detected 0^+ is particularly large in ^{168}Er. A high resolution experiment [56] has detected about 22 0^+ and 63 2^+ states up to 4 MeV. These findings have stimulated considerable interest and several theoretical approaches. Extended versions [57] of the IBM, including also p ($J = 1^-$) and f ($J = 3^-$) bosons, can account for a good fraction of the observed 0^+ states as long as several levels be associated to collective octupole two-phonon states. The most satisfactory and complete description of these states has been obtained within the QPM. This has been done in a series of papers [58] of the $Napoli - Dubna$ collaboration where an extensive study of low-spin states in ^{158}Gd and ^{168}Er has been performed. Energies, $E2$ and $E0$ transition strengths as well as two-nucleon transfer spectroscopic amplitudes have been computed and investigated. It comes out that while in ^{158}Gd an appreciable fraction of 0^+ states has a dominant two octupole phonon component, in agreement with the IBM2 results, their role is only marginal in the 0^+ states of ^{168}Er and other nuclei. The QPM predicts well the number and distribution in energy of both 0^+ and 2^+ levels in ^{168}Er and reproduces satisfactorily the reaction strengths for both kinds of states.

4. A fully microscopic treatment of collective excitations

Although adopted with success for studying collective states of complex nuclei, the QPM has some limitations. It applies in fact only to separable Hamiltonians and does not provide precise prescriptions for keeping under control the uncertainties due to the space truncation. The correlations in the ground state and the antisymmetrization of the multiphonon states are accounted for in the quasi boson approximation.

In principle a configuration interaction calculation is the most accurate approach to reproduce the energy spectrum of a nuclear system and evaluate its anharmonicities. A clear example is given by the microscopic analysis of the ground and excited states of the Na_8 metal cluster in the framework of the jellium model made by the *Catania* group [59]. They have performed a series of interaction configuration calculations in a Hartree-Fock basis and constructed the eigenstates of the Hamiltonian in a basis taking into account up to $4p-4h$ excitations. The analysis of the transition dipole strengths allows to single out the states which can be interpreted as the collective dipole plasmon and its double excitations and evaluate their degree of anharmonicity.

Shell model techniques are however very demanding as regards the numerical work and the above analysis cannot be easily extended to higher mass metal clusters. To give an answer to these contrasting requirements, an new equation of motion method has been proposed by the *Napoli* group [60]. In this approach, one solves exactly the nuclear many-body problem in a space which is the direct sum of n-phonon subspaces, with n ranging from 0 to an arbitrary value N . Starting from a general two-body Hamiltonian, one solves the following eigenproblem in each n-phonon subspace

$$\langle n;\beta_n \left| \left[H, a_p^\dagger a_h \right] \right| n-1;\alpha_{n-1} \rangle = \left(E_{\beta_n} - E_{\alpha_{n-1}} \right) X^{(ph)}_{\alpha_{n-1}\beta_n}, \tag{13}$$

where $a_p^\dagger$ and a_h create a particle and a hole, respectively, E_{β_n} are the unknown eigenvalues, with corresponding eigenvectors $|n;\beta_n\rangle$, of the Hamiltonian H in the n-phonon subspace and $E_{\alpha_{n-1}}$ and $|n-1;\beta_{n-1}\rangle$ are eigenvalues and eigenvectors in the $(n-1)$-phonon subspace. Once the commutator is unfolded one gets a set of equations in the unknown amplitudes

$$X^{(ph)}_{\alpha_{n-1}\beta_n} \equiv \langle n;\beta_n \left| a_p^\dagger a_h \right| n-1;\alpha_{n-1} \rangle \tag{14}$$

which establish recurrence relations between subspaces with different number of phonons. They can be solved iteratively, thereby generating the full multiphonon basis. The states so generated, being composed of products of microscopic *ph* phonons, are not antisymmetrized and therefore form an overcomplete set. One can however eliminate exactly and completely the redundant states in each n-phonon subspace by adopting a procedure, based on the Choleski decomposition of the overlap matrix of the basis, which allows to extract the correct number of linearly independent states out of a redundant set. Furthermore this procedure leaves unchanged also the phonon structure of the basis states and allows therefore to eliminate naturally and maximally the spurious admixtures induced by the center of mass motion. Once represented in the multphonon basis so constructed, the Hamiltonian matrix has a rather simple structure and can be easily, if necessary, reduced in size and diagonalized. A preliminary test on ^{16}O to explore the potential of the method is discussed in a contribution to these proceedings [61].

5. Concluding remarks

Though necessarily concise, the present review offers a quite encouraging and exciting picture of the research being pursued by the Italian community in the realm of nuclear structure, performed to a large extent under the Italian project *Teoria della struttura dei nuclei e della materia nucleare*, supported by the Italian Ministero dell'Istruzione, dell'Universitá e della Ricerca (MIUR).

Practically all subjects of current interest for the international community are touched. Often original solutions to long standing, still unsolved, problems have been proposed. Existing methods have been refined in order to account for the large variety of qualitatively new experimental data made available by experiments exploiting the new generation of detectors and accelerators. New algorithms and methods have been developed for facing problems of high complexity now at reach thanks to the spectacular advances in computer power. We can safely state that the Italian community can play a quite active role within the international community as testified by the strong interaction with research groups from all over the world.

References

1. N. Lo Iudice, in *Perspectives in Theoretical Nuclear Physics*, proceedings of the VII Convegno su Problemi di Fisica Nucleare Teorica, Cortona, 1998, edited by A. Fabbrocini, G. Pisent and S. Rosati (Edizioni ETS, Pisa, 1999), p. 125.
2. P. F. Bortignon in *Theoretical Nuclear Physics in Italy*, proceedings of the 8th Conference on Problems in Theoretical Nuclear Physics, Cortona, 2000, edited

by G. Pisent, S. Boffi, L. Canton, A. Covello, A. Fabbrocini, and S. Rosati (World Scientifio, Singapore, 2001), p. 3.
3. A. Gargano in *Theoretical Nuclear Physics in Italy*, proceedings of the 9th Conference on Problems in Theoretical Nuclear Physics, Cortona, 2002, edited by S. Boffi, A. Covello, M. Di Toro, A. Fabbrocini, G. Pisent and S. Rosati (World Scientifio, Singapore, 2003), p. 23.
4. G. Coló in *Theoretical Nuclear Physics in Italy*, proceedings of the 9th Conference on Problems in Theoretical Nuclear Physics, Cortona, 2004, edited by S. Boffi, A. Covello, M. Di Toro, A. Fabbrocini, G. Pisent and S. Rosati (World Scientifio, Singapore, 2005), p. 1.
5. V. G. J. Stoks *et al.* , Phys. Rev. C **49**,2950 (1994) ; R. B. Wiringa, V. G. J. Stoks, and R. Schiavilla, Phys. Rev. C **51**, 38 (1995).
6. R. Machleidt, Phys. Rev. C **63**, 024001 (2001).
7. R. Machleidt *et al.* Phys. Rep. **149**, 1 (1987); R. Machleidt, Adv. Nucl. Phys. **19**, 189 (1989).
8. S. Weinberg, Phys. Lett. **B251**, 288 (1990).
9. S. Weinberg, Nucl. Phys. **B363**, 3 (1991).
10. E. Epelbaum *et al.*, Nucl. Phys. **A671**, 295 (2000).
11. D. R. Entem and R. Machleidt, Phys. Rev. C **68**, 041001(R) (2003).
12. T. T. S. Kuo and E. Osnes,*Lecture Notes in Physics*, Vol. 364,(Springer-Verlag, Berlin, 1990).
13. S. Bogner, T.T.S. Kuo, and L. Coraggio, Nucl. Phys. **A684**, 432c (2001); S. Bogner, T.T.S. Kuo, L. Coraggio, A. Covello, and N. Itaco, Phys. Rev. C **65**, 051301 (2002).
14. F. Andreozzi, Phys.Rev. C **54**, 684 (1996).
15. L. Coraggio, A. Covello, A. Gargano, N. Itaco, T. T. S. Kuo, and R. Machleidt, Phys. Rev. C **71**, 014307 (2005); L. Coraggio, A. Covello, A. Gargano, N. Itaco, and T. T. S. Kuo, Phys. Rev. C **73**,014304 (2006).
16. L. Coraggio and N. Itaco, Phys. Lett. **B616**, 43 (2005).
17. L. Coraggio, A. Covello, A. Gargano, and N. Itaco, Phys. Rev. C **72**, 057302 (2005); L. Coraggio, A. Covello, A. Gargano, and N. Itaco, Phys. Rev. C **73**, 031302(R) (2006).
18. P. Guazzoni, L. Zetta, B. F.Bayman, A. Covello, A. Gargano, G Graw, R. Hertenberger, H. -F. Wirth, and M. Jaskola, Phys. Rev. C **72**, 044604 (2005).
19. G. Có, A. Fabrocini, S. Fantoni, and I. E. Lagaris, Nucl. Phys. **A549**, 439 (1992); F. Arias de Saavedra, G. Có, A. Fabrocini, and S. Fantoni, Nucl. Phys. **A605**, 359 (1996).
20. A. Fabrocini, F. Arias de Saavedra, G. Có, and P. Folgarait, Phys. Rev. C **57**, 1668 (1998); A. Fabrocini, F. Arias de Saavedra, and G. Có, Phys. Rev. C **61**, 044302 (2000).
21. C. Bisconti, F. Arias de Saavedra, G. Có, and A. Fabrocini, Phys. Rev. C **73**, 054304 (2006).
22. C. Bisconti, *these proceedings.*
23. S. Fracasso and G. Coló, Phys. Rev. C **72**, 064310 (2005).
24. J. Guillot, S. Galés, D. Beaumel, S. Fortier, E. Rich, N. Van Giai, G. Coló, A. M. van den Berg, S. Brandenburg, B. Davids, M. N. Harakeh, M. Hunyadi,

M. de Huu, S. Y. van der Werf, H. J. Wörtche, C. Bäumer, D. Frekers, E. -W. Grewe, P. Haefner, B. C. Junk, and M. Fujiwara, Phys. Rev. C **73**, 014616 (2006).
25. S. Péru, J. F. Berger, and P. F. Bortignon, Eur. Phys. J. A **28**,25 (2005).
26. H. Kurasawa, T. Suzuki, and P. F. Bortignon, Prog. Theor. Phys. **114**, 805 (2005).
27. S. Baroni, F. Barranco, P. F. Bortignon, R. A. Broglia, G. Coló, and E. Vigezzi, arXiv:nucl-th/0604018 (2006).
28. F. Barranco, R. A. Broglia, G. Coló, G. Gori, E. Vigezzi, and P. F. Bortignon, Eur. Phys. J. A **21**, 57 (2004).
29. P. Donati, G. Gori, F. Barranco, R. A. Broglia, and E. Vigezzi, J. Phys. G: Nucl. Part. Phys. **31**, 295 (2005).
30. G. Gori, F. Ramponi,F. Barranco, P. F. Bortignon, R. A. Broglia, G. Coló, and E. Vigezzi, Phys. Rev. C **72**, 011302 (2005).
31. F. Barranco, P. F. Bortignon, R. A. Broglia, G. Coló, P. Schuck, E. Vigezzi, and X. Viñas, Phys. Rev. C **72**, 054314 (2005).
32. J. Kvasil, N. Lo Iudice, R. G. Nazmitdinov, A. Porrino, and F. Knapp, Phys. Rev. C **69**, 064308 (2004).
33. N. Lo Iudice and F. Palumbo, Phys. Rev. Lett. **41**, 1532 (1978).
34. D. Bohle, A. Richter, W. Steffen, A. E. L. Dieperink, N. Lo Iudice, F. Palumbo, and O. Scholten, Phys. Lett. **137B**, 27 (1984).
35. J. Kvasil, N. Lo Iudice, F. Andreozzi, F. Knapp, and A. Porrino, Phys. Rev. C **73**, 034302 (2006).
36. M. Kneissl, H.H. Pitz, and A. Zilges, Prog. Part. Nucl. Phys. **37**, 439 (1996); M. Kneissl, N. Pietralla, and A. Zilges, J.Phys. G **32**, R217 (2006).
37. N. Frascaria, Nucl. Phys. **A482** , 245c (1988); T. Auman, P.F. Bortignon, and H. Hemling, Ann. Rev. Nucl. Part. Sc. **48**, 351 (1998)).
38. N. Pietralla, C. Fransen, D. Belic, P. von Brentano, C. Friessner, U. Kneissl, A. Linnemann, A. Nord, H. H. Pitz, T. Otsuka, I. Schneider, and I. Wiedenhover, Phys. Rev. Lett. **83**, 1303 (1999).
39. C. Fransen, V. Werner, D. Bandyopadhyay, N. Boukharouba, S. R. Lesher, M. T. McEllistrem, J. Jolie, N. Pietralla, P. von Brentano, and S. W. Yates, Phys. Rev. C **71**, 054304 (2005).
40. M. Kneissl, N. Pietralla, and A. Zilges, J. Phys. G: Nucl. Part. Phys. **32**, R217 (2006).
41. D. Gambacurta, M. Grasso, F. Catara, and M. Sambataro, Phys. Rev. C **73**, 024319 (2006).
42. D. Sarchi, P.F. Bortignon, and G. Coló, Phys. Lett. **B601**, 27 (2004).
43. S. T. Belyaev and V. G. Zelevinsky, Nucl. Phys. **39**, 582 (1962).
44. T. Marumori, M. Yamamura, and A. Tokunaga, Prog. Theor. Phys. **31**, 1009 (1964); T. Marumori, M. Yamamura, A. Tokunaga, and K. Takada, Prog. Theor. Phys. **32**, 726 (1964).
45. D. Gambacurta, M. Sambataro, and F. Catara, Phys. Rev. C **73**, 014310 (2006).
46. R. W. Richardson, J. Math. Phys. **18**, 1802 (1977).
47. F. Iachello, Phys. Rev. Lett. **85**, 3580 (2000); F. Iachello, Phys. Rev. Lett.

87, 052502 (2001); F. Iachello, Phys. Rev. Lett. **91**, 132502 (2003).
48. L. Fortunato, Eur. Phys. J. A **26**, s01, 1 (2005) ; S. De Baerdemacher, L. Fortunato, V. Hellemans, and K. Heide, Nucl. Phys. **A769**, 16 (2006).
49. L. Fortunato, S. De Baeremacker, and K. Heide, Phys. Rev. C **74**, 014310 (2006).
50. C. E. Alonso, J. M. Arias, L. Fortunato, and A. Vitturi, Phys. Rev. C **72**, 061302(R) (2005); C. E. Alonso, J. M. Arias, and A. Vitturi, Phys. Rev. C **74**, 027301 (2006).
51. F. Iachello, Phys. Rev. Lett. **95**, 052503 (2005).
52. A. Arima, T. Otsuka, F. Iachello, and I. Talmi, Phys. Lett. **B66**, 205 (1977).
53. V. G. Soloviev, *Theory of Atomic Nuclei: Quasiparticles and Phonons*, (Institute of Physics, Bristol, 1992).
54. N. Lo Iudice and Ch. Stoyanov, Phys. Rev. C **69**, 044312 (2004); N. Lo Iudice and Ch. Stoyanov, Phys. Rev. C **73**, 037305 (2006).
55. S. R. Lesher *et al.*, Phys. Rev. C **66**, 051305(R) (2002).
56. D. Bucurescu, G: Graw, R. Hertenberger, H. -F. Wirth,, N. Lo Iudice, A. V. Sushkov, N. Yu. Shirikova, Y. Sun, T. Fastermann, R. Krücken, M. Mahgoub, J. Jolie, P. von Brentano, N. Braun, S. Heinze, O. Möller, D. Mücher, C. Scholl, R. F. Casten, and D. A. Meyer, Phys. Rev. C **73**, 064309 (2006).
57. N. V. Zamfir, J. -Y. Zhang, and R. F. Casten, Phys. Rev. C **66**, 057303 (2002).
58. N. Lo Iudice, A. V. Sushkov, and N. Yu. Shirikova, Phys. Rev. C **70**, 064316 (2004); N. Lo Iudice, A. V. Sushkov, and N. Yu. Shirikova, Phys. Rev. C **72**, 034303 (2005).
59. F. Catara, D. Gambacurta, M. Grasso, and M. Sambataro, Phys. Lett. **A349**, 345 (2006).
60. F. Andreozzi, F. Knapp, N. Lo Iudice, A. Porrino, and J. Kvasil, *submitted to Phys. Rev. C for publication* .
61. F. Andreozzi, N. Lo Iudice, A. Porrino, F. Knapp, and J. Kvasil, *these proceedings.*

AN EQUATIONS OF MOTION METHOD FOR THE EXACT SOLUTION OF THE NUCLEAR EIGENVALUE PROBLEM IN A MULTIPHONON SPACE

F. ANDREOZZI, N. LO IUDICE, and A. PORRINO

Dipartimento di Scienze Fisiche, Universitá di Napoli "Federico II" and Istituto Nazionale di Fisica Nucleare, Monte S. Angelo, Via Cintia I-80126 Napoli, Italy
Andreozzi@na.infn.it,loiudice@na.infn.it,porrino@na.infn.it

F. KNAPP and J. KVASIL

Institute for Particle and Nuclear Physics MFF, Charles University, V Holesovickach 2, 180 00 Praha 8, Czech Republic
kvasil@ipnp.troja.mff.cuni.cz,knapp@ipnp.troja.mff.cuni.cz

A set of equations of motion are derived within a subspace spanned by states which are tensor products of n Tamm Dancoff phonons and solved iteratively starting from the particle-hole vacuum to generate a basis of microscopic multiphonon states. Such a basis is then used to diagonalize the full Hamiltonian. An numerical implementation of the method on ^{16}O is presented for illustrative purposes.

Keywords: Nuclear many-body; Eigenvalue problem; New methods.

1. Introduction

Recent experiments have considerably enriched the information on multiphonon collective modes in nuclei and their associated anharmonicity. Two- and three phonon multiplets have been observed. [1,2]. These include the ones described as proton-neutron (F-spin) mixed symmetry states. [3] At high energy, the double giant dipole resonance has been observed in a number of different reactions. [4,5]

While for a systematic of the low energy modes one may usefully resort to the interacting-boson model, (IBM) [6], for detailed and specific investigations we must adopt microscopic approaches, which extend the random-phase-approximation (RPA). Indeed, the double-giant dipole resonance was investigate using explicitly the particle-hole ($1p-1h$) and $2p-2h$ configurations [7] or adopting a Boson-Fermion mapping technique. [8] Widely adopted are the nuclear field theory, [9,10] especially suitable for characterizing the anharmonicities of the vibrational spectra and the spreading widths of the giant resonances, and the quasiparticle-phonon model (QPM), [11] extensively used for low and high energy multiphonon excitations, like the mixed-symmetry states [12] and the double giant dipole resonance. [13] Other microscopic methods have tried to go beyond the quasi boson approximation but with more limited success. [14,15]

We report here on an equation of motion method [16] which yields a basis of multiphonon states built out of phonons generated in the Tamm-Dancoff approximation (TDA) [17] and determines exact eigenvalues and eigenvectors of a Hamiltonian of general form.

2. Brief outline of the method

We intend to generate a set of n-phonon states $\mid n;\alpha\rangle$ such that the Hamiltonian H be diagonal within each separate n-phonon subspace. Under this request, we easily derive

$$< n;\beta|\,[H, b^{\dagger}_{ph}]\,|n-1;\alpha >= (E^{(n)}_{\beta} - E^{(n-1)}_{\alpha}) < n;\beta|b^{\dagger}_{ph}|n-1;\alpha >, \quad (1)$$

where $b^{\dagger}_{ph} = a^{\dagger}_{p} a_h$ is a bilinear form in the operators $a^{\dagger}_{p}$ and a_h which create respectively a particle (p) and a hole (h) with respect to the unperturbed ground state ($0p - 0h$ vacuum).

We then write the Hamiltonian in second quantized form and expand the commutator $[H, a^{\dagger}_{p} a_h]$ on the left-hand side of the equation. After a linearization procedure, we obtain for the n-phonon subspace the eigenvalue equation

$$\sum_{\gamma p' h'} A^{(n)}_{\alpha\gamma}(ph\ ;\ p'h')\ X^{(n)}_{\gamma\beta}(p'h') = E^{(n)}_{\beta}\ X^{(n)}_{\alpha\beta}(ph), \quad (2)$$

where

$$X^{(n)}_{\alpha\beta}(ph)\ \equiv < n;\beta|b^{\dagger}_{ph}|n-1;\alpha > \quad (3)$$

are the vector amplitudes and

$$\begin{aligned} A^{(n)}_{\alpha\gamma}(ph;p'h') &= \delta_{hh'}\delta_{pp'}\delta^{(n)}_{\alpha\gamma}\left[(\epsilon_p - \epsilon_h) + E^{(n-1)}_{\alpha}\right] + \\ &\sum_{h_1 p_1} V_{p'h_1 h' p_1}\left[\delta_{pp_1}\rho^{(n-1)}_{\alpha\gamma}(h_1 h) - \delta_{hh_1}\rho^{(n-1)}_{\alpha\gamma}(pp_1)\right] + \\ &\frac{1}{2}\left[\ \delta_{hh'}\sum_{p_1 p_2} V_{p' p_1 p p_2}\rho^{(n-1)}_{\alpha\gamma}(p_1 p_2) - \delta_{pp'}\sum_{h_1 h_2} V_{h h_1 h' h_2}\rho^{(n-1)}_{\alpha\gamma}(h_2 h_3)\right]. \quad (4) \end{aligned}$$

The symbols ϵ_p (ϵ_h) are single particle (hole) energies, V_{ijkl} the matrix elements of the two-body potential, and

$$\rho^{(n)}_{\alpha\gamma}(kl) =< n;\, \gamma | a^\dagger_k a_l | n;\, \alpha > \tag{5}$$

defines the density matrix with $a^\dagger_k a_l$ written in normal order with respect to the ph vacuum. This is a crucial quantity and is seen to weight the particle-hole, particle-particle and hole-hole interaction. For $n = 1$, Eqs. (2) and (4) yield the standard TDA equations. Our method is, therefore, nothing but the extension of TDA to multiphonon spaces.

The states $b^\dagger_{ph}|n-1;\, \alpha >$, being not completely antisymmetrized, are linearly dependent and, therefore, form an overcomplete set. We have to solve an eigenvalue problem of general form. To this purpose, we make the expansion

$$|n;\, \beta >= \sum_{\alpha\, ph} C^{(n)}_{\alpha\beta}(ph)\, b^\dagger_{ph}\ |n-1;\, \alpha > . \tag{6}$$

Upon insertion in Eqs. (2) and (3) , we get

$$X = \mathcal{D}C \tag{7}$$

$$A\mathcal{D}C = E\mathcal{D}C, \tag{8}$$

where $\mathcal{D}$ is the overlap or metric matrix

$$d^{(n-1)}_{\alpha\beta}(ph;p'h') = \langle n-1;\beta \mid b_{p'h'} b^\dagger_{ph} \mid n-1;\alpha\rangle. \tag{9}$$

Eq. (8) defines an eigenvalue equation of general form. This is ill-defined, however, since the determinant of $\mathcal{D}$ vanishes. The traditional prescription for curing such a disease is based on the straightforward diagonalization of $\mathcal{D}$ [18]. Moreover, one has to face the highly non trivial task of computing the metric matrix, for which complex procedures have been envisaged. [14,15]

We avoided the diagonalization of $\mathcal{D}$ by developing a method based on the Choleski decomposition [16], which extracts in a fast and efficient way a basis of linear independent states. As for the metric matrix $\mathcal{D}$, our method yields the simple formula

$$d^{(n-1)}_{\alpha\beta}(ph;p'h') = \sum_\gamma \left[\delta_{pp'}\delta_{\gamma\beta} - \rho^{(n-1)}_{\gamma\beta}(pp')\right] \rho^{(n-1)}_{\alpha\gamma}(hh'), \tag{10}$$

where the matrix densities are computed by using simple recursive relations. [16]

The generalized eigenvalue problem (8) can therefore be solved exactly and the basis for the n-phonon subspace generated. We then face the $(n+1)$-phonon subspace. We just need to evaluate the amplitudes $X^{(n)}_{\alpha\beta}(ph)$ and the density matrix $\rho^{(n)}_{\alpha\beta}(kl)$ within the $n-$phonon subspace.

The iterative process is clearly outlined. Starting with the lowest trivial 0-phonon subspace, the $0p-0h$ vacuum, we solve the equations of motions step by step up to a convenient n-phonon subspace obtaining a multiphonon basis which decomposes the Hamiltonian as follows

$$\begin{aligned} H = & \sum_{n\alpha} E_{\alpha}^{(n)} \mid n;\alpha\rangle\langle n;\alpha \mid \\ & + \sum_{n\alpha n'\beta} \mid n';\beta\rangle\langle n';\beta \mid H \mid n;\alpha\rangle\langle n;\alpha \mid . \end{aligned} \tag{11}$$

Here the off diagonal terms are given by recursive formulas and, therefore, easily computed.

The eigenvectors of this Hamiltonian have the phonon structure

$$\mid \Psi_\nu\rangle = \sum_{n,\alpha_n} C_{\alpha_n}^{(\nu)} \mid n;\alpha_n\rangle \tag{12}$$

which allows to compute easily the transition amplitudes.

3. A numerical illustrative application of the method for ^{16}O

For illustrative purposes, we apply the method to ^{16}O whose low-energy positive parity spectrum was studied in a shell model calculation which included up to $4p-4h$ and $4\hbar\omega$ configurations [19,20] and, more recently, up to $6\hbar\omega$. [21]

Our space is more restricted, including all ph configurations up to $3\hbar\omega$. This limits our phonon number at n. We used a modified harmonic oscillator one-body Nilsson Hamiltonian [22] plus a bare G-matrix deduced from the A-Bonn potential. [23]

To eliminate the spurious admixture induced by the center of mass motion we have adopted the method of Palumbo [24], applied to standard shell model by Glockner and Lawson [25]. This consists in adding an Harmonic Oscillator Hamiltonian in the center of mass coordinates multiplied by a large enough coupling constant so as to push the center of mass excited levels well above the intrinsic excitations.

The ground state is dominated by the unperturbed $0p-0h$ vacuum (80%). The 2-phonon components account for about 20%. The results obtained within a no-core and a symplectic shell model calculations are not very different [21]: About 60% for the $0p-0h$, 20% for $2p-2h$ and 20% for the other more complex configurations, excluded from our restricted space.

To investigate further the effects of the multiphonon configurations, we have studied the isoscalar and isovector dipole giant resonances. To this

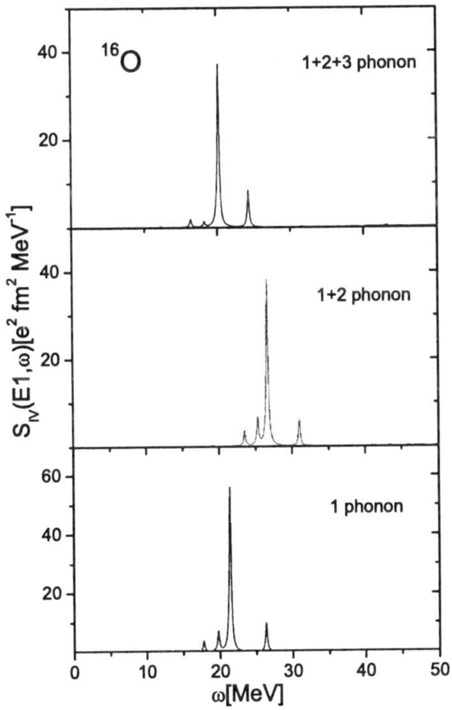

Fig. 1. Isovector $E1$ strength distributions in ^{16}O.

purpose, we have computed the strength function

$$\mathcal{S}(\lambda,\omega)=\sum_{\nu}B_{\nu}(\lambda)\,\delta(\omega-\omega_{\nu})\approx\sum_{\nu}B_{\nu}(\lambda)\,\rho_{\Delta}(\omega-\omega_{\nu}),\tag{13}$$

where ω is the energy variable and ω_{ν} the energy of the transition of multipolarity λ from the ground to the ν_{th} excited state $\Psi_{\lambda}^{(\nu)}$ of spin $J=\lambda$ (Eq. (12)). The δ function is replaced by a Lorentzian of width Δ as weight of the reduced strength $B^{(\nu)}(\lambda)$. For the isovector ($E1$) transitions, we adopt the multipole operator

$$\mathcal{M}_{IV}(E\lambda\mu)=\frac{e}{2}\sum_{i=1}^{A}\tau_{3}(i)r_{i}^{\lambda}Y_{\lambda\mu}(\hat{r_{i}}),\tag{14}$$

For the isoscalar one (squeezed dipole mode) we have instead

$$\mathcal{M}_{IS}(\lambda=1\mu)=\frac{1}{2}\sum_{i=1}^{A}r_{i}^{3}Y_{1\mu}(\hat{r_{i}}).\tag{15}$$

It is important to notice the absence of the corrective term generally included in order to eliminate the spurious contribution due to the center of mass excitation. Such a term is not necessary in our approach which guarantees a complete separation of the center of mass from the intrinsic motion.

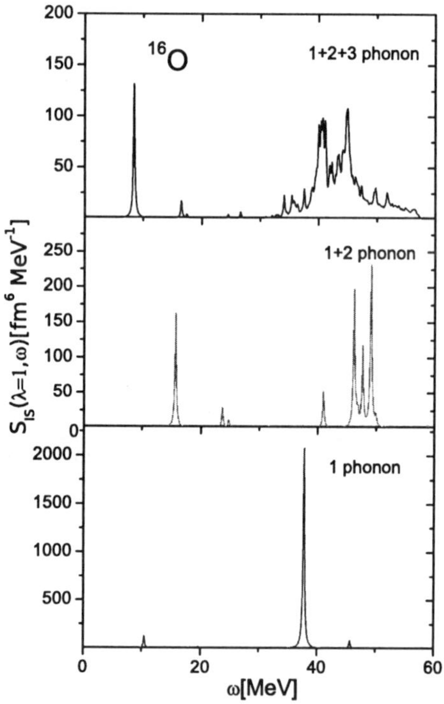

Fig. 2. Isoscalar $E1$ strength distributions in ^{16}O.

As shown in Fig. 1, the shape of the spectrum of the isovector $E1$ response $S_{IV}(E1, \omega)$ changes little as the number of phonon increases, while the main peak gets quenched. The full spectrum is shifted upward as we add the two-phonon configurations and, then, downward to the previous position when the three phonon configurations come into play.

More dramatic is the effect of the multiphonon excitations on the isoscalar $E1$ response. The strength gets spread over a much larger energy range as we increase the number of phonons (Fig. 2). Indeed, the isoscalar giant dipole resonance is due to ph excitations of $3\hbar\omega$. This is also the energy of many $2p - 2h$ as well as $3p - 3h$ configurations which are therefore to be included in a consistent description of the mode. The energy shift induced by the multiphonon configurations follows the same trend as in the isovector case.

Since our purpose is merely illustrative, we do not make any explicit comparison with experiments. Before doing this, we must investigate how sensitive each response is to the single particle energies and verify if the restricted space used here is adequate.

4. Conclusions

The equations of motion method we have proposed is exact and of easy implementation. Its exact numerical implementation in ^{16}O was confined to a space sufficient for our illustrative purposes, but too restricted to describe exhaustively all spectroscopic properties of this complex nucleus. Extending the calculation to a larger space is not straightforward. Indeed, the number of density matrices to be computed increases so rapidly with the number of phonons as to render the procedure unbearably slow. The method, however, generates a basis of correlated states. It is therefore conceivable that most of them are non collective and unnecessary. The selection of the relevant basis states may be done efficiently by an importance sampling algorithm developed recently [26], which allows a severe truncation while monitoring the accuracy of the solutions. The truncation procedure should render manageable the eigenvalue problem within a larger phonon space and, thus, may compete even with large scale shell model calculations.

The method can be extended in several ways. It can be reformulated so as to include RPA phonons. This extension, however, might be unnecessary since the method, already in its present TDA formulation, yields an explicitly correlated ground state.

A formulation of the method in terms of quasi-particle rather than particle-hole states is also straightforward and especially useful. It allows, indeed, to study anharmonicities and multiphonon excitations in open shell nuclei not easily accessible to shell model methods.

Acknowledgements

Work supported in part by the Italian Ministero della Istruzione Universitá e Ricerca (MIUR) and by the research plans MSM 0021620834 and GAUK 222/2006/B-FYZ/MFF of Czech Republic.

References

1. C. Fransen *et al.*, Phys. Rev. C **71**, 054304 (2005).
2. M. Kneissl, N. Pietralla, and A. Zilges, J. Phys. G: Nucl. Part. Phys. **32**, R217 (2006).
3. N. Pietralla *et al.*, Phys. Rev. Lett. **83**, 1303 (1999).
4. N. Frascaria, Nucl. Phys. **A482**, 245c (1988).
5. T. Auman, P. F. Bortignon, and H. Hemling, Annu. Rev. Nucl. Part. Sci. **48**, 351 (1998).
6. For a review see A. Arima and F. Iachello, Adv. Nucl. Phys. **13**, 139 (1984).
7. S. Nishizaki and J. Wambach, Phys. Rev. C **57**, 1115 (1998).
8. F. Catara, P. Chomaz, and N. Van Giai, Phys. Lett. **B233**, 6 (1989); Phys. Lett. **B277**, 1 (1992).
9. A. Bohr and B. R. Mottelson, *Nuclear Structure* Vol. II (Benjamin, New York, 1975).
10. P. F. Bortignon, R. A. Broglia, D. R. Bes, and R. Liotta, Phys. Rep. **30**, 305 (1977).
11. V. G. Soloviev, *Theory of Atomic Nuclei: Quasiparticles and Phonons* (Institute of Physics, Bristol, 1992).
12. N. Lo Iudice and Ch. Stoyanov, Phys. Rev. C **62**, 047302 (2000); Phys. Rev. C **65**, 064304 (2002).
13. V. Yu. Ponomarev, P. F. Bortignon, R. A. Broglia, and V. V. Voronov, Phys. Rev. Lett. **85**, 1400 (2000).
14. C. Pomar, J. Blomqvist, R. J. Liotta, and A. Insolia, Nucl. Phys. **A515**, 381 (1990).
15. M. Grinberg, R. Piepenbring, K. V. Protasov, and B. Silvestre-Brac, Nucl. Phys. **A597**, 355 (1996).
16. F. Andreozzi, N. Lo Iudice, A. Porrino, F. Knapp, and J. Kvasil, to be submitted to Phys. Rev. C.
17. D. J. Rowe, *Nuclear Collective Motion* (Methuen and Co. Ltd., London, 1970).
18. D. J. Rowe, J. Math. Phys. **10**, 1774 (1969).
19. G. E. Brown, A. M. Green, Nucl. Phys. **75**, 401 (1966).
20. W. C. Haxton and C. J. Johnson, Phys. Rev. Lett. **65**, 1325 (1990).
21. J. P. Draayer, private communication.
22. P. Ring and P. Schuck,*The Nuclear Many-Body Problem* (Springer-Verlag, New York, 1980).
23. R. Machleidt, Adv. Nucl. Phys. **19**, 189 (1989).
24. F. Palumbo, Nuc. Phys. **99**, 100 (1967).
25. D. H. Glockner and R. D. Lawson, Phys. Lett. **B53**, 313 (1974).
26. F. Andreozzi, N. Lo Iudice, and A. Porrino, J. Phys. G: Nucl. Part. Phys. **29**, 2319 (2003).

NEW RESULTS IN THE CBF THEORY FOR MEDIUM-HEAVY NUCLEI

C. BISCONTI

Dipartimento di Fisica, Università di Lecce, and I.N.F.N. sezione di Lecce, I-73100 Lecce, Italy
Dipartimento di Fisica, Università di Pisa, I-56100 Pisa, Italy

G. CO'

Dipartimento di Fisica, Università di Lecce, and I.N.F.N. sezione di Lecce, I-73100 Lecce, Italy

F. ARIAS DE SAAVEDRA

Departamento de Física Atómica, Molecular y Nuclear, Universidad de Granada, E-18071 Granada, Spain

Momentum distributions, spectroscopic factors and quasi-hole wave functions of medium-heavy doubly closed shell nuclei have been calculated in the framework of the Correlated Basis Function theory, by using the Fermi hypernetted chain resummation techniques. The calculations have been done by using microscopic two-body nucleon-nucleon potentials of Argonne type, together with three-body interactions. Operator dependent correlations, up to the tensor channels, have been used.

Keywords: Nuclear structure. Many-body theories. Closed shell nuclei

1. Introduction

The validity of the non relativistic description of the atomic nuclei has been well established in the last ten years. The idea is to describe the nucleus with a Hamiltonian of the type:

$$H = -\frac{\hbar^2}{2m}\sum_i \nabla^2 + \sum_{i<j} v_{ij} + \sum_{i<j<k} v_{ijk} \ , \tag{1}$$

where the two- and three-body interactions, v_{ij} and v_{ijk} respectively, are fixed to reproduce the properties of the two- and three-body nuclear systems.

About fifteen years ago, we started a project aimed to apply to the description of medium and heavy nuclei the Correlated Basis Function (CBF) theory, successfully used to describe the nuclear and neutron matter properties [1,2]. We solve the many-body Schrödinger equation by using the variational principle:

$$\delta E[\Psi] = \delta \frac{< \Psi|H|\Psi >}{< \Psi|\Psi >} = 0 \ . \tag{2}$$

The search for the minimum of the energy functional is carried out within a subspace of the full Hilbert space spanned by the A-body wave functions which can be expressed as:

$$\Psi(1, ..., A) = \mathcal{F}(1, ..., A)\Phi(1, ..., A) \ , \tag{3}$$

where $\mathcal{F}(1, ..., A)$ is a many-body correlation operator, and $\Phi(1, ..., A)$ is a Slater determinant composed by single particle wave functions, $\phi_\alpha(i)$. We use two and three-body interactions of Argonne and Urbana type, and we consider all the interaction channels up to the spin-orbit ones. The complexity of the interaction requires an analogously complex correlation:

$$\mathcal{F} = \mathcal{S}\left(\prod_{i<j=1}^{A} F_{ij}\right) \ , \tag{4}$$

where $\mathcal{S}$ is a symmetrizer operator and F_{ij} has the form:

$$F_{ij} = \sum_{p=1,6} f^p(r_{ij}) O^p_{ij} \ . \tag{5}$$

In the above equation we have adopted the nomenclature commonly used in this field, by defining the operators as:

$$O^{p=1,6}_{ij} = [1, \boldsymbol{\sigma}_i \cdot \boldsymbol{\sigma}_j, S_{ij}] \otimes [1, \boldsymbol{\tau}_i \cdot \boldsymbol{\tau}_j] \ , \tag{6}$$

where $S_{ij} = (3\hat{\mathbf{r}}_{ij} \cdot \boldsymbol{\sigma}_i \hat{\mathbf{r}}_{ij} \cdot \boldsymbol{\sigma}_j - \boldsymbol{\sigma}_i \cdot \boldsymbol{\sigma}_j)$ is the tensor operator.

The binding energies and the charge distributions of ^{12}C , ^{16}O , ^{40}Ca , ^{48}Ca and ^{208}Pb doubly closed shell nuclei have been obtained by solving the Fermi Hypernetted Chain (FHNC) equations in the Single Operator Chain (SOC) approximation [3]. These calculations have the same accuracy of the best variational calculations done in nuclear and neutron matter [1,2].

By using this FHNC/SOC computational scheme we have investigated the effects of the correlations on some ground state quantities which are related to observables. A first quantity we have studied is the momentum distribution. It is well known that the short–range correlations enhance by orders of magnitude the high-momentum components of the momentum distribution [4]. We have also studied the spectroscopic factors since, as we have already mentioned, their empirical values are always smaller than the independent particle model (IPM) predictions.

The present study has been done by using the many-body wave functions obtained in Ref. [3]. The Argonne v_8' two-nucleon potential, together with the Urbana XI three-body force has been used.

In the next sections we present the results of our study of the One-Body Density Matrix (OBDM) and of the momentum distribution and we evaluate the spectroscopic factors.

2. The momentum distributions

The OBDM, of a system of A nucleons is defined as:

$$\rho^{s,s';t}(\mathbf{r}_1,\mathbf{r}_1') = \frac{A}{<\Psi|\Psi>}\int dx_2 \ldots dx_A$$
$$\Psi^\dagger(x_1,x_2,\ldots,x_A)\chi_t(1)\chi_s(1)\chi_t^\dagger(1')\chi_{s'}^\dagger(1')\Psi(x_1',x_2,\ldots,x_A) \ . \qquad (7)$$

In the above expression, the variable x_i indicates the position ($\mathbf{r}_i$) and the third components of the spin (s) and of the isospin (t) of the single nucleon. With the integral sign we understand that also the sum on spin and isospin of all the particles, including $1'$, is performed. The OBDM of eq. (7) is characterized by the quantum numbers relative to the particle 1. In our calculations we are interested in the quantity:

$$\rho^t(\mathbf{r}_1,\mathbf{r}_1') = \sum_{s=\pm 1/2}\left[\rho^{s,s;t}(\mathbf{r}_1,\mathbf{r}_1') + \rho^{s,-s;t}(\mathbf{r}_1,\mathbf{r}_1')\right] \ , \qquad (8)$$

whose diagonal part represents the one-body density of neutrons or protons, this last one is related to the charge density distribution of the nucleus.

We define the momentum distributions of protons or neutrons as:

$$n^t(k) = \frac{1}{\mathcal{N}_t}\int d\mathbf{r}_1 d\mathbf{r}_1'\, e^{i\mathbf{k}\cdot(\mathbf{r}_1-\mathbf{r}_1')}\rho^t(\mathbf{r}_1,\mathbf{r}_1') \ , \qquad (9)$$

where we have indicated with $\mathcal{N}_t$ the number of protons or neutrons. We describe doubly closed shell nuclei, with different numbers of proton and neutrons, in a jj coupling scheme.

The correlated OBDM is obtained by using the ansatz (3) in Eq. (7). The calculation is done by following the lines indicated in Ref. [5] and considering, in addition, the presence of the antiparallel spin terms and distinguishing proton and neutron contributions. The diagonal part of the OBDM is the one-body density, normalized to the number of nucleons. We present in Fig.1 the proton momentum distributions, for the ^{12}C , ^{16}O ,

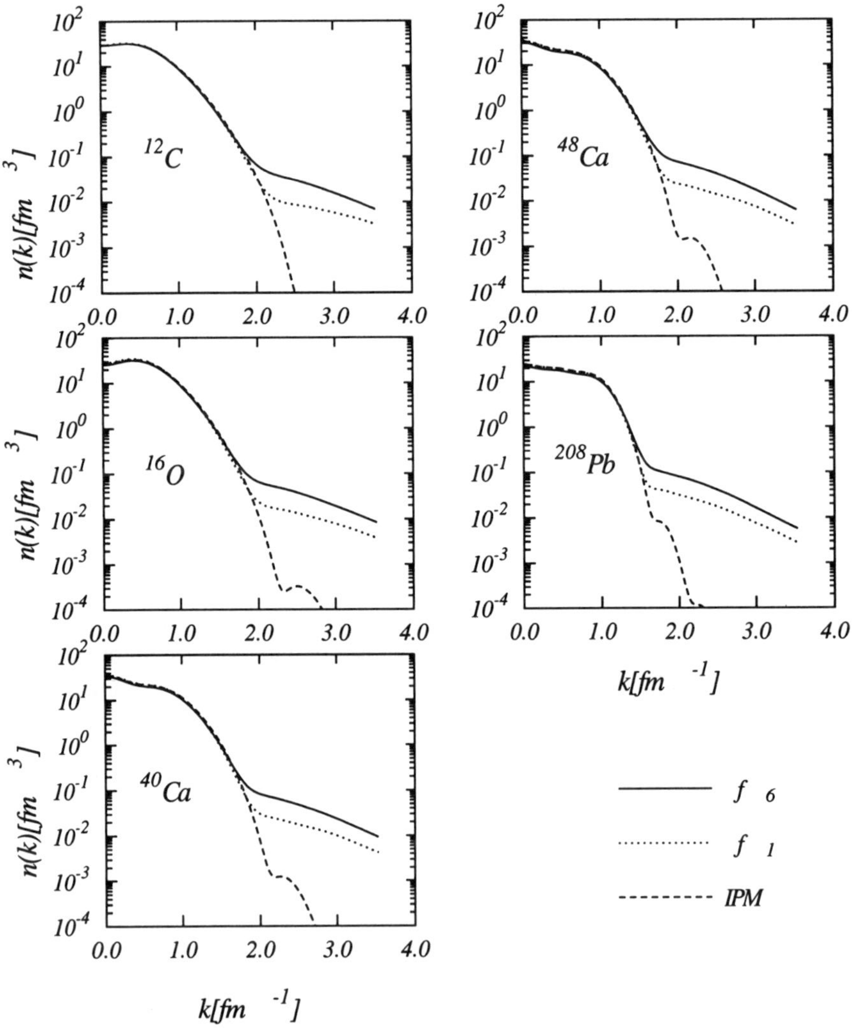

Fig. 1. The proton momentum distributions for the ^{12}C , ^{16}O , ^{40}Ca , ^{48}Ca , ^{208}Pb nuclei calculated in the IPM model (dashed lines), by using the scalar correlation only (dotted lines) and the full correlation operator in FHNC/SOC approximation (solid lines).

^{40}Ca , ^{48}Ca and ^{208}Pb nuclei. The dashed lines show the results obtained in the IPM model, the dotted lines those results calculated by using the scalar correlations only and, finally, the solid lines indicate the momentum distributions obtained by using the full operator dependent correlations.

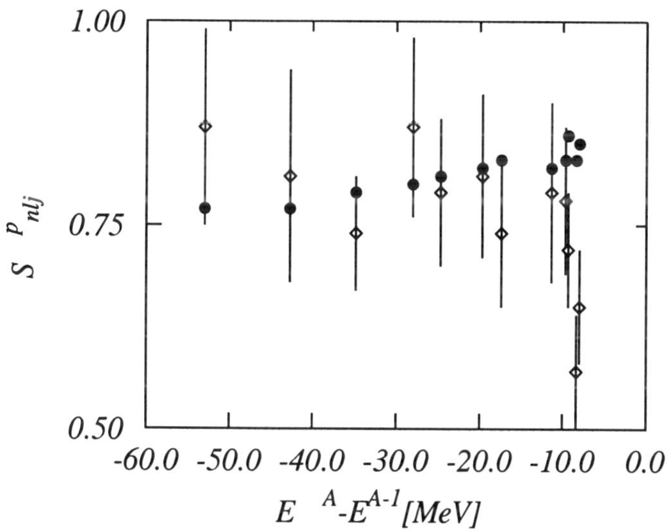

Fig. 2. Comparison between the theoretical, black points, and the experimental, empty diamonds, spectroscopic factors for some proton states of ^{208}Pb. In the x axis we indicate the separation energies.

The high momentum components of the correlated distributions are orders of magnitude larger than those produced by the IPM. The operator dependent terms of the correlations further increase this behavior.

3. The spectroscopic factors

The quasi-hole wave function is defined as:

$$\psi^t_{nljm}(x) = \sqrt{A}\frac{< \Psi_{nljm}(A-1)|\delta(x-x_A)P^t_A|\Psi(A) >}{< \Psi_{nljm}(A-1)|\Psi_{nljm}(A-1) >^{1/2}< \Psi(A)|\Psi(A) >^{1/2}} , \tag{10}$$

where we have indicated with $\Psi_{nljm}(A-1)$ and $\Psi(A)$ the wave functions describing the nuclei formed by $A-1$ and A nucleons respectively, and with P^t_A the isospin projector. The subindexes $nljm$ designate the quantum numbers of the odd-even nucleus.

We describe the wave function of the nucleus with $A-1$ nucleons by using an ansatz analogous to that of Eq. (3):

$$\Psi_{nljm}(A-1) = F(1,...,A-1)\,\Phi_{nljm}(1,...,A-1) , \tag{11}$$

where $\Phi_{nljm}(1,...,A-1)$ indicates a Slater determinant obtained by removing from the Slater determinant $\Phi(1,...,A)$ a single state characterized by the quantum numbers $nljm$. In the IPM the quasi-hole wave functions coincide with the mean-field wave functions.

In order to obtain the radial part of the quasi-hole wave function we multiply equation (10) by $< jm|$, we integrate over the angular coordinate Ω, and sum on the spin coordinates [5]:

$$\psi^t_{nlj}(r) = \sum_{\mu,s} < l\mu 1/2 s|jm > \int d\Omega \, Y^*_{l\mu}(\Omega) \chi^\dagger_s \, \psi^t_{nljm}(x) = \mathcal{X}^t_{nlj}(r)[\mathcal{N}^t_{nlj}]^{1/2} \; . \tag{12}$$

From the knowledge of the quasi-hole functions we obtain the spectroscopic factors as:

$$S^t_{nlj} = \int dr \, r_1^2 \, |\psi^t_{nlj}(r)|^2 \; . \tag{13}$$

In Fig. 2 we compare the theoretical spectroscopic factors calculated for the proton bound states of the ^{208}Pb nucleus with the experimental data of Ref. [6]. In abscissa we give the separation energies defined as the difference between the energy of a A-nucleon system and that of the $A-1$-nucleon system obtained by removing the $nljm$ state. The agreement between theory and experiment is better for the deeply bound shells than for those levels closer to the Fermi surface. This could be due to the strong coupling between the quasi-hole wave function and the low-lying surface vibrations. The effects of this coupling, usually called long-range correlations, not explicitly treated by our theory, are expected to be larger for the external shells than for the internal ones.

The effect of the correlations on the quasi-hole wave functions is presented in Fig. 3 where we have shown the squared quasi-hole $3s_{1/2}$ proton wave function calculated with increasing complexity in the correlation function. The full line indicates the IPM result, the other lines have been obtained by using only scalar correlations, f_1, operatorial correlations without the tensor channels, f_4, and correlations which include also the tensor dependent terms, f_6. The presence of the correlations produces a lowering of the quasi-hole wave function in the nuclear center. There is a consistent trend of the correlations effects: the more elaborated is the correlations the larger is the decreasing at the center of the nucleus.

4. Summary and Conclusions

In this work momentum distributions, spectroscopic factors and quasi-hole wave functions of medium-heavy doubly closed shell nuclei have been calculated by extending the FHNC/SOC computational scheme. The calcula-

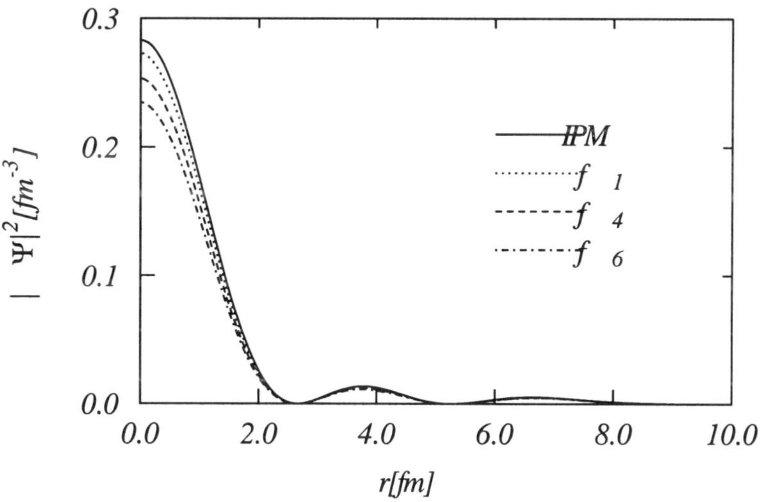

Fig. 3. The square of the quasi-hole wave function for the $3s_{1/2}$ proton state of the ^{208}Pb . The labels *unc*, *Jas*, f^4 and f^6 indicate the IPM,Jastrow, f^4 and f^6 models respectively.

tions have been done considering the different number of proton and neutrons and the single particle basis are given in a jj coupling scheme. A microscopic two-body interaction of Argonne type, implemented with the appropriated three-body force of Urbana type, have been used. The calculations have been done with operator dependent correlations which include, in addition to the four central channels, also tensor correlations.

The comparison between our results with those obtained in the IPM highlights the correlations effects. The correlated momentum distributions have high momentum tails which are orders of magnitude larger than the IPM results. The spectroscopic factors are always smaller than one, the IPM value, and in a reasonable agreement with the experimental values, especially for the more bounded states. The quasi-hole wave functions are depleted in the nuclear center by the correlations. We have found that the operator dependent terms emphasize the correlations effects.

AKNOWLEDGMENTS: This work has been partially supported by the agreement INFN-CICYT, by the Spanish Ministerio de Educación y Ciencia (FIS2005-02145) and by the MURST through the PRIN: *Teoria della struttura dei nuclei e della materia nucleare.*

References

1. R. B. Wiringa, V. Ficks, and A. Fabrocini, Phys. Rev. C **38**, 1010 (1988).
2. A. Akmal, V. Pandharipande, and D. G. Ravenhall, Phys. Rev. C **58**, 1804 (1998).
3. C. Bisconti, F. Arias de Saavedra, G. Co', and A. Fabrocini, Phys. Rev. C **73**, 054304 (2006).
4. A. N. Antonov, P. E. Hodgson, and I. Z. Petkov, *Nucleon momentum and density distributions* (Clarendon, Oxford, 1988).
5. A. Fabrocini and G. Co', Phys. Rev. C **63**, 044319 (2001).
6. M. F. van Batenburg, *Deeply bound protons in* ^{208}Pb, PhD thesis, Universiteit Utrecht (Nederlands) 2001, unpublished.

UNBOUND EXOTIC NUCLEI STUDIED BY PROJECTILE FRAGMENTATION *

G. BLANCHON

Scuola di Dottorato G. Galilei and Dipartimento di Fisica, Università di Pisa and INFN, Sez. di Pisa, Largo Pontecorvo 3, 56127 Pisa, Italy

A. BONACCORSO and A. GARCÍA- CAMACHO

INFN, Sezione di Pisa and Dipartimento di Fisica, Università di Pisa, Largo Pontecorvo 3, 56127 Pisa, Italy

D. M. BRINK

Department of Theoretical Physics, 1 Keble Road, Oxford OX1 3NP, U. K

N. VINH MAU

Institut de Physique Nucléaire, IN2P3-CNRS, F-91406, Orsay Cedex, France

We call *projectile fragmentation* of neutron halo nuclei the elastic breakup (diffraction) reaction, when the observable studied is the neutron-core relative energy spectrum. This observable has been measured in relation to the Coulomb breakup on heavy target and recently also on light targets. Such data enlighten the effect of the neutron final state interaction with the core of origin. Projectile fragmentation is studied here by a time dependent model for the excitation of a nucleon from a bound state to a continuum resonant state in a neutron-core complex potential which acts as a final state interaction. The final state is described by an optical model S-matrix so that both resonant and non resonant states of any continuum energy can be studied as well as deeply bound initial states. It turns out that due to the coupling between the initial and final states, the neutron-core free particle phase shifts are modified, in the exit channel, by an additional phase. Some typical numerical calculations for the relevant observables are presented and compared to experimental data. It is suggest that the excitation energy spectra of an unbound nucleus might reflect the structure of the parent nucleus from whose fragmentation they are obtained.

*In memory of Adelchi Fabrocini.

1. Introduction

All theoretical methods used so far to describe breakup rely on a basic approximation to describe the collision with only the three-body variables of nucleon coordinate, projectile coordinate, and target coordinate. Thus the dynamics is controlled by the three potentials describing nucleon-core, nucleon-target, and core-target interactions. In most cases the projectile-target relative motion is treated semiclassically by using a trajectory of the center of the projectile relative to the center of the target $\mathbf{R}(t) = \mathbf{b}_c + vt\hat{\mathbf{z}}$ with constant velocity v in the z direction and impact parameter $\mathbf{b}_c$ in the xy plane. This approximation makes our formalism applicable for incident energies above the Coulomb barrier. Along this trajectory the amplitude for a transition from a nucleon state ψ_i bound in the projectile, to a final continuum state ψ_f, is given by [1,2]

$$A_{fi} = \frac{1}{i\hbar}\int_{-\infty}^{\infty} dt\langle\psi_f(\mathbf{r},t)|V(\mathbf{r},\mathbf{R}(t))|\psi_i(\mathbf{r},t)\rangle, \tag{1}$$

where V is the interaction responsible for the transition which will be specified in the following. The probabilities for different processes can be represented in terms of the amplitude as $dP/d\xi = \sum |A_{fi}|^2\delta(\xi - \xi_f)$ where ξ can be momentum, energy or any other variable for which a differential cross section is measured. Direct one-particle re-arrangment reactions of the peripheral type in presence of strong core-target absorption can be described by an equation like [2–5]

$$\frac{d\sigma_{-n}}{d\xi_f} = C^2 S \int d\mathbf{b_c} \frac{dP_{b_{up}}(b_c)}{d\xi_f} P_{ct}(b_c), \tag{2}$$

(see Eq. (2.3) of [3]) and C^{2}S is the spectroscopic factor for the initial single particle state. The core survival probability is defined in terms of a S-matrix function of the core-target distance of closest approach b_c. A simple parameterisation is $P_{ct}(b_c) = |S_{ct}|^2 = e^{(-\ln 2exp[(R_s - b_c)/a])}$. It takes into account the peripheral nature of the reaction and naturally excludes the possibility of large overlaps between projectile and target. The strong absorption radius R$_s$ $\approx 1.4(A_p^{1/3} + A_t^{1/3})$ fm is defined as the distance of closest approach for a trajectory that is 50% absorbed from the elastic channel and a=0.6 fm is a diffuseness parameter. The values of R$_s$ thus obtained agree within a few percent with those of the Kox parameterization [6].

2. Projectile Fragmentation

Let us call *projectile fragmentation* the elastic breakup (diffraction dissociation) reaction, when the observable studied is the neutron-core relative energy spectrum. This kind of observable has been widely measured in relation to the Coulomb breakup on heavy target. Results on light targets have also been presented [7]. These data enlighten the effect of the neutron final state interaction with the core of origin, while observables like the core energy or momentum distributions enlighten the effect of the neutron final state interaction with the target.

Projectile fragmentation has also been used to study two neutron halo projectiles [8–23]. In this case it has been suggested that the reaction might proceed by the simultaneous emission of the two neutrons or by successive emissions [8]. The successive emission can be due to a mechanism in which one neutron is stripped by the interaction with the target, as in the one-neutron fragmentation case, while the other is left behind, for example in a resonance state, which then decays. This mechanism has been described by the sudden approximation [9] under the hypothesis that while the first neutron is stripped, the second neutron is emitted at large impact parameters with no final state interaction with the target. The emission can be expected sequential if the two neutrons are not strongly correlated.

If the two neutrons are strongly correlated they will preferentially be emitted simultaneously. If the neutron which is not detected is stripped while the other suffers an elastic scattering on the target, then in both cases to first order in the interaction the neutron ends-up in a plane wave final state [2]. It can then re-interact with the core which, for example, is going to be ^{10}Be in the case of the one-neutron halo projectile ^{11}Be, while it will be ^{12}Be in the case of the projectile fragmentation of ^{14}Be, since ^{13}Be is not bound. Experiments with a ^{14}B projectile [10] have also been performed, in which the n-^{12}Be relative energy spectra have been reconstructed by coincidence measurements. In such a nucleus the valence neutron is weakly bound, while the valence proton is strongly bound. Thus the neutron will probably be emitted in the first step and then re-scattered by the core minus one proton nucleus. The projectile-target distances at which this kind of mechanism would be relevant are probably not so large to neglect the effect of the neutron-target interaction.

2.1. *Inelastic excitation to the continuum.*

To first order the inelastic-like excitations can be described again by the time dependent perturbation amplitude Eq.(1) [1,2]. In this section also, the potential $V(\mathbf{r}, \mathbf{R}(t))$, which is the interaction responsible for the neutron transition, moves past on a constant velocity path as described in the previous sections. The radial part $\phi_i(\mathbf{r})$ of the single particle initial state wave function $\psi_i(\mathbf{r}, t)$ is calculated in a potential $V_{WS}(r)$ which is fixed in space. The coordinate system and other details of the calculations can be found in Ref. [11]. In the special case of exotic nuclei the traditional approach to inelastic excitations needs to be modified. For example the final state can be eigenstate of a potential V_1 modified with respect to V_{WS} because some other particle is emitted during the reaction process as discussed in the introduction. The final state interaction might also have an imaginary part which would take into account the coupling between a continuum state and an excited core. The first order time dependent perturbation amplitude then reads

$$A_{fi} = \frac{1}{i\hbar v} \int_{-\infty}^{\infty} dxdydz \; \phi_f^*(x, y, z)\phi_i(x, y, z)e^{iqz}\tilde{V}(x - b_c, y, q), \qquad (3)$$

where $\tilde{V}(x - b_c, y, q) = \int_{-\infty}^{\infty} dz V(x - b_c, y, z)e^{iqz}$, and we changed variables and put $z' = z - vt$ or $t = (z - z')/v$, $q = \varepsilon_f - \varepsilon_i/\hbar v$. Here ε_f is the neutron-core relative energy in the final state.

The rôle of the target represented by $\tilde{V}$ is just to perturb the initial bound state wave function and to allow the transition to the continuum by transferring some momentum to the neutron. Then it is enough to choose a simplified form of the interaction, such as a delta-function potential $V(r) = v_2\delta(x)\delta(y)\delta(z)$. The value of the strength $v_2 \equiv$ [MeV fm^3] used in the calculation is taken equal to the volume integral of the appropriate neutron-target interaction. It is clear that while in the sudden approach the initial and final state overlap is taken in the whole coordinate space, irrespective of the target and of the beam velocity, here the overlap of the initial and final wave functions depends on the core-target impact parameter. The neutron is emitted preferentially on the reaction plane and the z-component, being along the relative velocity axis is boosted by a momentum q.

Due of the strong core absorption discussed in Sec.1 these calculations are performed using the asymptotic form of the initial and final state wave functions. Introducing the quantization condition [2] the probability spectrum reads

$$\frac{dP_{in}}{d\varepsilon_f} = \frac{2}{\pi}\frac{v_2^2}{\hbar^2 v^2}C_i^2\frac{m}{\hbar^2 k}\frac{1}{2l_i+1}\Sigma_{m_i,m_f}|1-\bar{S}_{m_i,m_f}|^2|I_{m_i,m_f}|^2. \tag{4}$$

The generalization including spin is given in Appendix B of Ref. [11] and $|I_{m_i,m_f}|^2 = \left|\int_{-\infty}^{\infty} dz e^{iqz} i^{l_i}\gamma h_{l_i}^{(1)}(i\gamma r)Y_{l_i,m_i}(\theta,0)k\frac{i}{2}h_{l_f}^{(-)}(kr)Y_{l_f,m_f}(\theta,0)\right|^2$. The quantity $\bar{S} = e^{2i(\delta+\nu)}$ is an off-the-energy-shell S-matrix representing the final state interaction of the neutron with the projectile core. It depends on a phase which is the sum of δ, the free particle n-core phase shift, plus ν the phase of the matrix element $|I|$.

3. Applications

3.1. *The reaction* $^{11}Be \rightarrow n + ^{10}Be$

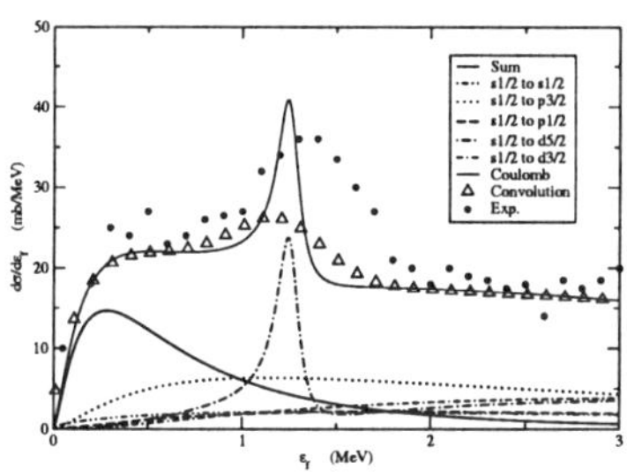

Fig. 1. n-^{10}Be relative energy spectrum, including Coulomb and nuclear breakup for the reaction ^{11}Be+^{12}C → n+^{10}Be+X at 69 A.MeV. Only the contributions from an s initial state with spectroscopic factor C^2S= 0.84 are calculated. The triangles are the total calculated result after convolution with the experimental resolution function. The dots are the experimental points from [7].

As a test of our model we calculate the relative energy spectrum n+^{10}Be obtained by the authors of Ref. [7] in the breakup reaction of ^{11}Be on ^{12}C at 70 A.MeV. The structure of ^{11}Be is well known: the valence neutron is bound by 0.503 MeV; the wave function is mainly a 2s state with a spectroscopic factor around 0.8 and there is also a small $d_{5/2}$ component. The main $d_{5/2}$ strength is in the continuum centered around 1.25. We have calculated the initial wave function for the s-state in a simple Woods-Saxon potential with strength fitted to the experimental separation energy and whose parameters are: r_0=1.25 fm, a=0.8 fm. As possible final states we have considered only the s, p and d partial waves calculated in the l-dependent potentials of [11]. The delta-function potential strength has been chosen as -4057.59 MeV fm^3. The authors of Ref. [7] have shown that the effect of Coulomb breakup is noticeable in their n+^{10}Be spectrum. We have also included this contribution, calculating it according to [4]. The spectrum of Fig.1 is very similar to the spectrum obtained in Ref. [24] by solving the time-dependent Schrödinger equation numerically, expanding the projectile wave function upon a three-dimensional spherical mesh. Similarly to the present model, a classical, straight line trajectory for the core-target scattering was used in Ref. [24]. Also our n-core potentials are very close to those of Ref. [24] and our δ-interaction strength is consistent with the volume integral of their neutron-target interaction. We have then folded the calculated spectrum througth the experimental resolution function of Fukuda *et al.* [7], as given in Ref. [24]. The result is shown in Fig.1 by the triangles. The full curve is the total spectrum, sum of Coulomb and nuclear breakup. Each individual transition, due to the nuclear interaction only, is also shown. The dots are the experimental points from [7]. The kind of discrepancy between our calculation and the data in the range 1-2 MeV is very similar to that of the calculations in Ref. [24].

3.2. *Structure of ^{14}Be and ^{14}B*

Uncertainties in the interpretation of experimental results as compared to structure calculations were at the origin of our motivations to try to understand whether the neutron-^{12}Be relative energy spectra obtained from fragmentation of ^{14}Be or ^{14}B would show differences predictable in a theoretical model. If differences will be found in the experimental results with ^{14}B and ^{14}Be beams they could be due to an interplay between structure and reaction effects.

The ground state of ^{14}Be has spin $J^\pi = 0^+$. In a simple model assuming two neutrons added to a ^{12}Be core in its ground state the wave function is:

$$|^{14}Be>= [b_1(2s_{1/2})^2 + b_2(1p_{1/2})^2 + b_3(1d_{5/2})^2] \otimes |^{12}Be, 0^+> . \qquad (5)$$

Then the bound neutron can be in a 2s, 1p$_{1/2}$ or 1d$_{5/2}$ state. However, as it has been discussed in the previous section, the situation is much more complicated [17–21] and in particular the calculations of Ref. [22] show that there is a large component $(2s_{1/2}, 1d_{5/2}) \otimes |^{12}Be, 2^+>$ with the core in its low energy 2$^+$ state which can modify the neutron distribution.

The ground state of ^{14}B has spin $J^\pi = 2^-$. In a model where it is described as a neutron-proton pair added to a ^{12}Be core in its 0$^+$ground

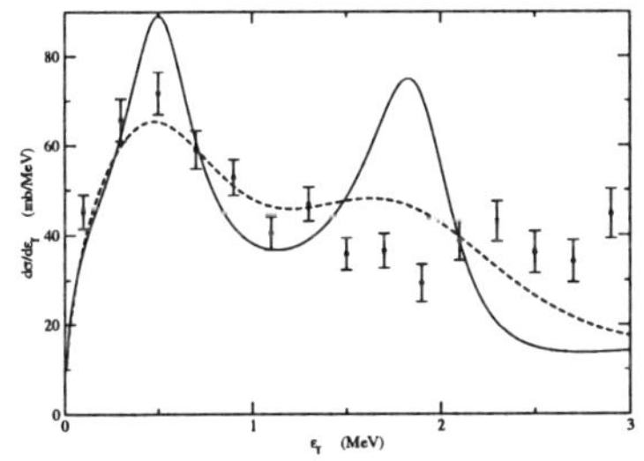

Fig. 2. Sum of all transitions from the s initial state for the reaction ^{14}Be+^{12}C $\to$ n+^{12}Be+X . Experimental points from [15]. Dashed line is the folding of the calculated spectrum with the experimental resolution curve.

state with the proton in the 1p$_{3/2}$ shell, its wave function may be written as:

$$|^{14}B> = [a_1(p_{3/2}, 2s_{1/2}) + a_2(p_{3/2}, d_{5/2})] \otimes |^{12}Be, 0^+ > . \qquad (6)$$

The present experimental information [25] on ^{14}B is that the neutron is in a state combination of s and d-components with weights 66% and 30% respectively, while shell model calculations show a similar mixture and no component with an excited state of the core. There are two possibilities for the reaction mechanism. One is that a proton is knocked out in the reaction with the target. The remaining ^{13}Be would be left in an unbound s-state with probability $|a_1|^2$, in a d$_{5/2}$-state with probability $|a_2|^2$. These unbound states would decay showing the s-wave threshold and d-wave resonance effects. As mentioned in the introduction, the second possibility is that the neutron is knocked out first due to its small separation energy and that the proton is stripped from the remaining ^{13}B.

To give another example of a possible comparison with available data, we show in Fig. 2 the experimental points from H. Simon *et al.* [15] for the reaction ^{14}Be+^{12}C $\to$ n+^{12}Be+X at 250 A.MeV. The normalization factor of the data to mb/MeV is 0.843. The solid line gives the sum of all transitions from the s initial state with ε_f=-1.85 MeV (solid line), renormalized with a factor 2.4. The dashed line is the folding of the calculated spectrum with the experimental resolution curve. Therefore the calculation underestimate the absolute experimental cross section by a factor of 2. In view of the incertitude in the strength of our n-target δ-potential and on the initial state spectroscopic factor which has been taken as unit, we can consider our absolute cross sections quite reasonable. A more detailed account of these calculations is given in [11].

4. Conclusions and Outlook

The field of Rare Isotopes Studies is very active, growing steadily and rapidly. Some recent achievements in the reaction theory for elastic breakup have been presented. From the structure point of view, in the search for the dripline position, a very important role is played by the study of nuclei unstable by neutron emission. This is one of the most important subjects which need to be adressed and further developed in the near future and for which some suggestions have been presented.

References

1. K. Alder and A. Winther, *Electromagnetic Excitation*, North-Holland, 1975.
2. A. Bonaccorso and D.M. Brink, Phys. Rev. C **38**, 1776 (1988); Phys. Rev. C **43**, 299 (1991); Phys. Rev. C **44**, 1559 (1991).
3. A. Bonaccorso, Phys. Rev. C **60**, 054604 (1999).
4. J. Margueron, A. Bonaccorso, and D.M. Brink, Nucl. Phys. **A703**, 105 (2002); Nucl. Phys. **A720**, 337 (2003).
5. R. A. Broglia and A. Winther, *Heavy Ion Reactions*, Benjamin, Reading, Mass, 1981.
6. S. Kox *et al.*, Phys. Rev. C **35**, 1678 (1987).
7. N. Fukuda *et al.*, Phys. Rev. C **70**, 054606 (2004).
8. F.M. Marqués *et al.*, Phys. Rev. C **64**, 061301(R) (2001). N.Orr, Prog. Theor. Phys. Suppl. **146**, 201 (2003).
9. G.F. Bertsch, K. Hencken, and H. Esbensen, Phys. Rev. C **57**, 1366 (1998).
10. J.L. Lecouey, Few-Body Syst. **34**, 21 (2004).
11. G. Blanchon *et al.*, Nucl. Phys. A (2006), in press.
12. G. Blanchon, A. Bonaccorso, and N. Vinh Mau, Nucl. Phys. **A739**, 259 (2004).
13. M. Thoennessen *et al.*, Phys. Rev. C **59**, 111 (1999); Phys. Rev. C **60**, 027303 (1999); Proceedings the Erice Int. School of Heavy-Ion Physics, 4th Course. Eds. R.A. Broglia and P.G. Hansen. World Scientific, Singapore 1998, p. 269.
14. M. Labiche *et al.*, Phys. Rev. Lett. **86**, 600 (2001).
15. H. Simon *et al.*, Nucl. Phys. **A734**, 323 (2004), and private communication.
16. A.A. Korsheninnikov *et al.*, Phys. Lett. **B343**, 53 (1995).
17. G.F. Bertsch and H. Esbensen, Ann. Phys. (N.Y.) **209**, 327 (1991).
18. I.J. Thompson and M.V. Zhukov, Phys. Rev. C **53**, 708 (1996).
19. N. Vinh Mau and J.C. Pacheco, Nucl. Phys. **A607**, 163 (1996).
20. J.C. Pacheco and N. Vinh Mau, Phys. Rev. C **65**, 044004 (2002).
21. P. Descouvemont, Phys. Lett. **B331**, 271 (1994); Phys. Rev. C **52**, 704 (1995).
22. T. Tarutina, I.J. Thompson, and J.A. Tostevin, Nucl. Phys. **A733**, 53 (2004).
23. M. Labiche, F.M. Marques, O. Sorlin, and N. Vinh Mau, Phys. Rev. C **60**, 027303 (1999).
24. P. Capel, D. Baye, Phys. Rev. C **70**, 064605 (2004).
25. V. Guimarães *et al.*, Phys. Rev. C **61**, 064609 (2000).

FULLY MICROSCOPIC CALCULATIONS FOR CLOSED-SHELL NUCLEI WITH REALISTIC NUCLEON-NUCLEON POTENTIALS

L. CORAGGIO, A. COVELLO, A. GARGANO, and N. ITACO

Dipartimento di Scienze Fisiche, Università di Napoli Federico II, and Istituto Nazionale di Fisica Nucleare, Complesso Universitario di Monte S. Angelo, Via Cintia - I-80126 Napoli, Italy

The ground-state energy of the doubly magic nuclei ^{4}He and ^{16}O has been calculated within the framework of the Goldstone expansion starting from modern nucleon-nucleon potentials. A low-momentum potential $V_{\rm low-k}$ has been derived from the bare potential by integrating out its high-momentum components beyond a cutoff Λ. We have employed a simple criterion to relate this cutoff momentum to a boundary condition for the two-nucleon model space spanned by a harmonic-oscillator basis. Convergence of the results has been obtained with a limited number of oscillator quanta.

1. Introduction

As is well known, the strong repulsive components in the high-momentum regime of a realistic nucleon-nucleon (NN) potential V_{NN} need to be renormalized in order to perform perturbative nuclear structure calculations. In Refs. [1,2] a new method to renormalize the NN interaction has been proposed, which consists in deriving an effective low-momentum potential $V_{\rm low-k}$ that satisfies a decoupling condition between the low- and high-momentum spaces. This $V_{\rm low-k}$ preserves exactly the on-shell properties of the original V_{NN} up to a cutoff momentum Λ, and is a smooth potential which can be used directly in nuclear structure calculations.

In the past few years, we have employed this approach to calculate the ground-state (g.s.) properties of doubly closed-shell nuclei within the framework of the Goldstone expansion [3,4], using a fixed value of the cutoff momentum.

Recently, we have investigated how the cutoff momentum Λ is related to the dimension of the configuration space in the coordinate representation [5], where our calculations are performed. We have shown how the choice of a cutoff momentum corresponds to fix a boundary for the two-nucleon model space.

In the present work, we calculate the g.s. energy of ^{4}He and ^{16}O in the framework of the Goldstone expansion with different NN potentials. To verify the validity of our approach, we compare the ^{4}He results with those obtained using the Faddeev-Yakubovsky (FY) method.

The paper is organized as follows. In Sec. 2 we give a brief description of our calculations. Sec. 3 is devoted to the presentation and discussion of our results for ^{4}He and ^{16}O. A summary of our study is given in Sec. 4.

2. Outline of calculations

As mentioned in the Introduction, the short-range repulsion of the NN potential is renormalized integrating out its high-momentum components through the so-called $V_{\rm low-k}$ approach (see Refs. [1,2]). The $V_{\rm low-k}$ preserves the physics of the two-nucleon system up to the cutoff momentum Λ. While this low-momentum potential is defined in the momentum space, we perform our calculations for finite nuclei in the coordinate space employing a truncated HO basis. This makes it desirable to map the cutoff momentum Λ, which decouples the momentum space into a low- and high-momentum regime, onto a boundary for the HO space [5].

If we consider the two-nucleon relative motion in a HO well in the momentum representation, then, for a given maximum relative momentum Λ, the corresponding maximum value of the energy is

$$E_{\rm max} = \frac{\hbar^2 \Lambda^2}{M} \quad , \tag{1}$$

where M is the nucleon mass.

We rewrite this relation in the relative coordinate system in terms of the maximum number $N_{\rm max}$ of HO quanta:

$$\left(N_{\rm max} + \frac{3}{2} \right) \hbar\omega = \frac{\hbar^2 \Lambda^2}{M} \quad , \tag{2}$$

for a given HO parameter $\hbar\omega$. The above equation provides a simple criterion to map out the two-nucleon HO model space. Let us write the two-nucleon states as the product of HO wave functions

$$|a\ b\rangle = |n_a l_a j_a,\ n_b l_b j_b\rangle \quad . \tag{3}$$

We define our HO model space as spanned by those two-nucleon states that satisfy the constraint

$$2n_a + l_a + 2n_b + l_b \leq N_{\max} \ . \tag{4}$$

In this paper, making use of the above approach, we have calculated the g.s. energies of ^{4}He and ^{16}O within the framework of the Goldstone expansion [6]. We start from the intrinsic hamiltonian

$$H = \left(1 - \frac{1}{A}\right) \sum_{i=1}^{A} \frac{p_i^2}{2M} + \sum_{i<j} \left(V_{ij} - \frac{\mathbf{p}_i \cdot \mathbf{p}_j}{MA}\right) \ , \tag{5}$$

where V_{ij} stands for the renormalized V_{NN} potential plus the Coulomb force, and construct the Hartree-Fock (HF) basis expanding the HF single particle (SP) states in terms of HO wave functions. The following step is to sum up the Goldstone expansion including contributions up to fourth-order in the two-body interaction. Using Padé approximants [7,8] one may obtain a value to which the perturbation series should converge. In this work, we report results obtained using the Padé approximant [2|2], whose explicit expression is

$$[2|2] = \frac{E_0(1+\gamma_1+\gamma_2) + E_1(1+\gamma_2) + E_2}{1+\gamma_1+\gamma_2} \ , \tag{6}$$

where

$$\gamma_1 = \frac{E_2 E_4 - E_3^2}{E_1 E_3 - E_2^2} \ , \qquad \gamma_2 = -\frac{E_3 + E_1 \gamma_1}{E_2} \ ,$$

E_i being the ith order energy contribution in the Goldstone expansion.

Our calculations are made in a truncated model space, whose size is related to the values of the cutoff momentum Λ and the $\hbar\omega$ parameter. The calculations are performed increasing the $N_{\max}$ value (and consequently Λ) and varying $\hbar\omega$ until the dependence on $N_{\max}$ (Λ) is minimized.

3. Results

We have calculated the binding energy of ^{4}He using different V_{NN}'s, and compared our results with those obtained by means of the FY method. This comparison is made in order to test the reliability of our approach. In Figs. 1, 2, and 3 the calculated ^{4}He g.s. energies obtained from the CD-Bonn [9],

N^3LO [10], and Bonn A [11] NN potentials are reported, for different values of $\hbar\omega$, as a function of the maximum number N_{max} of HO quanta. The FY result [12,13] is also shown.

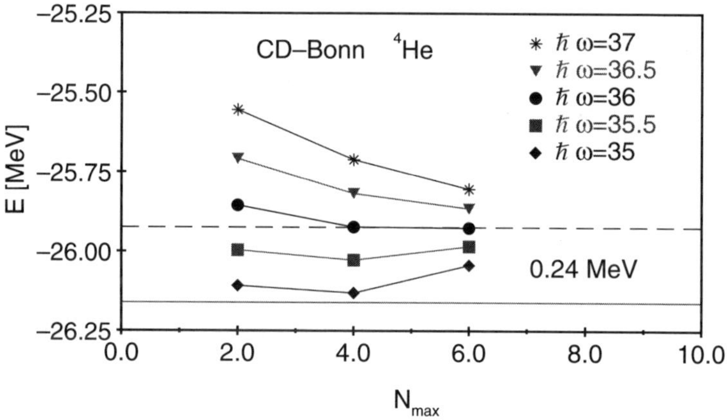

Fig. 1. Ground state energy of ^{4}He calculated with the CD-Bonn potential as function of N_{max}, for different values of $\hbar\omega$. The straight line represents the Faddeev-Yakubovsky result, while the dashed one our converged result. The difference in energy between the latter and the Faddeev-Yakubovsky result is also reported.

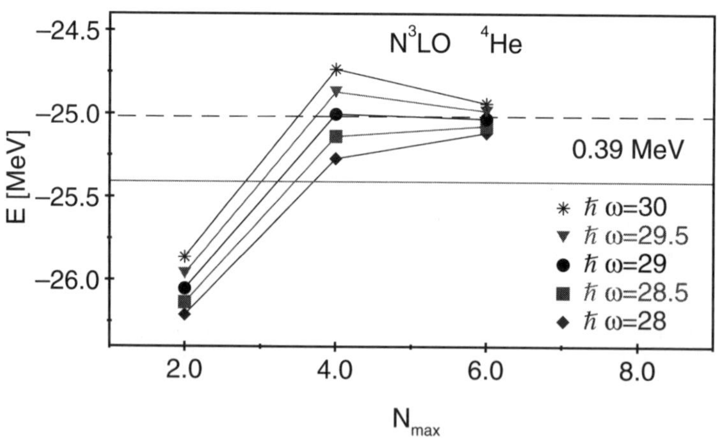

Fig. 2. Same as Fig. 1, but for the N^3LO potential.

For the sake of clarity, in Table 1 we report the numerical values obtained with the CD-Bonn potential. From the inspection of Table 1 it can be seen

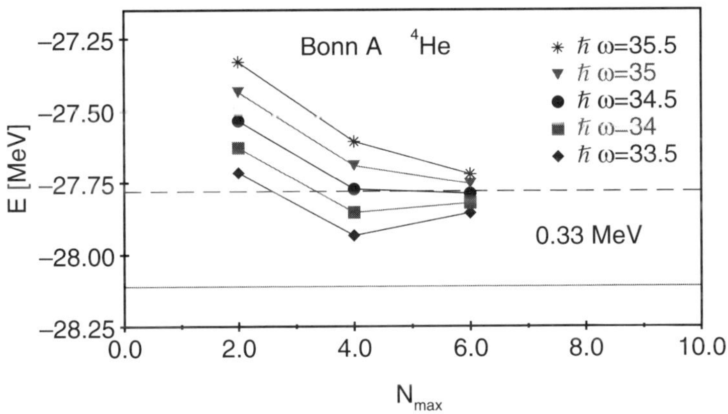

Fig. 3. Same as Fig. 1, but for the Bonn A potential.

that the g.s. energy does not change increasing $N_{\max}$ from 4 to 6 for $\hbar\omega = 36$ MeV. On these grounds, we choose as our final result that corresponding to the above $\hbar\omega$ value, i.e. -25.92 MeV. Moreover, we find it worthwhile to introduce a theoretical error due to the uncertainty when choosing $\hbar\omega_{\text{conv}}$, which corresponds to the one with the faster convergence with $N_{\max}$. We estimate this error as the largest difference in energy between the final result and those corresponding to the two $\hbar\omega$ values adjacent to $\hbar\omega_{\text{conv}}$. For the CD-Bonn potential, we see that this difference is 0.05 MeV for the largest $N_{\max}$.

Table 1. Ground state energy of ^{4}He (in MeV) calculated with the CD-Bonn potential for different values of $\hbar\omega$ and $N_{\max}$

$\hbar\omega$ (in MeV)	$N_{\max} = 2$	$N_{\max} = 4$	$N_{\max} = 6$
35.0	-26.11	-26.13	-26.04
35.5	-25.10	-26.03	-25.97
36.0	-25.86	-25.92	-25.92
36.5	-25.71	-25.83	-25.87
37.0	-25.56	-25.71	-25.80

Similarly, the results for the ^{4}He g.s. energy with the N^{3}LO and the Bonn A potentials are (-25.02 ± 0.05) and (-27.78 ± 0.03) MeV, respectively. These values are in good agreement with the FY results, the largest discrepancy being 0.39 MeV for N^{3}LO potential.

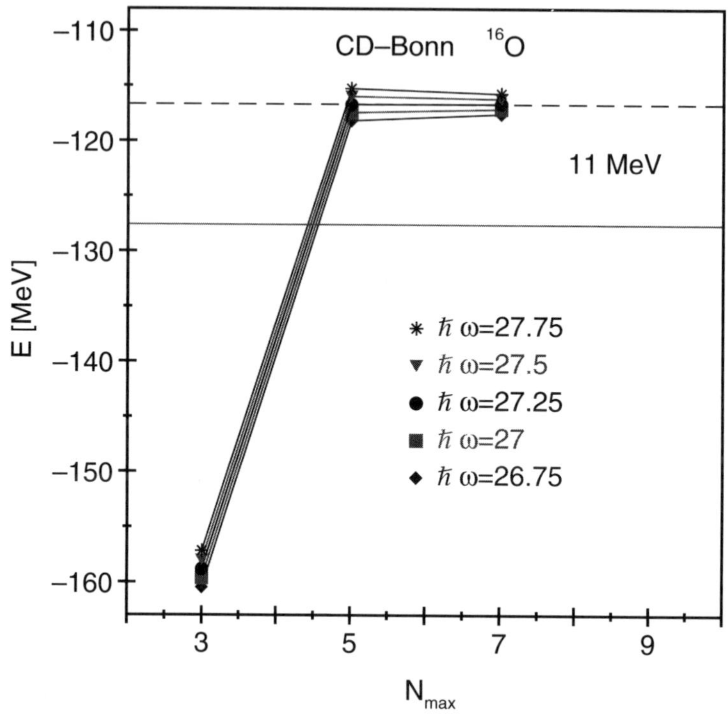

Fig. 4. Ground state energy of ^{16}O calculated with the CD-Bonn potential as function of $N_{\rm max}$, for different values of $\hbar\omega$. The straight line represents the experimental value [14], while the dashed one our converged result. The difference in energy between the latter and the experimental value is also reported.

We have also calculated the g.s. energy of ^{16}O starting from both the CD-Bonn and the Bonn A potential, as reported in Figs. 4 and 5, respectively. With the CD-Bonn potential, the converged value, obtained for $\hbar\omega =$ 27.25 MeV, is equal to (-117 ± 1) MeV, the discrepancy with the experimental value [14] being 11 MeV. This value is slightly different (≈ 1 MeV) from the one reported in our previous paper [5], because in the present work we have decreased by a factor 2 the spacings between the $\hbar\omega$ values. It is worth to point out that this result is consistent with those obtained by Fujii *et al.* using the unitary model-operator approach [15], and by Vary *et al.* in the no-core shell model framework [16].

A better agreement with experiment is obtained using the weaker tensor force NN potential Bonn A, our ^{16}O g.s. energy being (-130.0 ± 0.5) MeV.

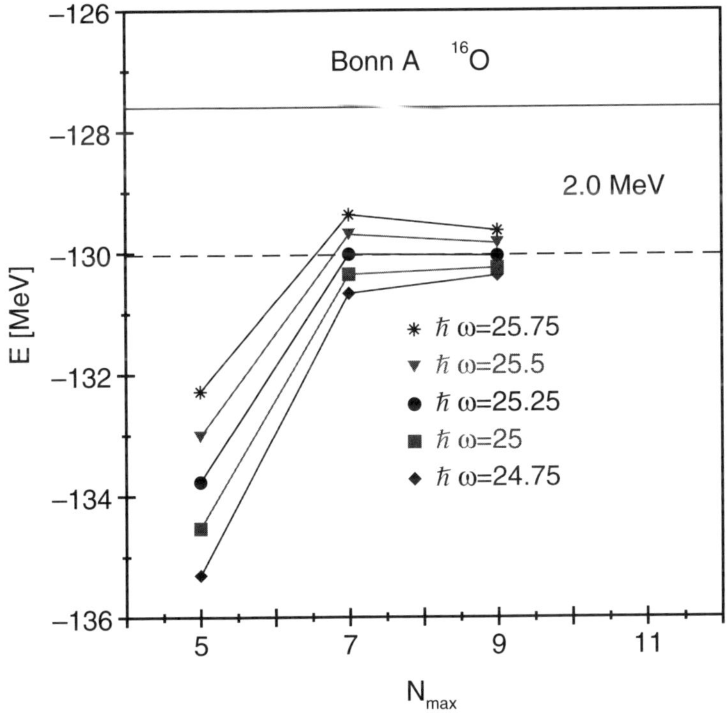

Fig. 5. Same as Fig. 4, but for the Bonn A potential.

4. Summary

In this work, we have calculated the g.s. energy of the doubly closed-shell nuclei ^{4}He and ^{16}O in the framework of the Goldstone expansion, starting from different realistic NN potentials. In order to renormalize their short-range repulsion, the high-momentum components of these potentials have been integrated out through the so-called $V_{\rm low-k}$ approach. We have employed a criterion to map out the model space of the two-nucleon states in the HO basis according to the value of the cutoff momentum Λ [5].

To show the validity of this procedure, we have calculated the g.s. energy of ^{4}He, with the CD-Bonn, N^{3}LO, and Bonn A potentials, comparing the results with the FY ones. We have found that the energy differences do not exceed 0.39 MeV. The limited size of the discrepancies evidences that our approach provides a reliable way to renormalize the NN potentials preserving not only the two-body but also the many-body physics.

On the above grounds, we have performed similar calculations for ^{16}O with the CD-Bonn and Bonn A NN potentials, and obtained converged results using model spaces not exceeding $N_{\max} = 9$. The rapid convergence of the results with the size of the HO model space makes it very interesting to study heavier systems employing our approach [17].

5. Acknowledgments

This work was supported in part by the Italian Ministero dell'Istruzione, dell'Università e della Ricerca (MIUR).

References

1. S. Bogner, T. T. S. Kuo and L. Coraggio, Nucl. Phys. **A684**, 432c (2001).
2. Scott Bogner, T. T. S. Kuo, L. Coraggio, A. Covello, and N. Itaco, Phys. Rev. C **65**, 051301(R) (2002).
3. L. Coraggio, N. Itaco, A. Covello, A. Gargano, and T. T. S. Kuo, Phys. Rev. C **68**, 034320 (2003).
4. L. Coraggio, A. Covello, A. Gargano, N. Itaco, T. T. S. Kuo, and R. Machleidt, Phys. Rev. C **71**, 014307 (2005).
5. L. Coraggio, A. Covello, A. Gargano, N. Itaco, and T. T. S. Kuo, Phys. Rev. C **73**, 014304 (2006).
6. J. Goldstone, Proc. R. Soc. London Ser. A **239**, 267 (1957).
7. G. A. Baker Jr and J. L. Gammel, *The Padé Approximant in Theoretical Physics*, Mathematics in Science and Engineering Vol. 71, edited by G. A. Baker Jr and J. L. Gammel (Academic Press, New York, 1970).
8. N. Y. Ayoub and H. A. Mavromatis, Nucl. Phys. **A323**, 125 (1970).
9. R. Machleidt, Phys. Rev. C **63**, 024001 (2001).
10. D. R. Entem and R. Machleidt, Phys. Rev. C **68**, 041001(R) (2003).
11. R. Machleidt, K. Holinde, and Ch. Elster, Phys. Rep. **149**, 1 (1987).
12. W. Glöckle and H. Kamada, Phys. Rev. Lett. **71**, 971 (1993).
13. A. Nogga, private communication.
14. G. Audi, A. H. Wapstra, and C. Thibault, Nucl. Phys. **A729**, 337 (2003).
15. S. Fujii, R. Okamoto, and K. Suzuki, Journal of Physics: Conference Series **20**, 83 (2005).
16. J. P. Vary *et al.*, Eur. Phys. J. A **25**, s01, 475 (2005).
17. L. Coraggio, A. Covello, A. Gargano, N. Itaco, and T. T. S. Kuo, in preparation.

SPATIAL DEPENDENCE OF PAIRING FIELD INDUCED BY THE EXCHANGE OF COLLECTIVE VIBRATIONS

A. PASTORE[1,2], F. BARRANCO[4], R.A. BROGLIA[1,2,3], G. POTEL[1,2], and E. VIGEZZI[2]

[1] *Dipartimento di Fisica, Università degli Studi, Milano, Italy*
Via Celoria 16, 20133 Milano, Italy
[2] *INFN, Sezione di Milano,*
Via Celoria 16, 20133 Milano, Italy
[3] *The Niels Bohr Institute, University of Copenhagen*
Blegdamsvej 17, 2100 Copenhagen Ø(Denmark)
[4] *Departamento de Fisica Aplicada III, Escuela Superior de Ingenieros,*
Camino de los Descubrimientos s/n, 41092 Sevilla, Spain

The interaction induced by the exchange of low-lying surface vibrations between pairs of orbitals close to the Fermi surface provides an important contribution to pairing correlations in superfluid nuclei. We study the spatial dependence of the pairing field produced by this induced interaction in ^{120}Sn.

1. Introduction

The superfluid properties of nuclear systems are strongly influenced by polarization phenomena. In particular, the exchange of low-lying surface vibrations between pairs of nucleons close to the Fermi surface gives rise to an attractive induced interaction, that accounts for about one half of the value of the pairing gap Δ_{exp} derived from the experimental odd-even mass differences in superfluid nuclei like ^{120}Sn [1,2]. Adding the matrix elements of the bare and of the induced interaction and then solving the BCS equations one obtains values of the pairing gaps in good agreement with Δ_{exp}, if one takes into account the other renormalization processes (self-energy and vertex corrections) [3]. Since the matrix elements of the induced interaction depend on specific properties of individual nuclei (in particular, on the features of their low-lying surface modes), it becomes possible to study detailed isotopic effects that can be obscured in phenomenological interactions like Gogny or Skyrme forces, which are devised to reproduce the pairing gaps on a global scale.

In this paper we study the induced interaction in coordinate space. The analysis will be performed for the case of ^{120}Sn. In the future we plan to use the method in the case of exotic, loosely-bound nuclei, where renormalization processes play an essential role.

2. The induced interaction

The matrix elements of the induced interaction can be written

$$\langle \nu'\bar{\nu}'|v_{ind}|\nu,\bar{\nu}\rangle = 2\sum_{LMn}\frac{<\nu|f_{Ln}Y_{LM}|\nu'><\bar{\nu}|f_{Ln}Y^*_{LM}|\bar{\nu}'>}{E_0-|e_\nu-e_F|-|e_{\nu'}-e_F|-\hbar\omega_{Ln}}, \qquad (1)$$

where L, M and n denote the quantum numbers of the exchanged collective vibrations, $f_{Ln}(r)=\beta_{Ln}R_0(dU/dr)$ is the associated radial formfactor (β_{Ln} being the deformation parameter associated with the n−th mode of multipolarity L, R_0 the ground-state radius, and U the average potential), $\hbar\omega_{Ln}$ is the vibrational energy of the n−th mode, while e_ν and e_F are the single-particle and Fermi energies [1,4]. E_0 is the pairing correlation energy per Cooper pair, which is of the order of -2Δ. In practice we have used E_0= - 2 MeV. The pre-factor 2 in Eq. (1) arises from the two possible time orderings associated with the one-phonon exchange. The matrix elements can then be introduced in the BCS or HFB gap equations, in order to obtain the associated pairing gap. We shall study the case of ^{120}Sn, using phonons calculated in the random phase approximation with a separable force, adjusted so as to reproduce the transition strength of the experimental phonons $2^+,3^-,4^+,5^-$ [5]. The calculation is essentially the same as in ref. [1], where a state-dependent pairing gap was obtained. In Fig. 1 we show the matrix elements (1) calculated as a function of the single particle energy, using a semiclassical approximation [4]. It is seen that they are strongly peaked around the Fermi surface. In the following we shall instead try to obtain an expression for the associated pairing field in coordinate space.

3. Pairing gap in coordinate space

We first collect some standard expressions, valid for the case of the $^1S_0$ pairing gap obtained in spherical nuclei with a local, finite range interaction, like the Argonne or the Gogny potentials [6]. The state dependent pairing gaps Δ_{nlj}, the quasiparticle energies E^q_{lj} and the occupation amplitudes U^q_{nlj}, V^q_{nlj} are obtained from the HFB equations expressed in a

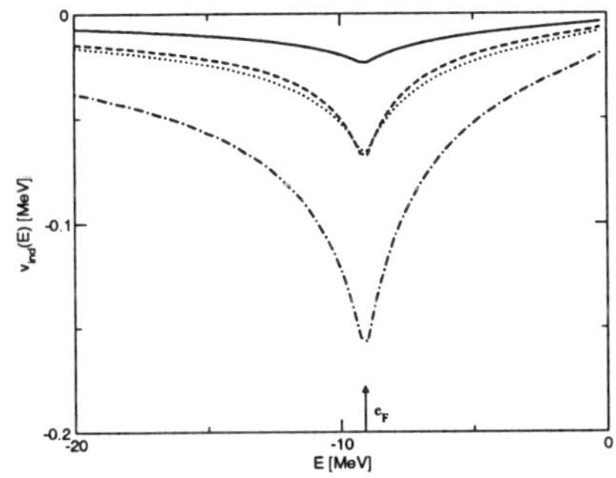

Fig. 1. Diagonal matrix elements of the induced interaction associated with the exchange of collective vibrations of multipolarities L =2,3 and 4 (dashed, dotted and solid line respectively). The sum of the three contributions is displayed by means of a dash-dotted line.

single-particle basis with eigenfunctions u_{nlj}:

$$(\epsilon_{nlj} - \epsilon_F)U^q_{nlj} + \sum_{n'} \Delta_{nn'lj} V^q_{n'lj} = E^q_{lj} U^q_{nlj}$$
$$-(\epsilon_{nlj} - \epsilon_F)V^q_{nlj} + \sum_{n'} \Delta_{nn'lj} U^q_{n'lj} = E^q_{lj} V^q_{nlj}. \quad (2)$$

where

$$\Delta_{nn'lj} = - < nn'lj|v(|\vec{r}_1 - \vec{r}_2|)|\Phi >, \quad (3)$$

with

$$|\Phi >= \sum_{nn'lj} \left(\sum_q U^q_{nlj} V^q_{n'lj} \right) \frac{2j+1}{2} |nn'lj > . \quad (4)$$

Actually, we shall solve the HFB equations selfconsistently only in the pairing channel, while in the particle-hole channel we shall use a fixed Woods-Saxon potential, parameterized as in ref. [7]. Let us now introduce the spatial representation of the pairing field, $\Delta(\vec{r}_1, \vec{r}_2)$, which is equal to

$$\Delta(\vec{r}_1, \vec{r}_2) = -v(r_{12}) \Phi_{S=0}(\vec{r}_1, \vec{r}_2), \quad (5)$$

where $v(r_{12})$ is the ($S = 0$ channel) two-body interaction, and $\Phi_{S=0}(\vec{r}_1, \vec{r}_2)$ is the anomalous density in coordinate space:

$$\Phi_{S=0}(\vec{r}_1, \vec{r}_2) = \sum_{nn'lj} \left(\sum_q U^q_{nlj} V^q_{n'lj} \right) \frac{2j+1}{2} \psi_{nn'lj}, \quad (6)$$

expressed in terms of the $S = 0$ part of the two-particle wavefunction $< \vec{r}_1, \vec{r}_2 | nn'lj >$

$$\psi_{nn'lj} = \frac{1}{4\pi} u_{nlj}(r_1) u_{n'lj}(r_2) P_l(\cos\theta_{12}). \tag{7}$$

In the following it will be useful to express the pairing gap and the anomalous density as a function of the moduli of the center of mass and of the relative distance, R_{cm} and r_{12}, and of the angle α between them. Using the HFB gap equation, one can show that the projection of $\Delta(\vec{r}_1, \vec{r}_2)$ onto the state $\psi_{nn'lj}$ is precisely the pairing field matrix element $\Delta_{nn'lj}$(cf. (3)). At this point, we observe that the basis $\psi_{nn'lj^\uparrow}$, with $j^\uparrow = l + 1/2, l = 0, 1, ...$ and the basis $\psi_{nn'lj^\downarrow}$, with $j^\downarrow = 1/2, l = 0$ and $j^\downarrow = l - 1/2, l = 1, 2, ...$ both represent a complete basis on which it is possible to expand the function $\Delta(\vec{r}_1, \vec{r}_2)$. Correspondingly, we can write

$$\Delta(\vec{r}_1, \vec{r}_2) = \sum_{nn'lj^\uparrow} (2l+1) \Delta_{nn'lj} \psi_{nn'lj^\uparrow} = \sum_{nn'lj^\downarrow} (2l+1) \Delta_{nn'lj} \psi_{nn'lj^\downarrow}, \tag{8}$$

whose projections onto the states $\psi_{nn'lj^\uparrow}$ and $\psi_{nn'lj^\downarrow}$ are $\Delta_{nn'lj^\uparrow}$ and $\Delta_{nn'lj^\downarrow}$ respectively (the $2l+1$ weighting factor accounts for the normalization of the $\psi_{nn'lj}$ waves).

In Fig.2 we show as an example the pairing field (5) associated with the simple attractive Gaussian interaction [8]:

$$v(r_{12}) = -V_0 \cdot \exp\left(-(r_{12}/r_0)^2\right), \tag{9}$$

with $V_0 = -28$ MeV, $r_0 = 1.6$ fm. The pairing field is shown for a fixed value of the angle α between $\vec{R}_{cm}$ and $\vec{r}_{12}$; this angular dependence in practice turns out to be very weak. We have verified that including states up to E_{cut}= 200 MeV we can reproduce very well the pairing field (5) with the expression (8). This would not be the case for repulsive, short-range interactions, where one should sum over a much larger basis.

Turning now to the case of the induced interaction, in which we do not have Eq. (5) at our disposal, we can use Eq. (8) to define two pairing fields:

$$\Delta^\uparrow = \sum_{nn'lj^\uparrow} (2l+1) \Delta_{nn'lj} \psi_{nn'lj^\uparrow}; \Delta^\downarrow = \sum_{nn'lj^\downarrow} (2l+1) \Delta_{nn'lj} \psi_{nn'lj^\downarrow}, \tag{10}$$

where the quantities $\Delta_{nn'lj}$ are obtained from the solution of the HFB equation with the matrix elements (1). Correspondingly, we can introduce the two anomalous densities $\Phi^\uparrow_{S=0}$ and $\Phi^\downarrow_{S=0}$, restricting the sum over j in Eq. (6) to $j^\uparrow$ and to $j^\downarrow$, respectively.

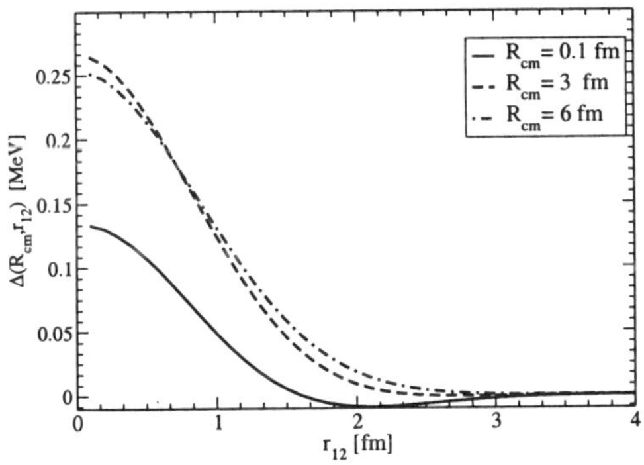

Fig. 2. The pairing field associated with the interaction (9) is shown as a function of the relative distance r_{12} for various values of the center of mass R_{cm}.

The pairing energy can be obtained from $E_{pair} = E^{\uparrow}_{pair} + E^{\downarrow}_{pair} - E_{l=0}$, where

$$E^{\uparrow}_{pair} = \frac{1}{2}\int d\tau \Delta^{\uparrow}(\vec{r}_1,\vec{r}_2)\Phi^{\uparrow}(\vec{r}_1,\vec{r}_2)\ ;\ E^{\downarrow}_{pair} = \frac{1}{2}\int d\tau \Delta^{\downarrow}(\vec{r}_1,\vec{r}_2)\Phi^{\downarrow}(\vec{r}_1,\vec{r}_2), \tag{11}$$

and one subtracts the contribution from the sum over $l = 0$ which enters into both $\Delta^{\uparrow}$ and $\Delta^{\downarrow}$ and otherwise would be counted twice.

The fields $\Delta^{\uparrow}$ and $\Delta^{\downarrow}$ can be quite different, because the matrix elements of the induced interaction are strongly peaked around the Fermi surface, and are much smaller if the involve a spin flip. In the present case, the level $1h_{11/2}$, which is the only level above the Fermi surface plays a key role, so that $\Delta^{\uparrow}$ dominates, and $E^{\uparrow}_{pair}$ and $E^{\downarrow}_{pair}$ are equal respectively to -4.56 MeV and to -1.49 MeV while $E_{l=0} = -0.5$ MeV, so that $E_{pair} = -5.55$ MeV. The pairing field $\Delta^{\uparrow}$ is shown in Fig. 3 as a function of the relative distance r_{12} for different values of the center of mass of the pair, R_{cm}. As expected, the largest contribution comes from values of R_{cm} close to the surface of the nucleus, $R_{cm} = 6.2$ fm, with a range of the order of 2 fm. Although their contribution to the pairing energy is very small, it is interesting to observe that there are also contributions from large values of r_{12}, when $R_{cm} \approx 0$. They arise from nucleons which are on opposite sides of the surface. In Fig. 4 we show the function $\Phi^{\uparrow}_{S=0}$. We can also try to devise an approximate expression for the induced interaction in coordinate space, fitting a function $v_{ind}(r_{12})$ so that the relation $\Delta(r_{12}) \approx v_{ind}(r_{12})\Phi_{S=0}(r_{12})$ (cf. Eq. (5)) is fulfilled for given values of R_{cm} (as we have already mentioned, the angular dependence plays a little role, and we have put $\alpha = 0$).

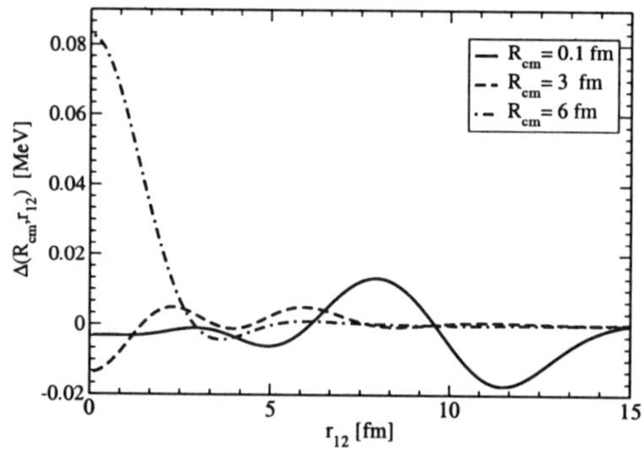

Fig. 3. The pairing field $\Delta^{\uparrow}$ is shown as a function of r_{12} for various values of R_{cm}.

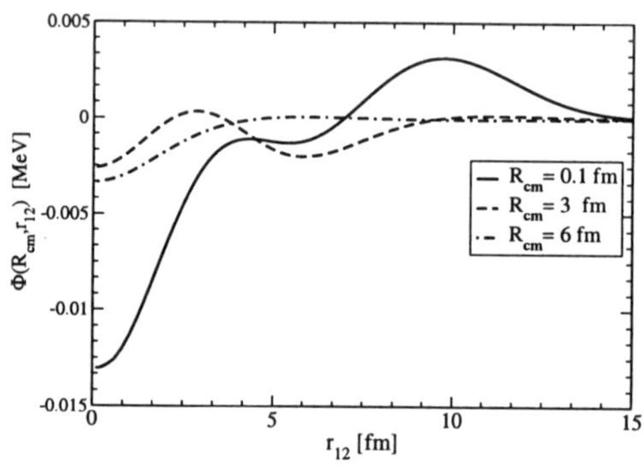

Fig. 4. The anomalous density $\Phi^{\uparrow}_{S=0}$ is shown as a function of r_{12} for various values of R_{cm}.

More precisely, we have performed two separate fits, one for $\Delta^{\uparrow}$ and one for $\Delta^{\downarrow}$, and have obtained correspondingly two interactions $v^{\uparrow}_{ind}$ and $v^{\downarrow}_{ind}$. The functional form we have adopted for v_{ind} is a simple gaussian

$$v_{ind}(r_{12}) = -b \cdot \exp\left(-\left((r_{12}-c)/a\right)^2\right), \tag{12}$$

where the a, b, c parameters are determined for every value of R_{cm}. In practice we have found that an overall fit can be obtained using the fixed value $a = 1.8$ fm for the width parameter, and the simple and intuitive function $c = 2|R_{nuc} - R_{cm}|$ for the centroid parameter, where R_{nuc} is the radius of the Woods-Saxon potential, so that at least one of the two nucleons is close to the surface . The values obtained for the strength parameter as a function of R_{cm} are presented in Fig. 5.

Note the smaller intensity of $v^{\downarrow}_{ind}$. The surface character of v_{ind} is clearly

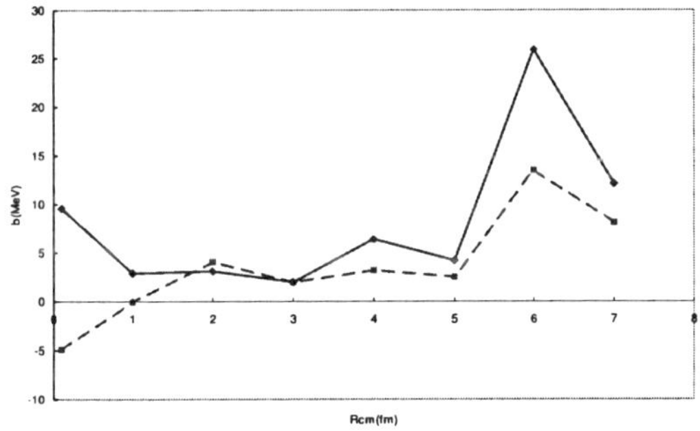

Fig. 5. Strength parameter of the gaussian fit to the induced interaction, both for $v^{\uparrow}_{ind}$(solid line) and $v^{\downarrow}_{ind}$ (dashed line).

seen in the figure, in keeping with Eq. (1) and its connection with the coupling to surface vibrations. In fact it should be possible to establish a link between the fitted gaussian parameters and the basic phonon data. As a first attempt one can try to parametrize the gaussian strength with the standard $R_{cm}dU/dR_{cm}$ surface peaked function. A good description for the most relevant region ($R_{cm} > 3$fm) can be obtained with

$$b^{\uparrow} = 0.5R_{cm}dU/dR_{cm}; \qquad b^{\downarrow} = 0.3R_{cm}dU/dR_{cm}. \tag{13}$$

The scale of the prefactors in Eq. (12) lies within the range of the usual zero-point deformation parameters associated to the collective low-lying states. One can also relate the width of the gaussians to zero point motion. In fact the ratio $a/R_{nuc} = 0.3$ is again in the range of the usual zero-point deformation parameter associated with collective low-lying states.

We remark that in several applications, and in particular when HFB equations are solved in coordinate space, a zero-range interaction is used, leading to a local pairing field $\Delta(r)$. We can obtain an approximate local expression from the complete $\Delta(R_{cm}, r_{12})$, in two steps. We first perform a Fourier transform in the variable r_{12}, obtaining the pairing field $\Delta(k, R_{cm})$ as a function of the relative momentum and of the center of mass. We then use a semiclassical approximation, $\Delta(r) \approx \Delta(k_F(R_{cm} = r), R_{cm})$, where $k_F(R_{cm})$ denotes the local Fermi momentum [6,8]. The pairing fields obtained in this way for the case of the Gaussian interaction (9) and for the induced interaction are shown in Fig. 6. Solving HFB equations in coordinate space [9] with the approximate pairing field in the case of the induced interaction, we obtained a total pairing energy of -4.65 MeV instead of the exact value -5.55 MeV reported above. We also note that the form of the pairing gap in this last case is similar to that obtained with phenomenological density-dependent, zero range pairing forces, using the values of the parameters $V_0 = 400$ MeV fm^3, $\alpha = 1.2$, $\eta = 1$ and cut off energy E_{cutoff}=60 MeV, in the expression [10] $v(\vec{r}_1, \vec{r}_2) = V_0\,(1 - \eta\,(\rho(R_{cm})/\rho_0)^{\alpha})\,\delta(\vec{r}_1 - \vec{r}_2)$, where ρ_0 denotes the saturation density.

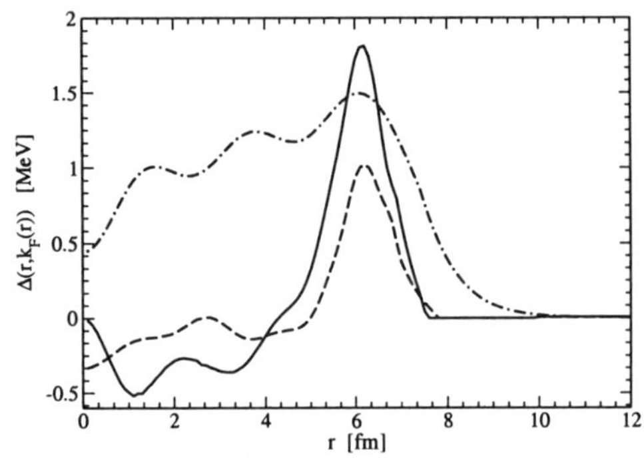

Fig. 6. The semiclassical approximation to the pairing gap is shown for the pairing fields associated with the interaction (9) (dash-dotted line) and for the pairing fields $\Delta^{\uparrow}$ and $\Delta^{\downarrow}$ associated with the parameterizations $v^{\uparrow}_{ind}$ (solid line) and $v^{\downarrow}_{ind}$ (dashed line).

References

1. F. Barranco *et al.*, Phys. Rev. Lett. **83**, 2147 (1999).
2. R.A. Broglia *et al.*, Phys Lett. **B37**, 159 (1971).
3. F. Barranco *et al.*, Eur. Phys. J. A **21**, 57 (2004).
4. F. Barranco, P.F. Bortignon, R.A. Broglia, G. Coló, P. Schuck, E. Vigezzi, and X. Viñas, Phys.Rev. C **72**, 054314 (2005).
5. N.G. Jonsson *et al.*, Nucl. Phys. **A417**, 326 (1981).
6. F.Barranco, R.A. Broglia, H. Esbensen, and E. Vigezzi, Phys. Rev. C **58**, 1257 (1998).
7. A.Bohr and B.R. Mottelson, *Nuclear Structure*, Vol. I, Benjamin (1965).
8. P.Ring and P.Schuck, *The Nuclear Many-Body Problem*, Springer-Verlag (2000), p. 551.
9. I. Hamamoto and B.R. Mottelson, Phys. Rev. C **68**, 034312 (2003).
10. G.F. Bertsch and H. Esbensen, Ann. Phys. (N.Y.) **209**, 327 (1991); E. Garrido, P. Sarriguren, E. Moya de Guerra, and P. Schuck, Phys. Rev. C **60**, 064312 (1999).

Nuclear Astrophysics

NUCLEAR ASTROPHYSICS

E. VIGEZZI
INFN, Sezione di Milano
Via Celoria 16, 20133 Milano, Italy
E-mail: vigezzi@mi.infn.it

The activity of the Italian community in the period 2004-2006 in the field of compact stars is reviewed.

1. Introduction

In this report I shall summarize the activity in the field of theoretical nuclear astrophysics, carried out in the period 2004-2006 by the Italian groups which regularly meet at Cortona. The present contribution is the logical continuation of two previous analogous reports [1,2].

As a matter of fact, during this period the work of this community has largely concentrated on various aspects of the physics of compact stars, and I shall limit myself to this topic. Section 2 deals with the inner crust of neutron stars, and especially its superfluid properties, also considering some natural connections with finite nuclei. Section 3 deals with the core of compact stars, which reaches densities up to several times the saturation density. In this case, the main issue has concerned the possible transition from hadronic to quark matter. Some of the connections between theoretical models and observables are briefly sketched in Section 4.

2. The inner crust of neutron stars

Neutron stars are created in a supernova explosion caused by the gravitational collapse of a massive star. The resulting protoneutron star reaches temperatures of the order of 50 MeV, but it cools very rapidly, becoming transparent to neutrinos. Subsequently the star becomes isothermal, with a temperature of the order of 10-100 keV. Most of the work referred below deals with this regime of cold neutrons stars; a few exceptions will be explicitly indicated.

The density of the star increases as one proceeds from the surface towards the interior. One first encounters the outer crust, consisting of a Coulomb lattice of heavy nuclei, in β−equilibrium with a relativistic electron gas. The neutron chemical potential increases in order to maintain β-equilibrium, until it becomes positive, and neutrons start to drip out. This defines the outer boundary of the inner crust. The presently accepted theoretical description of the inner crust is essentially the one proposed in 1973 by Negele and Vautherin [3], who performed a microscopic self-consistent calculation based on the density matrix expansion method (close to a Hartree-Fock calculation with a Skyrme functional). According to their results, for baryon densities $0.001\rho_0 \lesssim \rho \lesssim 0.3\rho_0$, where ρ_0 is the saturation density, the structure of a neutron star consists of a lattice of spherically symmetric nuclear clusters, accomodating a little more than a hundred bound neutrons, immersed in a sea of delocalized neutrons and electrons. Proton single-particle levels are deeply bound and localized inside the clusters. Negele and Vautherin determined the lattice step and the number of protons in the clusters at a given density, imposing the condition of β−stability and minimizing the energy. They worked within the Wigner-Seitz cell approximation, that is, replacing the crystal by a sum of identical spherical cells with a nuclear cluster at the center.

They found that the step decreases going deeper into the crust, and that the favoured number of protons is determined by proton shell effects, being $Z = 40$ (at lower and higher densities) or $Z = 50$ (at intermediate densities), or $Z = 32$ (at the inner boundary).

More recently, other studies [4] have shown that at large densities ($\rho \gtrsim 0.3\rho_0$) non-spherical clusters of various shapes (like rod and plate geometries) are energetically favoured, until for $\rho \gtrsim 0.5\rho_0$ (the outer boundary of the crust) the structures merge in uniform, very asymmetric nuclear matter (with a proton fraction of the order of a few per cent).

The properties of the neutron sea in the inner crust can be deduced by studies of uniform matter. The Equation of State (EoS) for uniform neutron and nuclear matter in the density range $0.01\ \text{fm}^{-3} \lesssim \rho \lesssim 0.12\ \text{fm}^{-3}$ has been recently studied within the Bethe-Brueckner-Goldstone approach with microscopic realistic potentials [5]. This calculation is expected to be quite accurate, because three-body forces play a negligible role at these densities. The results have been compared with phenomenological Skyrme and Gogny forces, finding good agreement for nuclear matter, but discrepancies of the order of 1 MeV/particle for neutron matter, similar to previous studies [6]. It is interesting to observe that the derived functional, supplemented with a gradient and a Coulomb term, and fitting only two parameters, could reproduce fairly well the binding energy and the charge radius of several magic nuclei (see also ref. [7] for a related approach).

According to BCS calculations with realistic forces [8], homogeneous neutron matter at the densities typical of the inner crust should be superfluid in the $^1S_0$ channel: the pairing gap Δ displays a bell-shape dependence on Fermi momentum, reaching its maximum value (2.5-3 MeV) around $k_F = 0.8$ fm^{-1}, and vanishing for k_F larger than about 1.4 fm^{-1}. Detailed Hartree-Fock-Bogoliubov (HFB) calculations have also been performed, in order to assess the role of the inhomogeneities in the inner crust, taking into account the coexistence of the nuclear clusters with the neutron sea [9–11]. According to the local density approximation, one would expect that the pairing gap were suppressed in the region of the nuclear cluster, due to the large value of the local Fermi momentum, as compared to the outer region. The HFB results show that such differences tend to be smoothed out, due to proximity effects (the coherence length being of the order of 10 fm). The most relevant quantity is the value of the pairing gap close to the Fermi energy, which is affected by a few hundred keVs compared to the value in homogeneous neutron matter. This influences the specific heat of the neutrons, and can be important for the cooling processes of the star.

Negele and Vautherin did not take pairing correlations into account. Recently, Baldo and collaborators have updated their calculations for the region of spherical clusters, adopting modern energy functionals which include pairing correlations [12–15]. At first they have used an energy functional developed by Fayans *et al.* [16] to describe the whole Wigner-Seitz cell, while later they have performed a more elaborate calculation, using the Fayans functional to describe the nuclear cluster, matching it to the microscopic EoS for uniform neutron matter described above, and to a BCS calculation in uniform matter. The favoured proton number turns out to be sensitive to the adopted functional and also to pairing, and it often assumes values different from those determined by Negele and Vautherin. For example, at $k_F = 1$ fm^{-1} the equilibrium value is $Z \approx 20$. Reducing the pairing gap in the neutron sea, in order to simulate many-body effects (cf. below), can produce changes in the optimal values of Z up to 6 units [17]. On the other hand, one must also consider that when pairing is included differences between minima are quite small, of the order of 10 keV/nucleon (cf. Fig. 1). Thermal effects have not been included.

A source of uncertainty, besides the matching procedure itself, arises from the use of boundary conditions in the isolated Wigner-Seitz cell, which

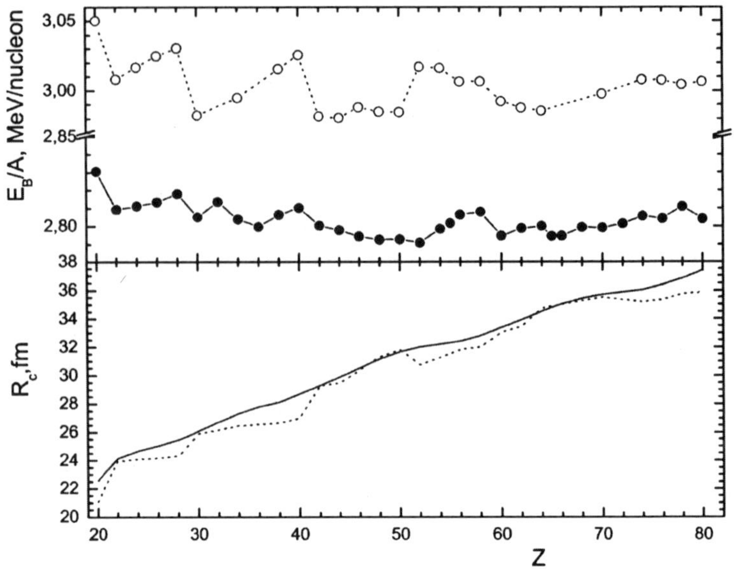

Fig. 1. Energy per particle (upper panel) and radius of the Wigner-Seitz cell as a function of the number of protons in the cell, calculated [13] with pairing (bold circles, solid lines) and without pairing (open circles, dotted lines) at an average density equal to 0.012 fm^{-3} (k_F=0.7 fm^{-1}).

are not consistent with the lattice periodicity. In ref. [3] mixed boundary conditions were used for the single-particle wavefunctions at the cell boundary, where wavefunctions of even parity, and derivatives of odd parity wavefunctions vanish. In this way the density is constant up to the edges of the cell, rather than being zero or twice the average density, as it would be the case if either condition were applied to all wavefunctions. In ref. [15] it was found that exchanging the parity condition can lead to quite different results for the pairing gaps for small cells (large densities): in fact in this case the boundary conditions can influence the single-particle spectrum close to Fermi energy in a strong way, affecting the value of the pairing gap. This is a signature of the fact that in such cases one should take the actual periodicity of the lattice into account.

The calculations mentioned above have been performed within mean field theory, and therefore do not take into account many-body effects, associated with self-energy processes and with screening effects due to the polarization of the medium, in particular with the exchange of density and spin fluctuations. In neutron matter, according to most studies (with one recent exception based on Monte Carlo many-body calculation [18]) many-body effects lead to a decrease of the gap by a factor 2 or 3 [19,20]. Unfortunately, however, there is no detailed quantitative agreement among different calculations. Recently, a comprehensive calculation of the neutron $^1S_0$ pairing gap was performed within Brueckner theory based on the $G-$matrix, taking the main effects into account [21]. In this work, the reduction of the quasi-particle strength is found to be the main factor which weakens the pairing gap. In neutron matter, this effect is reinforced by the interaction induced by the exchange of fluctuations, calculated summing up the RPA series. The particle-hole interaction is evaluated in terms of Landau parameters, derived by a Skyrme functional that reproduces the EoS obtained with Brueckner theory [7]. However, it is important to correct the values of the Landau parameters, taking into account the so-called particle-hole induced interaction [22]: otherwise F_0 would be close to -1 in neutron matter and even smaller than -1 in nuclear matter at the densities relevant for the inner crust, and a mechanical instability would appear. In neutron matter, repulsive spin fluctuations prevail over attractive density fluctuations. In nuclear matter, the effect of fluctuations on the neutron pairing gap is reversed because of the role played by the proton-neutron interaction, and the gap decreases only slightly. The same basic elements are at work in finite superfluid nuclei: also in this case, the reduction of quasiparticle strength tends to decrease the gap. On the other hand, the attractive interaction induced by the coupling to low-lying surface vibrations plays a key role, so that the pairing gap is enhanced, as compared to the result with the bare interaction [23–25], bringing it in overall agreement with the values deduced from experimental odd-even mass differences.

How to extend the previous studies to the case of the inhomogeneous inner crust in a consistent way is an open question. A first calculation of the screening interaction in the crust has been performed [26], taking the vibrations of the inhomogeneous system explicitly into account, calculating spin and density phonons in the RPA, up to high multipolarity ($l \approx 30$). It was found that the presence of the nuclear clusters tends to counteract the gap reduction typical of neutron matter. A much simpler approach, already mentioned above, is to reduce the pairing gap in the neutron sea by an ad hoc factor, simulating the reduction found in uniform matter.

One of the most important consequences of superfluidity in rotating bodies is the existence of vortices. In an inhomogeneous system, the dynamics of vortices can be strongly influenced by pinning, if it is energetically favourable for a vortex line to be anchored to the impurity - in the present case, to the nuclear clusters. About 30 yers ago, Anderson and Itoh [27] proposed that the phenomenon of glitches - sudden spinups observed in the period of pulsars [28], which is otherwise extremely regular - could be related to a sudden unpinning of vortices, and to the subsequent release of angular momentum to the crust. In fact, if the vortex lines are pinned to the crystal lattice, the angular velocity of the superfluid is essentially fixed. The angular velocity of the crust slowly decreases due to magnetic braking, becoming smaller than the angular velocity of the superfluid. This creates a centrifugal force, the Magnus force, which tends to pull the vortex lines away from the pinning sites. A glitch could then be originated by the simultaneous unpinning from many sites, when the Magnus force reaches a critical value.

Great observational advances have allowed systematic statistical analyses on glitches, leading to constraints on the different models and pinning sites that have been proposed (cf. for example [29,30]). On the other hand, theoretical studies of the basic parameters entering into the glitch model have been scarce, and the connection between observations and detailed models of the inner crust is still at a very early stage. The most important quantity in the Anderson-Itoh model is the pinning energy, namely, the difference between the energy cost to create a vortex at a distance from the nuclear cluster, and on top of it. Since one can estimate that the distance between vortices is much larger than the lattice spacing in the crust, it is possible to study the interaction of a single vortex with a nuclear cluster in a Wigner-Seitz cell. The first calculations were based on the Ginzburg-Landau theory [31], and found that vortex lines can pin to nuclei well inside the crust, while they are repelled in the outer zone. Recently, the vortex structure has been calculated in a more detailed microscopic treatment based on the local density approximation, determining the configurations with the lowest energy, ensuring thermodynamical and mechanical equilibrium [32,33]. It was found that vortices present a core made of normal matter, whose radius is determined to a very good approximation by the local value of the coherence length. Pinning on nuclei takes place only in a limited range, at high density in the crust. This semiclassical model has been improved, including several other effects, like the reduction of the pairing gap due to the many-body effects described above, and a more realistic description of the vortex flow, which lead to a reduction of the pinning energy [34].

The first fully quantum mean-field HFB calculations for the inner crust have been performed recently [35], following similar calculations in the case

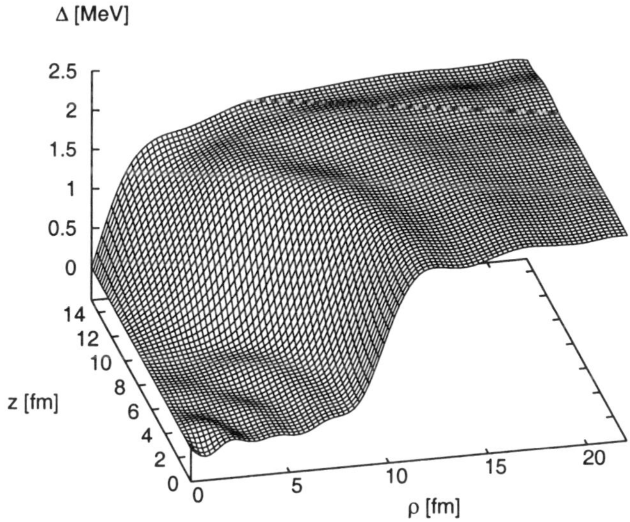

Fig. 2. Pairing gap associated with a vortex pinned on a nuclear cluster in the Wigner-Seitz cell, resulting from a HFB calculation at an average density equal to 0.012 fm^{-3} ($k_F = 0.7\ \text{fm}^{-1}$) [35]. The gap is shown as a function of the cylindrical coordinates z and ρ. The vortex axis is the $z-$axis, on which the pairing gap vanishes. The origin of the nucleus is at $z = 0, \rho = 0$. For values of z far from the nucleus, the dependence of Δ on ρ is the same as the one found for a vortex in uniform neutron matter. It is seen that decreasing z the vortex widens and encircles the nuclear region.

of homogeneous neutron matter [36]. A Skyrme interaction has been used in the particle-hole sector, while a density-dependent force which reproduces the pairing gap obtained with a bare interaction in neutron matter has been adopted as the pairing force. In the quantum calculation the core of the pinned vortex is much larger than in the semiclassical calculation (cf. Fig. 2). This has been attributed to quantal finite-size effects, which hinder the formation of Cooper pairs carrying one unit of angular momentum. Therefore, the gain in condensation energy found in the semiclassical approach (due to the fact that the vortex destroys a smaller pairing gap when is pinned on the nucleus, than in the outer neutron sea) is not present in the quantum calculation, because in this case the vortex destroys pairing also on the surface of the nucleus, where the gap reaches a large value in the absence of the vortex. This leads to a quite different density dependence of the pinning energy: pinning is favoured in the outer layers of the crust, contrary to previous estimates. It remains to be seen, to what extent this behaviour depends on the adopted microscopic forces in the particle-hole and pairing channels. For more details, I refer to the contribution by P. Avogadro *et al.* in this volume.

3. The core of compact stars

For $\rho \gtrsim 0.5\rho_o$ nuclei merge at the crust-core interface and then the star consists of asymmetric nuclear matter. This is the beginning of the core of the star, where about 99% of the mass is found, and which can reach densities of the order of 10 times ρ_0, depending on the theoretical model. The theoretical description of the core is fascinating but is at the same time subject to many uncertainties, because it involves unique conditions and phenomena [37]. To indicate a few basic issues, one can mention that the EoS of nuclear matter depends on little known three-body forces, leading in particular to large differences in the density dependence of the symmetry energy among various models, at densities larger than twice ρ_0 [38,39]. With increasing density and chemical potentials, other hadron species can appear, hyperons in particular, so that little known hyperon-hyperon interactions need to be introduced for a consistent theoretical study. Strangeness in neutron stars may exist also in deconfined form, in keeping with the idea, proposed a long time ago, that a phase transition to deconfined quark matter composed of up, down and strange quarks can occur, leading to hybrid stars, and that even stars entirely composed of quark matter may exist (strange stars), possibly surrounded by a thin nuclear crust [40].

3.1. *Hyperonic stars*

For a review of recent advances in the study of the nucleonic EoS, I refer to the paper by M. Baldo in this volume. When the neutron chemical potential becomes sufficiently large, it is possible to satisfy the equations of chemical equilibrium including Σ^-, and Λ hyperons [41]. The onset of hyperons, which are accomodated in lower momentum states and have a large bare mass, leads to a decrease of the kinetic energy and to a considerable softening of the EOS, as compared with the nucleonic case. The resulting matter composition at a given density depends strongly on the adopted potentials. The maximum mass of the star is instead rather insensitive to them, being close to 1.6 $M_\odot$ in relativistic models, compared to the value of 2 $M_\odot$ calculated for nucleonic matter. Studies based on the Brueckner-Hartree-Fock (BHF) approach lead to an even softer EoS and to even smaller values of the

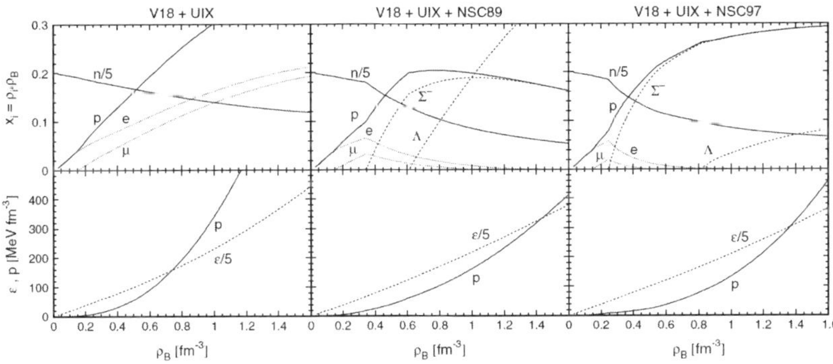

Fig. 3. Nucleon and hyperon compositions of β−stable matter obtained [42] with different forces (upper panels) and corresponding EoS (lower panels). Hyperons are taken into account in the central and in the right panels, using respectively the potentials NSC89 and NSC97e.

maximum mass. It has been argued that these low values can be related to the poor knowledge of hyperon-hyperon potentials, which are particularly uncertain, because of the lack of hyperon-hyperon scattering data. A recent study of hyperonic stars [42], based on BHF theory, has investigated the effect of two different three-body nucleonic forces (Urbana UIX or GLMM), and of two of the soft-core Nijmegen nucleon-hyperon and hyperon-hyperon potentials. These results confirm a previous analysis based on a different nucleonic EoS [43]. In Fig. 3 one can see that the inclusion of hyperons (center and right panels) leads to a much softer EoS, compared to the nucleonic case (left panels). The larger population of Σ^- and the retarded onset of Λ obtained with the NSC97e hyperon potential is related to the presence of very attractive $n\Sigma^-$ and $\Sigma^-\Sigma^-$ forces, and repulsive $\Lambda\Lambda$, $\Lambda\Sigma^-$ forces, compared to the NSC89 potential. These different compositions reflect themselves in the resulting mass-radius relations (cf. Fig. 4). However, the maximum mass is similar in the two cases, being close to 1.4 $M_\odot$. One can also observe that the introduction of hyperons reduces the differences existing between the various EoS at the nucleonic level. The differences between hyperon-nucleon potentials are seen very strongly in the pairing channel: in particular, it has been shown [44] that the NSC89 hyperon-nucleon potential, produces a large $n\Sigma^-$ pairing gap in the $^3PF_2$ channel at baryon densities $\rho > 1$ fm^{-3}. This could have important consequences for the internal structure of the star.

This kind of study has been extended to finite temperature [45], in order to deal with the case of protoneutron stars, formed just after a successful

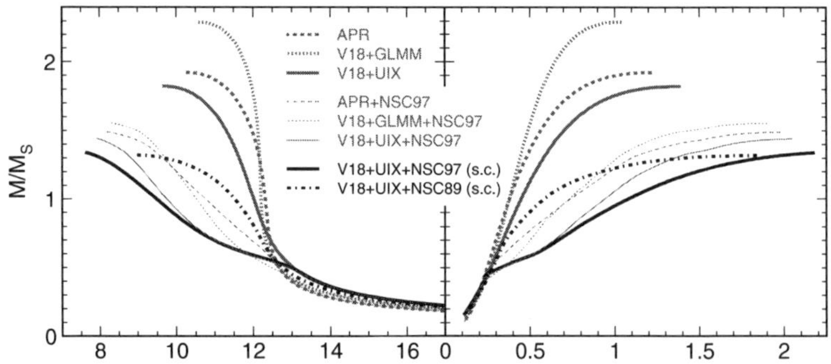

Fig. 4. Relation between mass and radius (left) and between mass and central density for hyperonic and nucleonic stars (right) [42]. The two lower dashed and solid thick curves refer to hyperonic stars, the three upper curves to nucleonic stars.

supernova explosion, leading either to neutron star or to a black hole. In this situation, one has to consider temperatures of the order of 30-40 MeV, and a finite value for the neutrino chemical potential, because neutrinos are trapped in the star. The nucleon-hyperon interaction has been considered only at T=0, while the hyperon-hyperon interaction has been neglected. One finds that matter composition is relatively little influenced by thermal effects, especially at high densities, although they tend to reduce the maximum mass of purely nucleonic stars. On the other hand, the presence of neutrinos plays an important role, retarding the appearance of hyperons. As a consequence, the effects of hyperons on the EoS, which, as we saw above, lead to a strong reduction of the maximum mass, are less pronounced in the presence of neutrinos. The hadron-quark phase transition has also been investigated in the same framework [46].

For more details, I refer to the paper by O. Nicotra in this volume.

3.2. *Quark stars*

It has been proposed a long time ago that at large densities cold baryonic matter can make a phase transition to deconfined quark matter [47]. The first studies ignored the strange quark, and found that the transition (determined by the condition that the chemical potentials of the two phases are equal for a given pressure) would occur at densities higher than those found in the center of neutron stars [48]. The inclusion of the strange quark, and the constraint of β−equilibrium, lowered the transition density substantially [49]. In these initial studies the quark phase has been described with the bag model, which continues to be widely employed because of its simplicity, and because first-principle calculations of the transition are not feasible. Besides the bag model, one can rely on effective models of various complexity, like the Nambu-Jona Lasinio (NJL) model, or the color dielectric model (CDM).

Quark matter can become superconducting, due to the attractive interaction among quarks in some channels [50]. The interplay of color and flavor degrees of freedom, produces a complex pattern of possible superconducting phases, which is being actively investigated. In particular, estimates of the value of the pairing gaps have been made [51], and the consequences of color superconductivity in neutron stars, where the constraints of $\beta-$stability and charge neutrality play an essential role, started to be investigated [52].

Recent studies of the quark phase transition have been carried out comparing the properties of the EoS associated with models of the quark phase, with hadronic EoS based on Brueckner theory including hyperons, discussed in Section 3.1 [53–57]. The transition has often been modeled as a sharp first-order transition, occurring when the chemical potential of the two phases is the same for a given value of the pressure. Some calculations consider mixed phases, caused by the simultaneous conservation of the baryon and of charge number [49]. The parameters of the models have been constrained, imposing that for symmetric matter at saturation density the hadronic phase is favoured. In particular, this puts a lower limit to the value of the bag constant B, $B \gtrsim 55\text{MeV fm}^{-3}$ (low values of B favour the quark transition); on the contrary, for $B \gtrsim 90$ MeV fm^{-3} no transition is found in symmetric matter at any density [53]. Nevertheless, the values of the density at which the transition to quark matter can occur, and the composition of the star, strongly depend on the models adopted to describe the quark phase. For example, in the CDM the onset of the quark phase occurs at low density, and the deconfined phase occupies most of the star. In the bag model, the the quark phase is also dominant, but a hadronic phase is present in the lower density region. In the case of NJL, the transition to quark matter can take only place only if the hyperonic phase is described in a relativistic model, rather than within Brueckner theory, and if color superconductivity is taken into account in the quark phase [54]. On the other hand, the maximum mass is always close to 1.6 $M_{\odot}$ in these calculations; the presence of the quark phase increases or decreases the mass, depending on whether the hadronic phase is described by Brueckner theory, or by relativistic models. The maximum mass turns out to be higher if hyperons are not considered in the hadronic phase, reaching values around 1.8

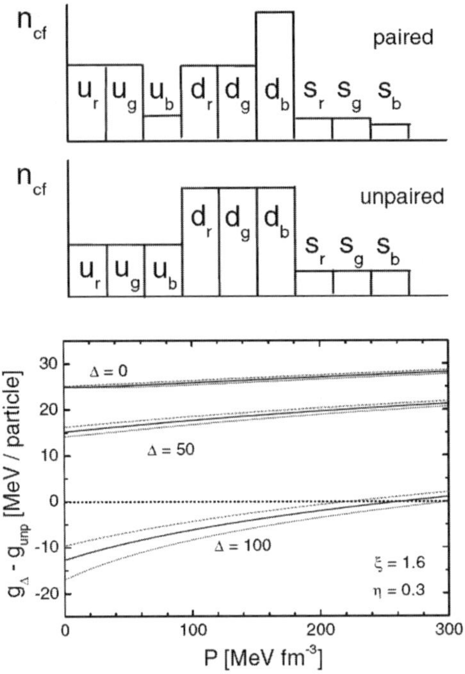

Fig. 5. Upper panel: number densities of u, d, s quarks of the three colors in paired or unpaired quark matter, following the deconfinement of hyperonic β−stable matter, imposing flavor and charge conservation [59]. In the paired case, pairing occurs between u_r and d_g and u_g and d_r quarks. Bottom panel: difference in the Gibbs energy between the paired and unpaired phase, as a function of the pressure in the star for selected values of the pairing gap Δ. The various curves associated with the same value of Δ correspond to different values of the bag constant B.

$M_\odot$ or 2 $M_\odot$, respectively with or without superconductivity in the quark phase. The effects of neutrino trapping on the phase transition have also been considered [58].

The features of the transition from hadronic to superconducting quark matter in the interior of compact stars have also been studied intensively, especially within the bag model. In this case, pairing provides a positive contribution to the pressure which is of order Δ^2/μ^2 respect to the contribution from the quark kinetic energy (typical values of the pairing gap and of the chemical potential are $\Delta \sim 100$ MeV and $\mu \sim 400$ MeV), and effectively lowers the value of the bag constant, favouring the transition to quark matter.

A detailed study of the conditions for the phase transition, treating the bag constant B and the pairing gap Δ as free parameters, has been performed [59], using a relativistic EOS for the hyperonic phase (cf. also refs. [60,61]). It has been supposed that flavor is conserved in the transition, which takes place due to the strong interaction. The resulting quark phase is not $\beta-$stable, and the conversion to the β-stable phase of lower energy takes place on a longer time scale due to the weak interactions. Calculating the pressure and the composition of the hyperonic star, based on a given EoS, it is then possible to determine the abundancies of each quark species in the deconfined phase, supposed to be colorless. These abundancies depend on whether the quark phase is superconducting or not (cf. Fig. 5).

Considering that pairing is favoured between species of different flavor and color having the same Fermi energy, it turns out that the flavor-conserving phase transition can lead to a two flavor paired quark phase, with pairing between up and down quarks of two different colors; up and down quarks of the third color and strange quarks are not paired. For values of Δ smaller than about 100 MeV, however, the unpaired phase is favoured, because the condensation energy does not compensate for the loss caused by equating the Fermi energy of the paired species. This can be seen in the phase diagram of Fig. 6, in which the deconfinement is studied as a function of Δ and B for a 1.6 $M_\odot$ star. For $\Delta \lesssim 70$ MeV, either there is no deconfinement (for $B > 130$ MeV fm^{-3}) or there is deconfinement to an unpaired phase. It can be seen that for $\Delta \gtrsim 70$ MeV pairing helps deconfinement. In Fig. 6 one can also see that for small values of B the final

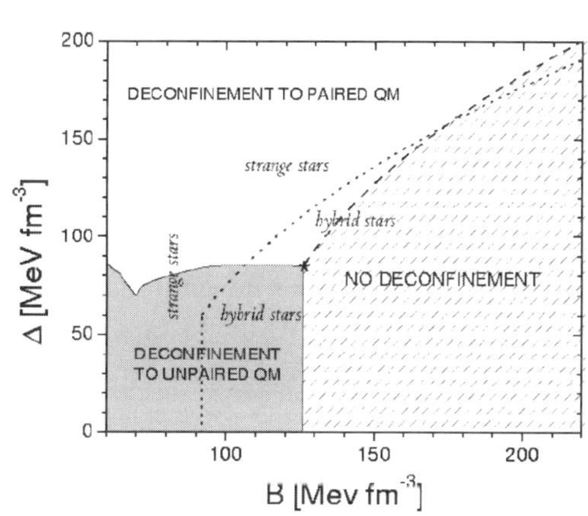

Fig. 6. Phase space for 1.6 $M_\odot$ star, indicating the values of the bag constant B and of the pairing gap Δ for which deconfinement can occur [59] .

$\beta-$stable configuration correspond to absolutely stable quark matter, leading to the formation of strange stars. The boundaries between phases move as a function of the mass of the star; increasing the mass, deconfinement can take places also for larger values of B.

The analysis described above does not consider finite-size effects which have to be taken into account in a detailed theory of the nucleation process, leading to the formation of a drop of quark matter [62–64]. Finite-size effects have been considered in detail both for unpaired [65,66] and paired [67] quark matter. The main effects can be expressed through the value the surface tension, although its value is highly uncertain ($\sigma \sim 10-100$ MeV fm^{-2}). In this way one can obtain an estimate of the nucleation time τ. When the system is close to the static transition point, τ decreases exponentially with increasing mass of the star [68,69]. It is then possible that a cold hyperonic or nucleonic star makes a transition to a quark star a long period after its birth in a supernova explosion, having for example reached its critical mass by accretion. The transition would be associated with an energy release of the order of $10^{52}-10^{53}$ erg [70]. It has been proposed that the conversion of part of this energy may be at the origin of at least some of the observed Gamma-Ray-Bursts (see Section 4.2).

The nature of the transition from hadronic to quark matter (described by the bag model) has been recently studied in detail, based on a fluido-dynamical description [71]. The main result is that the transition occurs as a subsonic process, corresponding to a strong deflagration. It has been found that the process is accelerated by the presence of instabilities, reaching velocities much larger than the laminar velocity, so that the transition should take place with a velocity of 10^3-10^4km/s. This can be contrasted with another kind of process, namely detonation, which would take place on an even faster time scale, but actually does not occur according to the analysis of ref. [71]. For more details concerning the nucleation process I refer to the contribution by I. Parenti *et al.* in this volume.

It is interesting to observe that according to a recent study [72], the transition to a mixed phase of quark and hadrons at relatively low temperatures (of the order of tens of MeV) might be approached in heavy-ion reactions at energies of a few GeV/A. In fact, for not too large values of the bag constant, the transition density to $u-d$ deconfined matter (no strangeness would be produced in these reactions) decreases rapidly with the ratio Z/A, and it would be around 2-3 ρ_0 for $Z/A \approx 0.4$: such densities can be reached in semicentral collisions with beams of neutron-rich species like $^{132}Sn +^{132} Sn$ or $^{238}U +^{238} U$.

4. Connection with observables

It is clearly very important to assess the possible observational signatures associated with the various theoretical proposed models for matter in compact stars. This is a very broad field, and I will outline only a few topics which have been examined in the works of members of the Italian community: for example, transport properties have not been much investigated. The discussion will be brief, because some of the main points have already been covered in the previous reviews [1,2].

4.1. *Mass-radius relation*

The different EoS mentioned in Section 3 can lead to quite different dependences between the mass and the radius of compact stars. An example has been shown in Fig. 4: the EoS for hyperonic stars is much softer than the EoS for a nucleonic star, leading to much smaller maximum masses (typically 1.4-1.6 $M_\odot$ against 2.-2.2 $M_\odot$). On the other hand, the transition to quark matter can lead to larger masses (cf. Section 3.1), but generally even quark matter tends to soften the EoS as compared to two-flavor nuclear matter. Studies based on the transition from a relativistic nucleonic EoS to the EoS of superconducting color flavor locked quark matter have produced quark stars with mass close to 1.7 $M_\odot$ [73,74]. Some care should be taken in the choice of parameters, in order to guarantee that symmetric nuclear matter is not unstable already at saturation density, as discussed above in Section 3.1 [53,55].

Accurate determinations of neutron star masses are rare, but recently masses larger than 1.7 $M_\odot$ with relatively small errors have been reported [37]. In particular, a determination of the radius and mass of the neutron star source EXO 0748-676 has led to a radius of 13.8 ± 1.8 km and to a mass of $2.1 \pm 0.28 M_\odot$ [75]. These values clearly constitute a severe challenge for many of the current theoretical models. However, it has been claimed that even if these high mass values were confirmed, they would not exclude the possibility of quark matter, based on more elaborate theoretical descriptions [76]. In particular, it has been shown that QCD corrections can stiffen the EoS of quark matter, making it resemble closely to that of purely nucleonic matter [77]. The treatment is however still highly phenomenological, as it is based on several essentially free parameters, and has not included hyperons in the hadronic equation of state.

On the other hand, there are regions of the $M - R$ plane (in particular small radii and small masses), that cannot be reached neither by hadronic, nor by hybrid stars, but only by strange stars, whose radii increase with increasing mass. Some strange star candidates have indeed been discovered [1]. In general, the search for strange matter is currently a very active field, and the mass-radius dependence of compact stars is only one of several possible signatures [78].

4.2. *Gamma-Ray-Bursts*

Data collected in the last decade have clarified many basic features of long-soft Gamma-Ray-Bursts (GRB), and in particular their complex times structure and their connection with Supernova events [79]. These data are usually interpreted according to the collapsar model [80], being generated by relativistic jets from the collapse of the core of massive stars.

It has been proposed that an alternative explanation is possible at least for a fraction of the observed GRB, associating them with the transition from a neutron to a quark star discussed in Section 3 [68,81,82]. As we have seen above, the energy release associated with transition would be compatible with GRB. The GRB would be generated by neutrino-antineutrino annihilation. It is important that the transition takes place as a deflagration, because the energy released goes mainly into heat. This avoids an excessive baryonic contamination of the region surrounding the star, which would occur in a detonation, making it impossible to accelerate the plasma to the very large Lorentz factors associated with GRB [83]. In this way, a plasma of electron-positron pairs can be created, powering the GRB [84].

The model implies in a natural way a time delay between the supernova explosion leading to a neutron star of subcritical mass, and the subsequent burst associated with the hadron-quark transition, taking place after the star has accreted mass beyond the critical mass. The pattern can be complicated by the presence of different quark phases [85]. Given the uncertainty associated with the values of the parameters of the model, the comparison with the observed GRB must rely on a statistical analysis of the expected and observed time intervals [85,86]. For more details concerning the nucleation process and its possible connection with GRB, I refer to the contribution by A. Drago and G. Pagliara in this volume.

4.3. *Gravitational waves*

Under a perturbation a neutron star can be set into non-radial oscillations and emit gravitational waves at the characteristic frequencies of its quasi-normal modes. By observing these stellar pulsation modes, one can hope to

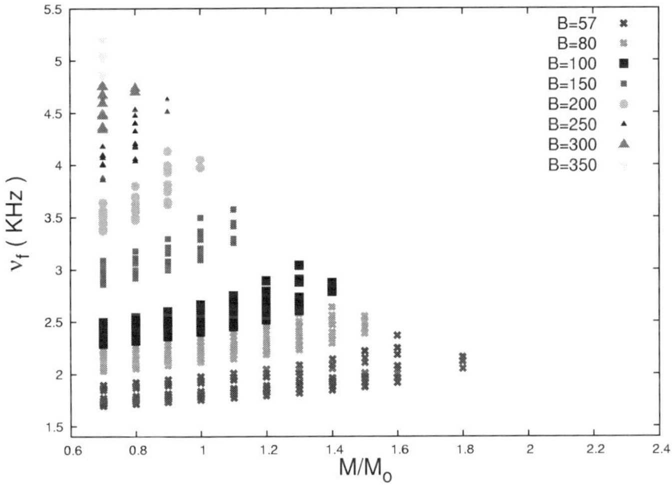

Fig. 7. The calculated frequency of the fundamental mode of strange stars is plotted for various values of the bag constant, as a function of the gravitational mass B [90]. The points referring to the same value of B differ in the adopted value of the mass of the strange quark and of the color coupling constant.

probe the interior of the star: this the field of asteroseismology. Although it is not likely that such waves can be revealed by the present generation of detectors, it is particularly interesting to investigate the relation between different EoS and the frequencies and damping times of the various modes in a star of a given mass [87–90]. For example, the frequency of the fundamental mode, ν_f, scales as the square root of the average density of the star: this has been explicitly verified calculating the star properties with various EoS. This means in particular that strange stars, which are the most compact, are associated with the largest frequencies. The results shown in Fig. 7, display the dependence of ν_f on the mass of strange stars. They refer to a calculation [90] of the structure of strange stars based on the bag model, in which also residual interactions between quarks have been taken into account in perturbation theory, in terms of a color coupling constant. It is seen that value of the frequency of strange stars essentially depends on the value of the bag constant. According to these results, the determination of the frequency of the fundamental mode of a compact star, together with its mass, could allow one to discriminate between hadronic and strange stars, and even to constrain the value of the bag constant.

Discussions with M. Baldo, I. Bombaci and A. Drago are gratefully acknowledged.

References

1. I. Bombaci, in *Theoretical Nuclear Physics in Italy*, Proceedings of the 9th Conference on Problems in Theoretical Nuclear Physics in Italy, Cortona, 2002, G. Pisent, S. Boffi, L. Canton, A. Covello, A. Fabrocini, and S. Rosati eds. (World Scientific, Singapore, 2003), p. 47.
2. A. Drago, in *Theoretical Nuclear Physics in Italy*, Proceedings of the 10th Conference on Problems in Theoretical Nuclear Physics in Italy, Cortona, 2004, S. Boffi, A. Covello, M. Di Toro, A. Fabrocini, G. Pisent, and S. Rosati eds. (World Scientific, Singapore, 2005), p. 89.
3. J.W. Negele and D. Vautherin, Nucl. Phys. **A207**, 298 (1973).
4. C.J. Pethick and D.G. Ravenhall, Ann. Rev. Nucl. Part. Sci. **45**, 429 (1995).
5. M. Baldo, C. Maieron, P. Schuck, and X. Viñas, Nucl. Phys. **A736**, 241 (2004).
6. C.J. Pethick, D.G. Ravenhall, and C.P. Lorenz, Nucl. Phys. **A584**, 675 (1995).
7. L.G. Cao, U. Lombardo, C.W. Shen, and N. Van Giai, Phys. Rev. C **73**, 014313 (2006).
8. U. Lombardo and H.-J. Schultze, in *Physics of Neutron Star Interiors*, D. Blaschke, N.K. Glendenning, and A. Sedrakian eds., Springer-Verlag (2001), p. 30.
9. F. Barranco, R.A. Broglia, H. Esbensen, and E. Vigezzi, Phys. Rev. C **58**, 1257 (1998).
10. P.M. Pizzochero, F. Barranco, R.A. Broglia, and E. Vigezzi, Astroph. J. **569**, 381 (2002).
11. N. Sandulescu, N. Van Giai, and R.J. Liotta, Phys. Rev. C **69**, 045802 (2004); N. Sandulescu, Phys. Rev. C **70**, 025801 (2004).
12. M. Baldo, U. Lombardo, E.E. Saperstein, and S.V. Tolokonnikov, JETP Lett. **80**, 523 (2004).
13. M. Baldo, U. Lombardo, E.E. Saperstein, and S.V. Tolokonnikov, Nucl. Phys. **A750**, 409 (2005).
14. M. Baldo, U. Lombardo, E.E. Saperstein, and S.V. Tolokonnikov, Phys. Atom. Nucl. **68**, 1812 (2005).
15. M. Baldo, E.E. Saperstein, and S.V. Tolokonnikov, Nucl. Phys. **A775**, 235 (2006).
16. S.A. Fayans, S.V. Tolokonnikov, E.L. Trykov, and D. Zawischa, Nucl. Phys. **A676**, 49 (2000).
17. M. Baldo, E.E. Saperstein, and S.V. Tolokonnikov, nucl-th/0609031.
18. A. Fabrocini, S. Fantoni, A.Y. Illarionov, and K.E. Schmidt, Phys. Rev. Lett. **95**, 192501 (2005).
19. M. Baldo, E.E. Saperstein, and S.V. Tolokonnikov, Nucl. Phys. **A749**, 42c (2005).
20. A. Sedrakian and J.W. Clark, nucl-th/0607028.
21. L. G. Cao, U. Lombardo, and P. Schuck, Phys. Rev. C **74**, 064301 (2006).
22. S. Babu and G.E. Brown, Ann. Phys. (N.Y.) **78**, 1 (1973).

23. F. Barranco, R.A. Broglia, G. Colò, G. Gori, E. Vigezzi, and P.F. Bortignon, Eur. Phys. J. **A21**, 57 (2004).
24. G. Gori, F. Ramponi, F. Barranco, P.F. Bortignon, R.A. Broglia, G. L. Colò, and E. Vigezzi, Phys. Rev. C **72**, 011302 (2005).
25. F. Barranco, P.F. Bortignon, R.A. Broglia, P. Schuck, E. Vigezzi, and X. Viñas, Phys. Rev. C **72**, 054314 (2005).
26. G. Gori, F. Ramponi, F. Barranco, R.A. Broglia, G. L. Colò, D. Sarchi, and E. Vigezzi, Nucl. Phys. **A731**, 401c (2004).
27. P.W. Anderson and N. Itoh, Nature **256**, 25 (1975).
28. A.G. Lyne, S.L. Shemar, and F. Graham Smith, Mon. Not. R. Astron. Soc. **315**, 534 (2000).
29. B. Link, Phys. Rev. Lett. **91**, 101011 (2003).
30. J.E. Horvath, Int. Jou. Mod. Phys. **D13**, 1327 (2004).
31. R.I. Epstein and G. Baym, Astroph. J. **328**, 680 (1988).
32. P. Donati and P.M. Pizzochero, Phys. Rev. Lett. **90**, 211101 (2003).
33. P. Donati and P.M. Pizzochero, Nucl. Phys. **A742**, 363 (2004).
34. P. Donati and P.M. Pizzochero, Phys. Lett. **B640**, 74 (2006).
35. P. Avogadro, F. Barranco, R.A. Broglia, and E. Vigezzi, nucl-th/060208.
36. Y. Yu and A. Bulgac, Phys. Rev. Lett. **90**, 161101 (2003).
37. J.M. Lattimer and M. Prakash, Science **304**, 536 (2004).
38. X.R. Zhou, G.F. Burgio, U. Lombardo, and H.-J. Schulze, Phys. Rev. C **69**, 018801 (2004).
39. A.W. Steiner, M. Prakash, J.M. Lattimer, and P.J. Ellis, Phys. Rep. **411**, 325 (2005).
40. C. Alcock and A.V. Olinto, Annu. Rev. Nucl. Part. Sci. **38**, 161 (1988).
41. I. Bombaci, Nucl. Phys. **A754**, 335c (2005).
42. H.-J. Schulze, A. Polls, A. Ramos, and I. Vidaña, Phys. Rev. C **73**, 056601 (2006).
43. I. Vidaña, A. Polls, A. Ramos, L. Engvik, and M. Hjorth-Jensen, Phys. Rev. C **62**, 035801 (2000).
44. X-R. Zhou, H.-J. Schulze, F. Pan, and J.P. Draayer, Phys. Rev. Lett. **95**, 051101 (2005).
45. O.E. Nicotra, M. Baldo, G.F. Burgio, and H.-J. Schulze, Astr. Astroph. **451**, 213 (2006).
46. O.E. Nicotra, M. Baldo, G.F. Burgio, and H.-J. Schulze, astro-ph/0608021.
47. J.C. Collins and M.J. Perry, Phys. Rev. Lett. **34**, 1353 (1975).
48. G. Baym and S.A. Chin, Phys. Lett. **62B**, 241 (1976).
49. N.K. Glendenning, Phys. Rev. D **46**, 1274 (1992).
50. I.A. Shovkovy, Found. Phys. **35**, 1309 (2005).
51. K. Rajagopal and F. Wilczek, hep-ph/0011333.
52. M. Alford and S. Reddy, Phys. Rev. D **67**, 074024 (2003).
53. G.F. Burgio, M. Baldo, P.K. Sahu, and H.-J. Schulze, Phys. Rev. C **66**, 025802 (2002).
54. M. Baldo, M. Buballa, G.F. Burgio, F. Neumann, M. Oertel, and H.-J. Schulze, Phys. Lett. **B562**, 153 (2003).
55. M. Baldo, Eur. Phys. J. **A19**, 189 (2004).

56. C. Maieron, M. Baldo, G.F. Burgio, and H.-J. Schulze, Phys. Rev. D **70**, 043010 (2004).
57. G.F. Burgio, Nucl. Phys **A749**, 337c (2005).
58. I. Vidaña, I. Bombaci, and I. Parenti, J. Phys. **G31**, S1165 (2005).
59. G. Lugones and I. Bombaci, Phys. Rev. D **72**, 065021 (2005).
60. A. Lavagno and G. Pagliara, Int. Jou. Mod. Phys. **D15**, 1515 (2006).
61. A. Lavagno and G. Pagliara, Eur. Phys. J. **A27**, s01, 289 (2006).
62. H. Heiselberg, C.J. Pethick, and E.F. Staubo, Phys. Rev. Lett. **70**, 1355 (1993).
63. K. Iida and K. Sato, Phys. Rev. C **58**, 2538 (1998).
64. D. Voskresensky, M. Yasuhira, and T. Tatsumi, Nucl. Phys. **A723**, 291 (2003).
65. I. Bombaci, I. Parenti, and I. Vidaña, Astroph. J. **614**, 314 (2004).
66. I. Vidaña, I. Bombaci, and I. Parenti, Nucl. Phys. **A754**, 345c (2005).
67. I. Bombaci, G. Lugones, and I. Vidaña, astro-ph/0603644.
68. Z. Berezhiani, I. Bombaci, A. Drago, F. Frontera, and A. Lavagno, Astrophys. J. **586**, 1250 (2003).
69. I. Bombaci, J. Phys. **G31**, S825 (2005).
70. A. Drago, A. Lavagno, and G. Pagliara, Phys. Rev. D **69**, 057505 (2004).
71. A. Drago, A. Lavagno, and I. Parenti, astro-ph/0512652.
72. M. Di Toro, A. Drago, T. Gaitanos, V. Greco, and A. Lavagno, nucl-th/0602052.
73. A. Drago, A. Lavagno, and G. Pagliara, Eur. Phys. J. **A19**, 197 (2004).
74. A. Drago, A. Lavagno, and G. Pagliara, Nucl. Phys. **B138**, 522 (2005).
75. F. Özel, astro-ph/0605106.
76. M. Alford, D. Blaschke, A. Drago, T. Klähn, G. Pagliara, and J. Schaffner-Bielich, astro-ph/0606524.
77. M. Alford, M. Braby, M. Paris, and S. Reddy, Astroph. J. **629**, 969 (2005).
78. F. Weber, Progr. Part. Nucl. Sci. **54**, 193 (2005).
79. M. Della Valle, astro-ph/0604110.
80. S.E. Woosley and A. Heger, Astrophys. J. **637**, 914 (2006).
81. K.S. Cheng and Z.G. Dai, Phys. Rev. Lett. **77**, 1210 (1996).
82. B. Paczyński and P. Haensel, Mon. Not. R. Astron. Soc. **363**, L4 (2005).
83. C.L. Fryer and S.E. Woosley, Astrophys. J. **501**, 780 (1988).
84. A. Drago, A. Lavagno, and I. Parenti, astro-ph/0608224.
85. A. Drago, A. Lavagno, and G. Pagliara, Nucl. Phys. **A774**, 823 (2006).
86. A. Drago and G. Pagliara, astro-ph/0512602.
87. N. Andersson and K.D. Kokkotas, Mon. Not. R. Astron. Soc. **299**, 1059 (1998).
88. O. Benhar, V. Ferrari, and L. Gualtieri, Phys. Rev. D **70**, 124015 (2004).
89. O. Benhar, Mod. Phys. Lett. **A31**, 2335 (2005).
90. O. Benhar, V. Ferrari, L. Gualtieri, and S. Marassi, astro-ph/0603464.

HARTREE FOCK BOGOLIUBOV CALCULATION OF THE PINNING ENERGY OF VORTICES ON NUCLEI IN THE INNER CRUST OF NEUTRON STARS

P. AVOGADRO

Dipartimento di Fisica, Università degli Studi di Milano, via Celoria 16, 20133 Milano, Italy
and INFN, Sezione di Milano, via Celoria 16, 20133 Milano, Italy.

F. BARRANCO

Departamento de Fisica Aplicada III, Escuela Superior de Ingenieros, Camino de los Descubrimientos s/n, 41092 Sevilla, Spain.

R.A. BROGLIA

The Niels Bohr Institute, University of Copenhagen, Blegdamsvej 17, 2100 Copenhagen Ø, Denmark
and Dipartimento di Fisica, Università degli Studi di Milano, via Celoria 16, 20133 Milano, Italy
and INFN, Sezione di Milano, via Celoria 16, 20133 Milano, Italy

E. VIGEZZI

INFN, Sezione di Milano, via Celoria 16, 20133 Milano, Italy.

We address the problem of vortex nucleus interaction in the inner crust of neutron stars. The relevance of this problem is connected with the observed glitches in pulsars that can be explained with an exchange of angular momentum from the superfluid neutrons in the inner crust to the outer crust. The problem is solved within the Hartree-Fock-Bogoliubov theory. We find that the properties of the vortex are strongly influenced by finite size effects, leading to qualitative differences with respect to results based on semiclassical approximations.

1. Introduction

Pulsars are fast rotating neutron stars. They are studied in the whole electromagnetic spectrum but in particular they have been discovered in the radio waves. The period of the signals we receive from a pulsar is the period of rotation of that star. In the models this happens because the emission is collimated and not aligned with the rotation axis. For this reason there is a "lighthouse" effect and every time the beam is directed towards the earth we receive that pulse. Typical periods of rotation vary from milliseconds to a few seconds. The emitted signals are very stable, but since the star slows down due to electromagnetic torque, a slow decrease of the frequency can be observed. The most important irregularities observed in the pulsed signal of neutron stars are sudden spinups, called "glitches", in which the frequency of rotation increases, typically [1] by $\frac{\Delta\nu}{\nu} \sim 10^{-9} - 10^{-6}$. It has been shown that glitches can be considered self regulating instabilities [2] of the rotating neutron star. The first model attributed the origin of this phenomenon to starquakes [3] but this model has been ruled out because it can't explain the observed timing of the glitches. Anderson and Itoh [4] suggested instead that glitches are due to the interaction of the superfluid present in the inner crust with the outer crust. The superfluid present in the rotating neutron star develops vortices. Vortices interact with the nuclei which form a Coulomb lattice in the crust. If the interaction is attractive and sufficiently strong, vortices can be anchored, or *pinned*, to the nuclei. As the star slows down, a difference develops between the velocity of the vortices, pinned to the crust, and the velocity of the superfluid. This produces a Magnus force, a hydrodynamical force whose value per unit length reads $\overrightarrow{F}_m = n_s\overrightarrow{K} \times (\overrightarrow{V_v} - \overrightarrow{V_s})$ (n_s being the superfluid density, $|K| = \oint v \cdot dr$ the modulus of the circulation of the velocity field produced by the vortex, its direction equal with that of the vortex line, and $\overrightarrow{V_v}$ and $\overrightarrow{V_s}$ are the velocity of the vortex line and the superfluid respectively). The modulus of the Magnus force increases as the star slows down. When the velocity of the superfluid respect to the crust is large enough, pinned vortices can be moved from their site. This creates a new velocity field for the superfluid neutrons and produces a transfer of angular momentum to the crust. In order to assess the validity of the Anderson-Itoh model, and to compare it with existing data, it is essential to have a quantitative estimate of the energy associated with vortex pinning.

2. The Neutron star structure

It is useful to distinguish four different zones of the neutron star in order of increasing density. The external shell is called outer crust. It is formed by a lattice of heavy nuclei surrounded by relativistic electrons. The density range of the outer crust is from $10^7 g/cm^3$ to $4 \cdot 10^{11} g/cm^3$. The outer crust should have a thickness of few hundred meters. At a density of $4 \cdot 10^{11} g/cm^3$ the neutrons start to drip out of the nuclei, occupying states in the continuum; this represents the beginning of the inner crust. Calculations show that up to densities of the order of 0.3 n_0, where $n_0 = 2.8 \cdot 10^{14} g/cm^3$ is the nuclear saturation density, the energetically favored phase in the inner crust is a Coulomb lattice of spherical nuclei surrounded by a sea of superfluid neutrons and relativistic electrons [5–7]. In the present paper we focus on this region. The distance between nuclei decreases for increasing density. In the region 0.3-0.6 n_0 other phases associated with different nuclear shapes are predicted, until for larger densities nuclei totally merge and the core of the star is reached. Even if the inner crust is a rather thin shell (of the order of a few hundred meters), it plays a significant role in the thermal properties of the star. The inner crust has been studied in detail in ref. [5] and ref. [6] dividing it into Wigner-Seitz cells each containing a single nucleus at their center, and determining the favored number of protons and the radius of the cell at a given average neutron density. The inner crust ends where nuclei can no more be considered separated and the outer core begins. The outer core consist in a uniform system of protons, neutrons, electrons. When the average density exceeds two times the nuclear saturation density the inner core begins, where densities up to 10-15 n_0 can be reached. The lattice of the crust and the core should form a rather rigid body, thus the decoupled superfluid matter that doesn't slow down in the inner crust accounts for less than 1% of the total mass of the star.

3. Hartree-Fock-Bogoliubov equations

To study the interaction of vortices with nuclei we divide the inner crust in cells. In keeping with the Wigner-Seitz approximation and with the fact that the distance between vortices is orders of magnitude larger than the size of the cells, we shall study the interaction of an isolated vortex with a single nucleus placed at the center of a cylindrical cell. We shall also assume that the presence of the nucleus affects the vortex only within a distance that is smaller than the typical internuclear distance (the diameter of the Wigner-Seitz cell). In other words this means that far enough from the nucleus, the properties associated with the vortex (density, pairing gap, velocity field) are similar to those calculated for uniform matter. We shall calculate the properties of different systems (*uniform* neutron matter,*vortex* in uniform matter, cell with a *nucleus* and vortex *pinned* on a nucleus) making use of quantum mean field theory. For this purpose, we solve the Hartree Fock Bogoliubov (HFB) equations, commonly used in nuclear physics to calculate the structure of finite nuclei [8,9]. In the study of neutron stars, HFB equations have been used previously to calculate microscopically the coexistence of finite nuclei and the sea of free neutrons of the inner crust [10–12]. They have also been used before by Elgarøy and De Blasio [13] to calculate the structure of vortices. However, they assumed cylindrical symmetry and a fixed Woods-Saxon potential. We shall instead perform a fully self-consistent calculation, assuming only axial symmetry. Our calculation for the vortex in uniform matter essentially reduces to that of ref. [14], except for the use of a different pairing interaction. We have used the Skyrme SII interaction in the particle-hole channel [15]. For the pairing interaction between neutrons we have used a contact interaction of the form [16]:

$$V_{pair}(\vec{x},\vec{x}') = V_0 \cdot \left(1 - 0.7 \cdot \left(\frac{n(\vec{x})}{0.08}\right)^{0.45}\right) \delta(\vec{x}-\vec{x}'), \tag{1}$$

where V_0 = -481 MeV fm^3, and $n(\vec{x})$ denotes the neutron density. The use of a contact interaction requires the introduction of a cutoff. We adopt $E_{cut} \approx 60$ MeV. This choice for the pairing interaction reproduces the gap obtained with a bare interaction (Argonne) in uniform neutron matter with the effective mass associated with the SII interaction. With the assumed zero-range interactions, the HFB equations for the quasiparticle amplitudes U_i and V_i can be written as:

$$\begin{cases} (H(\vec{x})-\lambda)U_i(\vec{x}) + \Delta(\vec{x})V_i(\vec{x}) &= E_iU_i(\vec{x}) \\ \Delta^*(\vec{x})U_i(\vec{x}) - (H(\vec{x})-\lambda)V_i(\vec{x}) &= E_iV_i(\vec{x}) \end{cases} \tag{2}$$

where λ denotes the chemical potential, $H(\vec{x})$ and $\Delta(\vec{x})$ are the single particle Hartree-Fock Hamiltonian and the pairing field, while E_i is the quasi particle energy. The selfconsistent pairing field is local and is given by $\Delta = -V_{pair}\sum_i U_iV_i^*$. According to the characteristic ansatz for the pairing field of a vortex [17] and choosing the z axis as the vortex axis, one writes:

$$\Delta(\rho,z,\phi) = \Delta(\rho,z)e^{i\nu\phi}, \tag{3}$$

where ν is the number of quanta of angular momentum carried by each Cooper pair along the vortex axis. Correspondingly, we write for the quasi-particle amplitudes

$$\begin{aligned} U_i(\rho,z,\phi) &= U_i(\rho,z)e^{im\phi} \quad , \\ V_i(\rho,z,\phi) &= V_i(\rho,z)e^{i(m-\nu)\phi}, \end{aligned} \tag{4}$$

where m is an integer, associated to the single-particle orbital angular momentum along the $z-$axis. In particular if we set $\nu = 0$ we obtain the usual

HFB equations, which can describe uniform neutron matter, the isolated nucleus or the nucleus embedded in the neutron sea, depending on the inclusion of protons and on the value of the Fermi energy. Setting instead $\nu = 1$, we can describe a vortex configuration in which each Cooper pair carries one unit of angular momentum along the $z-$axis. We consider pairs whose relative wavefunction is in the $^1S_0$ state, and the angular momentum of the pair comes from the motion of the center of mass around the vortex axis. The system has mirror symmetry respect to the $x-y$ plane, so that $\Delta(\rho, z) = \Delta(\rho, -z)$ and the parity of the function (3) is given by

$$\pi = (-1)^{\nu}. \tag{5}$$

In particular, we shall consider vortices with $\nu = 1$, in keeping with the fact that experiments on superfluids and theoretical considerations indicate that the energetically most favorable configuration is an array of $\nu = 1$ vortices instead of a few vortices carrying many quanta of angular momentum. In the case $\nu = 1$ the Cooper pairs are made of single particle levels of opposite parity. The velocity field obtained for the superfluid neutrons is the superfluid current divided by the density:

$$Vel = -\frac{i\hbar}{m\rho n(\rho, z)} \sum_{\alpha} V_{\alpha}^{*}(\rho, z, \phi) \frac{\partial V_{\alpha}(\rho, z, \phi)}{\partial \phi}. \tag{6}$$

Eq.(2) is expanded on a single particle basis:

$$\Phi_{nmk}(\rho, \phi, z) = c_{nmk} J_m(k_{nm}\rho) \sin(kz + \delta) e^{im\phi} \tag{7}$$

where $k_{nm} = \frac{\alpha_{nm}}{\rho_{box}}$, ρ_{box} is the radius of the cylindrical box, α_{nm} is the n-th zero of the Bessel function $J_m(x)$ and c_{nmk} is a normalization factor. The functions vanish on the edges of the cylinder.

The proton density is supposed to keep spherical symmetry. This is justified by the fact that the protons are deeply bound so that they essentially do not interact with the vortex. Furthermore, we have neglected the spin-orbit interaction for simplicity, assuming a number of protons $Z = 40$, associated with a closed shell configuration. The HFB equations for the protons then reduce to HF.

4. Results

In this section we shall discuss results obtained solving the HFB equations in a cylinder of radius $\rho_{box} = 30$ fm and height $h_{box} = 40$ fm (the nucleus, when present, is at the center of the cell). Four cells with different average densities have been calculated. The pinning energy, (the parameter used to study the interaction of the vortex with a nucleus) is defined as the difference between the energy cost to create a vortex on the top of a nucleus and the energy cost to create it in uniform matter. The cost to build a vortex on a nucleus is the total energy of the cell in the *pinned* case minus the total energy of the *nucleus* case (similarly the total cost to create a vortex in uniform matter is the total energy of the *vortex* cell minus the energy of the *uniform* cell). In order to compare the cells we imposed that the number of particles has to be the same in alike cells (*pinned* and *nucleus*; *vortex* and *uniform*). In fig.1 we compare the results obtained for the *pinned*, *vortex*

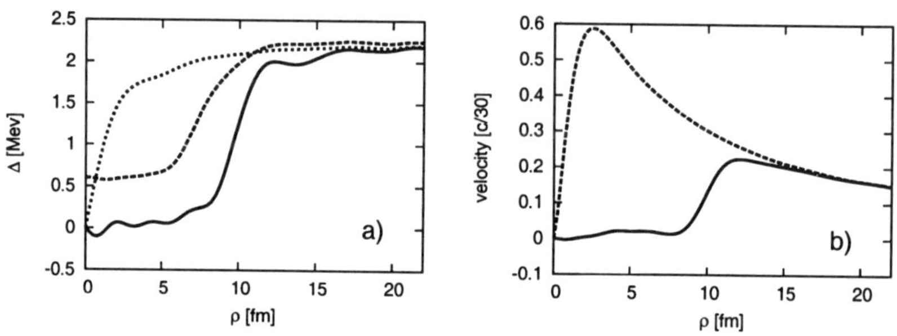

Fig. 1. (a) Pairing gap associated with the *vortex* (dotted line), with *nucleus* (dashed line) and with the *pinned* configuration. (b) Velocity field associated with the *vortex* (dashed line) and with the *pinned* configuration, in units of 1/30th of the speed of light.

and *nucleus* configurations as a function of the distance from the vortex axis at $z = 0$. In the pinned configuration the pairing gap displays a strong suppression in the nuclear region (cf. fig.1 (a)) and it is almost zero even in the nuclear surface. Also the velocity field is suppressed in the same region where the pairing goes to zero (fig.1(b)). The difference between the *pinned* and the *nucleus* density indicates that a depletion happens outside the nuclear volume, and in contrast the density is slightly increased on the top of the nucleus (cf fig.2 (b)). On the other hand in a vortex in uniform matter the density depletion can be observed on the axis (this feature was first observed in ref. [14]). These results indicate that a $\nu = 1$ vortex can hardly form within the nuclear volume. In this case, in fact, Cooper pairs are formed with single particle levels with opposite parity. In the nuclear region due to shell effects these levels should be too distant to allow Cooper pair formation, thus leading to a gap suppression.

The quantal effects discussed above lead to important differences compared to the semiclassical approximation [18], in which one assumes a sharp

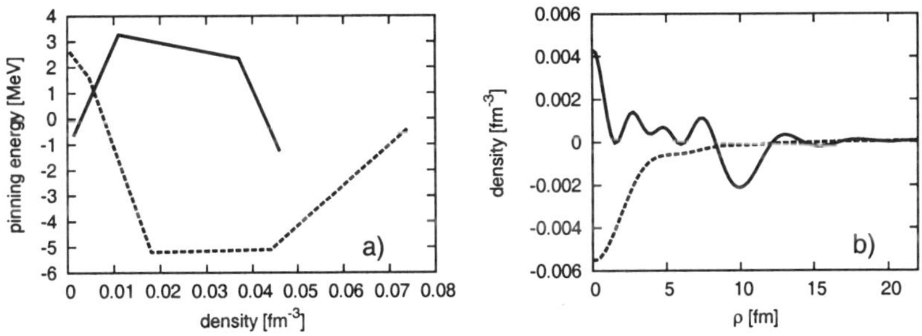

Fig. 2. (a) Pinning energies obtained at different asymptotic densities: our results (solid line), semiclassical results by Donati and Pizzochero (dashed line). (b) The density differences: *pinned-nucleus* (solid line), *vortex-uniform* (dashed line).The density depletion happens outside the nucleus in the pinned configuration, while on the top of the nucleus ($\rho = 0$) there is a slight increase.

boundary, corresponding to a vortex core $\rho_{core}(z)$ separating the normal zone close to the vortex axis, from the superfluid. The value of ρ_{core} is essentially given by the value of the coherence length, which is only slightly affected by the presence of the nucleus.

In the semiclassical model the pairing gap is calculated accoarding to the local density approximation (LDA), ignoring the shell effects produced by the nuclear potential. The only effect of the nucleus is to change the local Fermi momentum and thus the associated LDA pairing gap. This is a minor effect compared to the drastic quantal shell effects. The change of the local k_F may even lead to a decrease of the vortex core radius in the *pinned* case compared to the *uniform* case, what is completely at variance from what we find in the quantal calculations. The differences in the pairing gap lead to different values of the condensation energy of the system. More specifically the condensation energy of a pinned vortex is very much overstimated in the LDA semiclassical approximation. In fact, due to shell effects, our quantal calculations give rise to a strong suppression of the gap in a rather large volume around the nuclear surface (cf. fig.1(a)) and as a consequence to a much smaller condensation energy than the semiclassical estimate. As a rule there is no gain (in condensation energy) for the vortex in being pinned to the nucleus. In clear contrast, the semiclassical results give rise as a rule to an energy gain due to the overlap between the (almost unperturbed) pinned vortex and nuclear volumes, which reduces the net volume where the pairing gap is small. Thus the quantal picture of pinning is rather different than the semiclassical one. In fact, pinning, when it is energetically favoured ($E_{pinning} < 0$) is due to the gain in vortical kinetic energy, which because of quantal shell effects is also rather different from that obtained in the semiclassical models. It is not strange that this qualitative differences between both models lead also to quite different dependences on the density. In fig.2(a) the resulting pinning energy of the two models as a function of the density is shown. The astrophysical consequences of these new results remain to be explored, but it is quite likely that they will give rise to quite different vortex dynamics and to qualitative differences in the behaviour of the neutron star angular velocity during glitches.

References

1. *The structure and evolution of neutron stars*, D. Pines, R. Tamagaki, S. Tsuruta (eds.), Addison-Wesley, New York (1992).
2. B. Link, R.I. Epstein, and J.M.Lattimer, Phys. Rev. Lett. **83**, 3362 (1999).
3. M. Ruderman, Nature **223**, 597 (1969).
4. P.W. Anderson and N. Itoh, Nature **256**, 25 (1975).
5. J.W. Negele and D. Vautherin, Nucl. Phys. **A207**, 298 (1973).
6. M. Baldo, E.E. Saperstein, and S.V. Tolokonnikov, Nucl. Phys. **A775**, 235 (2006).
7. C.P. Lorenz, D.G. Ravenhall, and C. J. Pethick, Phys. Rev. Lett **70**, 379 (1993).
8. P. Ring and P. Schuck, *The Nuclear Many Body Problem*, Springer Verlag, Berlin-Heidelberg (1980).
9. A. Bulgac, nucl-th/9907088.
10. P.M. Pizzochero, F. Barranco, R.A. Broglia, and E. Vigezzi, ApJ. **569**, 381 (2002).
11. N. Sandulescu, Nguyen Van Giai, and R.J. Liotta, Phys. Rev. C **69**, 045802 (2004).
12. F. Montani, C. May, and H. Müther, Phys. Rev. C **69**, 065801 (2004).
13. Ø. Elgaroy and F.V. De Blasio, Astron. Astrophys. **370**, 939 (2001).
14. Y. Yu and A. Bulgac, Phys. Rev. Lett. **90**, 161101 (2003).
15. D. M. Brink and D. Vautherin, Phys. Rev. C **5**, 626 (1972).
16. E. Garrido, P. Sarriguren, E. Moya de Guerra, and P. Schuck, Phys. Rev. C **60**, 064312 (1999).
17. J.B. Ketterson and S.N. Song, *Superconductivity*, Cambridge University Press (1999).
18. P.M. Donati and P.M. Pizzochero, Nucl. Phys. **A742**, 363 (2004).

GAMMA-RAY-BURSTS AND QUARK PHASES

A. DRAGO
Dipartimento di Fisica - Università di Ferrara and INFN Sez. di Ferrara, 44100 Ferrara, Italy

G. PAGLIARA
Institut für Theoretische Physik, Goethe Universität, D-60438, Frankfurt am Main, Germany

We discuss the temporal structure of the Gamma-Ray-Bursts (GRBs) light curves and we analyse the occurrence of quiescent times which are long periods within the prompt emission in which the inner engine is not active. We show that if a long quiescent time is present, it is possible to divide the total duration of GRBs into three periods: the pre-quiescence emission, the quiescent time and the post-quiescence emission. We then discuss a model of the GRBs inner engine based on the formation of quark phases during the life of an hadronic star. Within this model the pre-quiescence emission is interpreted as due to the deconfinement of quark inside an hadronic star and the formation of 2SC quark matter or unpaired quark matter (UQM). The post-quiescence emission is due to the conversion of 2SC (or UQM) into the Color-Flavor-Locking (CFL) phase. The temporal delay between these two processes is connected with the nucleation time of the CFL phase in the 2SC (UQM) phase and it can be associated with the observed quiescent times in the GRBs light curves.

Keywords: Gamma-Ray-Burst, Compact stars, Quark matter

1. Introduction

The time structure of GRBs is usually complex and it often displays, during the phase of the prompt emission, several short pulses separated by time intervals lasting from fractions of second to several ten of seconds. In some cases are present very long period of vanishing signal, at least in the more energetic channels, the so called *quiescent times* (QTs), which can have durations comparable with the durations of the emission periods. Within the internal shocks model QTs shorter than few tens of seconds can be explained by the modulation of a continuous shells emission from the Inner Engine (IE) [1]. When a long quiescent time is present in the light curve, as we will discuss in the following, it is more plausible to assume that the inner engine is dormant.

Interestingly, in some bursts, it is also possible to find periods of quiescence separating the so-called *precursor* from the prompt emission. and in these cases the periods of quiescence can last up to hundreds of seconds.

The possibility that the inner engine switches off both during the prompt emission and during the time interval separating the precursor from the main event, poses strong constraints to the inner engine models. In this paper we discuss the quark deconfinement model of the GRBs inner engine. We will propose an unique interpretation of QTs (both in the prompt emission or after a precursor) as the time intervals between readjustments of the structure of a compact star. We suggest that the GRB emissions are due to first order phase transitions occurring between different phases of the strong interacting matter and we associate the periods of dormancy with the nucleation times needed to trigger the phase transitions.

2. Quiescent times in the GRBs prompt emission

A previous statistical analysis [2] has shown that there are three time-scales in the GRB light curves: the shortest one is the variability scale determining the pulses durations and the intervals between pulses; the largest one describes the total duration of the bursts and, finally, an intermediate time scale is associated with long periods within the bursts having no activity, the quiescent times. The origin of these periods of quiescence is still unclear.

We have recently [3] performed a new statistical analysis of the time intervals Δt between adjacent peaks in the light curve of GRBs using the algorithm introduced in Ref. [4]. We have applied this analysis to all the light curves of the BATSE catalogue. In a first investigation we have merged all the bursts of the catalogue into one sample from which we compute the cumulative probability $c(\Delta t)$ of finding time intervals Δt which are not QTs i.e. we compute the distribution of the time intervals within each active period. In Fig. 1a, we show that $c(\Delta t)$ is well described by a log-normal distribution. In Fig. 1b, the histogram of QTs is displayed together with a log-normal distribution. As already observed by previous authors [2], there is an evident deviation of the data points respect to the log-normal distribution for time intervals longer than a few seconds, indicating an excess of long Δt. In Fig. 1c we show a power law fit of the tail of the QTs distribution which displays a very good agreement with the data, as already observed by [5]. The physical interpretation of this distribution will be discussed later. Finally, in Fig. 1d we show a correlation function, indicating the probability of finding at least 2 QTs longer than ΔT in a same GRB. As shown in the figure this probability rapidly decreases and it essentially vanishes for $\Delta T > 40$ s.

We can now define a subsample of the BATSE catalogue composed of all the bursts having a QT longer than 40 s and study its properties. From the result of Fig. 1d, the bursts of the subsample contain only one long QT and it is therefore possible to divide each burst into a pre-quiescence emission (PreQE) and a post-quiescence emission (PostQE) of which we will compare the temporal and spectral structure.

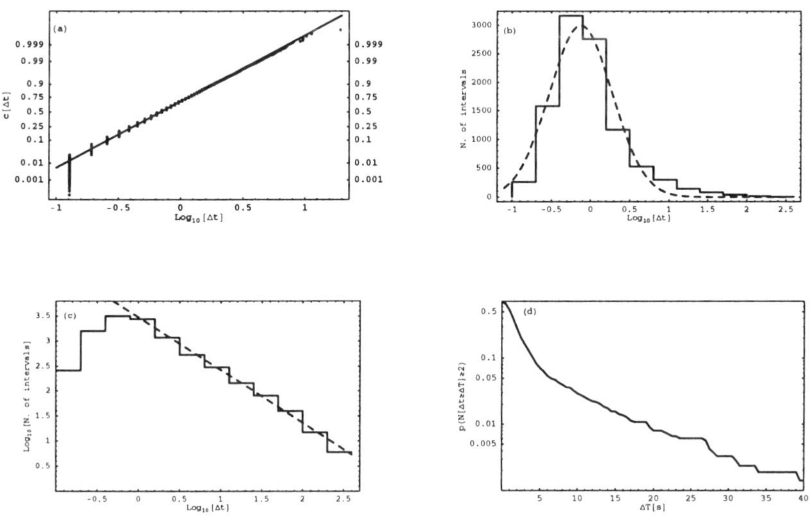

Fig. 1. **Analysis of time intervals between peaks a** The cumulative distribution of time intervals Δt which are not QTs (black point), is compared with its best fit log-normal distribution (solid line). **b** Histogram of the QTs and its log-normal fit (dashed line). **c** Histogram of QTs and power-low fit of its tail (dashed line). The fit is based only on QTs longer than 40s. **d** Frequency of bursts containing at least two QTs longer than ΔT.

In Fig. 2 we display the cumulative distributions $c_1(\Delta t)$ and $c_2(\Delta t)$ within each of the two emission periods. The two distributions are very similar. The χ^2-test provides a significance of 34% for that the two data sets are drawn from the same distribution function. Let us remind that within the internal-external-shocks model [6,7], external shocks produce emissions lacking the short time scale variability produced by internal shocks [8]. The result of Fig. 2 rules out a scenario in which PostQE is dominated by external shocks and PreQE by internal shocks. This in turn excludes the

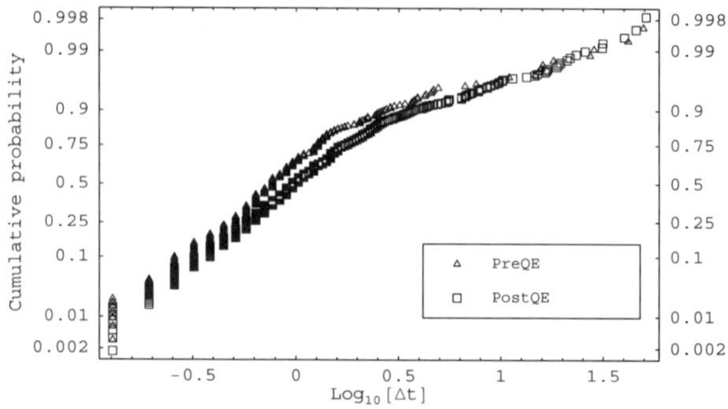

Fig. 2. **Analysis of time intervals between peaks within the two emission periods** The cumulative distributions of Δt are shown for the two emission episodes including QTs.

possibility of associating the QTs with the time needed to the jet to reach and interact with the interstellar medium.

We perform now a statistical analysis of the durations $D1$ and $D2$ of the two emission periods. As shown in Fig. 3a, the two data sets are well fitted by two log-normal distributions (the Kolmogorov-Smirnov test provides a significance of $\sim 90\%$). The two distributions have different mean values ($D1_{ave} \sim 21s$, $D2_{ave} \sim 41s$) and almost identical standard deviations ($\sigma_1 = 36s$, $\sigma_2 = 33s$). We have repeated the previous analysis by dividing PreQE and PostQE each in two parts, using the longest QT within each emission as a divider. The distributions of the duration of all parts are shown in Fig. 3b. The durations of the two parts within each emission period share the same distribution (the χ^2-test provides significances larger than 50% in both cases) but, in agreement with the previous findings, the average durations of the two parts of PostQE are longer than the two parts of PreQE. Therefore, the longer duration of PostQE cannot be attributed to a continuous modification of the emission but is a specific feature of the second part of the GRB.

To estimate the emitted energy during PreQE and PostQE we have analysed the hardness ratios, defined as the ratios between the photon counts in two BATSE channels (the second and the third in our case). The average hardness of PostQE turns out to be only marginally smaller ($\sim 20\%$) than the average hardness of PreQE. Also the power emitted during PreQE and PostQE are on average the same.

Let us now discuss the implications of this analysis on the origin of QTs. As observed by [1], within the internal shocks model it is possible to explain the QTs either as a turn-off of the IE or as a modulation of a continuous relativistic wind emitted by the IE (Wind Modulation Model WMM). Both hypothesis are consistent with the result of Fig. 2.

The main difference between the WMM and the dormant engine scenario is that in the WMM the inner engine has to provide a constant power during the whole duration of the burst. In our subsample, we have several bursts whose total duration (including the QT) approaches 300 s. These durations have to be corrected taking into account the average redshift of the BATSE catalogue, $z_{ave} \sim 2$ [9], but even after this renormalization, durations of a hundred seconds or more are not too rare. This time scale has to be compared with the typical duration of the emission period of the inner engine, as estimated in various models. For instance, in all numerical investigations of the collapsar model [10] the IE remains switched-on during some 20s. Also in the quark deconfinement model which will be discussed in the next section the inner engine remains active during periods of the order of a few ten seconds corresponding to the cooling time of the compact stellar object.

We conclude therefore that the energy requirement within the WMM scenario is too large respect to the results of the theoretical models. It is instead more plausible to assume that during a QT the inner engine switches off, sometimes for very long periods, and then restart producing a PostQE very similar to the PreQE.

3. Phase transitions between QCD phases

Let us now discuss how to generate dormancy periods in the prompt emission within the quark deconfinement model. In this model the energy source powering the GRB is the transition from a star containing only hadrons to a star composed, at least in part, of deconfined quarks [11,12]. An important feature of this model is that it accounts for the SuperNova(SN)-GRBs connection which has been confirmed several times by the observations. Moreover the two explosive events can be temporally separated, with the SN preceding the GRB by a delay which can vary from minutes to years. This time delay in the model is due to the existence of a finite value of the surface tension separating the hadronic phase from the quark phase as discussed in Ref. [11]. In the first calculations within the quark deconfinement model, the equation of state of quarks was computed using the MIT bag model. Actually, in the last years the possibility of forming a diquark condensate at the center of a compact star has been widely discussed in the literature [13]. It was shown that the formation of a color superconducting

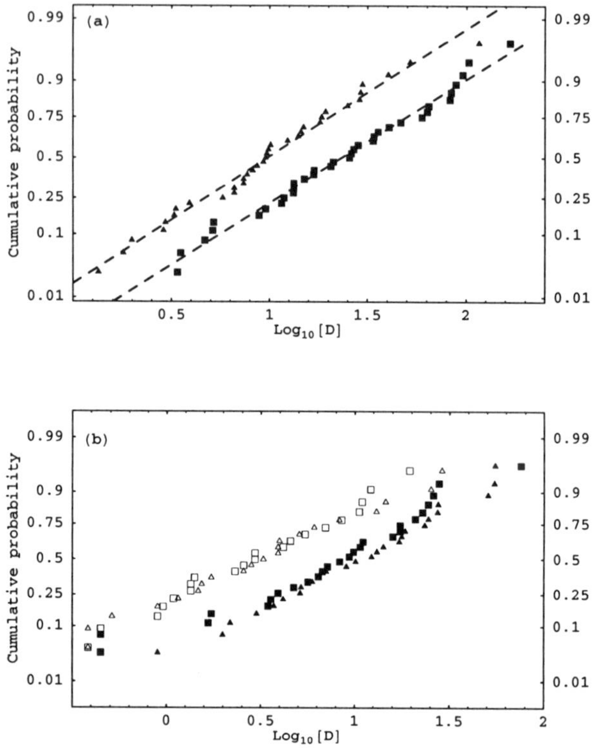

Fig. 3. **Analysis of the durations D of the two emission periods a** Cumulative distributions of durations of PreQE (filled triangles) and of PostQE (filled boxes) and their best-fit log-normal distributions (dotted lines). **b** Cumulative distributions of durations of the first and second part of PreQE (empty triangles and boxes) and of PostQE (filled triangles and boxes).

quark core can increase the energy released by a significant amount [14]. In particular, many calculations indicate that Color-Flavor-Locked (CFL) quark matter is the most stable configuration at large density and that the transition from normal quarks to CFL matter can take place as a first order if the leptonic content of the newly formed normal quark matter phase is not too small [15]. It is therefore tempting to associate PreQE with the transition from hadronic to normal quark matter and PostQE with the formation of the superconducting phase [16]. In this scenario the two dimensional scales regulating the durations of PreQE and PostQE are the energies released in the two transitions. Finally, also the power-law distribution of long QTs can have in this model a plausible explanation: after the first phase transition powering PreQE, the resulting hybrid or quark star (composed partially or totally of 2SC or UQM phase) have a mass ranging with a narrow mass interval. Then the nucleation time of the CFL phase has, in each star, a different numerical value. This in turns implies that the temporal distribution of the second emission is not simply an exponential distribution as one would expect if all the stars (after the first transition) have the same mass. It is instead a superposition of exponential distributions with different decay times which can look like be a power law distribution [17].

Finally, concerning the conversion process between different QCD phases, it turns out that the conversion always takes place as a strong deflagration and never as a detonation [18]. This is important because in the case of a detonation the region in which the electron-photon plasma forms (e.g. via neutrino-antineutrino annihilation near the surface of the compact star) would be contaminated by the baryonic load and it would be impossible to accelerate the plasma up to the enormous Lorentz factors needed to explain the GRBs.

4. Conclusions

We can combine the information provided in the previous sections and formulate a model for the GRBs based on the following scheme:

i) a compact star forms after a SN explosion. The explosion can be entirely successful or marginally failed, so that in both cases the mass-fallback is moderate (fraction of a solar mass);

ii) the compact star is now metastable respect to the formation of quark matter (if deconfinement at finite density takes place as a first order transition) and after a short delay the formation of deconfined quarks takes place as a deflagration. A hot hybrid or quark star remains, and it cools-down through neutrino-antineutrino emission;

iii) the cooling of the hybrid-quark stars formed after the first transition triggers a new first order phase transition from the 2SC or UQM to the CFL phases. In this way, after a quiescent time, a second GRB emission can be powered.

v) the neutrino-antineutrino emitted by the compact star can annihilate near the surface with an efficiency of order percent. Electrons and positrons add to the photons directly emitted. The energy deposited in the electron-positron-gamma plasma can be large enough to power a GRB. The typical duration of the cooling of the compact star is of the order of a few ten seconds. The emissions generated by the various cooling periods of the compact star can explain the main event.

Finally in the quark deconfinement model also the origin of precursors could be addressed: after the SN explosion the newly born neutron star starts readjusting its internal structure. A first event could be associated with the formation of kaon condensation (or of hyperons if it goes through a first order transition [19]). This first structural modification could be relatively small, involving only a modification of the central region of the star, but the presence of strangeness can trigger the instability respect to the formation of strange quark matter. The precursors could be due to this process.

It is interesting to observe that in the quark deconfinement model all the complex structures of the GRBs light curve are interpreted in a unique scheme in which a compact stars releases a huge amount of energy every time it readjusts its internal structure. The complex time structure of the GRBs light curves would reflect therefore the rich structure of the QCD phase diagram.

References

1. E. Ramirez-Ruiz, A. Merloni, and M.J. Rees, Mon. Not. Roy. Astron. Soc. **324**, 1147 (2001).
2. E. Nakar and T. Piran, Mon. Not. Roy. Astron. Soc. **331**, 40 (2002).
3. A. Drago and G. Pagliara, astro-ph/0512602 (2005).
4. E. Nakar and T. Piran, Mon. Not. Roy. Astron. Soc. **330**, 920 (2002).
5. F. Quilligan, B. McBreen, L. Hanlon, S. McBreen, K.J. Hurley, and D. Watson, Astron. Astrophys. **385**, 377 (2002).
6. T. Piran, Rev.Mod.Phys. **76**, 1143 (2004).
7. B. Zhang and P. Meszaros, Int. J. Mod. Phys. **A19**, 2385 (2004).
8. R. Sari and T. Piran, Astrophys. J. **485**, 270 (1997).
9. T. Piran, Phys. Rep. **314**, 575 (1999).
10. A.L. MacFadyen and S.E. Woosley, Astrophys. J., **524**, 262 (1999).
11. Z. Berezhiani, I. Bombaci, A. Drago, F. Frontera, and A. Lavagno, Astrophys. J. **586**, 1250 (2003).
12. B. Paczynski and P. Haensel, Mon. Not. Roy. Astron. Soc. Lett. **362**, L4 (2005).
13. K. Rajagopal and F. Wilczek, hep-ph/0011333 (2000).
14. A. Drago, A. Lavagno, and G. Pagliara, Phys. Rev. D **69**, 057505 (2004).
15. S.B. Ruster, V. Werth, M. Buballa, I.A. Shovkovy, and D.H. Rischke, Phys. Rev. D **73**, 034025 (2005).
16. A. Drago, A. Lavagno, and G. Pagliara, astro-ph/0510018, Proceedings QM2005, August 2005 Budapest.
17. M.S. Wheatland, Astrophys. J. **536**, L109 (2000).
18. See I. Parenti contribution, these proceedings.
19. J. Schaffner-Bielich *et al.*, Phys. Rev. Lett. **89**, 171101 (2002).

HYBRID PROTONEUTRON STARS WITHIN A STATIC APPROACH

O.E. NICOTRA

Dipartimento di Fisica e Astronomia, Università di Catania and INFN, Sezione di Catania, Via Santa Sofia 64, 95123 Catania, Italy

We study the hadron-quark phase transition in the interior of protoneutron stars. For the hadronic sector, we use a microscopic equation of state involving nucleons and hyperons derived within the finite-temperature Brueckner-Bethe-Goldstone many-body theory, with realistic two-body and three-body forces. For the description of quark matter, we employ the MIT bag model both with a constant and a density-dependent bag parameter. We calculate the structure of protostars within a static approach. In particular we focus on a suitable temperature profile, suggested by dynamical calculations, which plays a fundamental role in determining the value of the minimum gravitational mass. The maximum mass instead depends only upon the equation of state employed.

Keywords: dense matter; equation of state; stars:interiors; stars:neutron.

1. Introduction

After a protoneutron star (PNS) is successfully formed in a supernova explosion, neutrinos are temporarily trapped within the star (Prakash *et al.* 1997). The subsequent evolution of the PNS is strongly dependent on the stellar composition, which is mainly determined by the number of trapped neutrinos, and by thermal effects with values of temperatures up to 30-40 MeV (Burrows and Lattimer 1986; Pons *et al.* 1999). Hence, the equation of state (EOS) of dense matter at finite temperature is crucial for studying the macrophysical evolution of protoneutron stars.

The dynamical transformation of a PNS into a NS could be strongly influenced by a phase transition to quark matter in the central region of the star. Calculations of PNS structure, based on a microscopic nucleonic equation of state (EOS), indicate that for the heaviest PNS, close to the maximum mass (about two solar masses), the central particle density reaches values larger than $1/\text{fm}^3$. In this density range the nucleon cores (dimension

≈ 0.5 fm) start to touch each other, and it is likely that quark degrees of freedom will play a role.
In this work we will focus on a possible hadron-quark phase transition. In fact, as in the case of cold NS, the addition of hyperons demands for the inclusion of quark degrees of freedom in order to obtain a maximum mass larger than the observational lower limit. For this purpose we use the Brueckner-Bethe-Goldstone (BBG) theory of nuclear matter, extended to finite temperature, for describing the hadronic phase and the MIT bag model at finite temperature for the quark matter (QM) phase. We employ both a constant and a density-dependent bag parameter B. We find that the presence of QM increases the value of the maximum mass of a PNS, and stabilizes it at about 1.5–1.6 $M_{\odot}$, no matter the value of the temperature.
The paper is organized as follows. In sec. 2 we present a new static model for PNS. Sec. 3 is devoted to the description of the hadron-quark phase transition within the EOS mentioned above. In sec. 4 we present the results about the structure of hybrid PNS and, finally, we draw our conclusions.

2. A static model for PNS

Calculations of static models of protoneutron stars should be considered as a first step to describe these objects. In principle the temperature profile has to be determined via dynamical calculations taking into account neutrino transport properly [1]. Many static approaches have been developed in the past decade [2,3], implementing several finite temperature EOS and assuming an isentropic or an isothermal [4] profile throughout the star.
In our model we assume that a PNS in its early stage is composed of a hot, neutrino opaque, and isothermal core separated from an outer cold crust by an isentropic, neutrino-free intermediate layer, which will be called the envelope throughout the paper. For a PNS core in which the strongly interacting particles are only baryons, its composition is determined by requirements of charge neutrality and equilibrium under weak semileptonic processes, $B_1 \rightarrow B_2 + l + \bar{\nu}_l$ and $B_2 + l \rightarrow B_1 + \nu_l$, where B_1 and B_2 are baryons and l is a lepton (either an electron or a muon). Under the condition that neutrinos are trapped in the system, the beta equilibrium equations read explicitly

$$\mu_i = b_i\mu_n - q_i(\mu_l - \mu_{\nu_l}) \,, \tag{1}$$

where b_i is the baryon number, and q_i the electric charge of the species i. Because of trapping, the numbers of leptons per baryon of each flavour ($l = e, \mu$), $Y_l = x_l - x_{\bar{l}} + x_{\nu_l} - x_{\bar{\nu}_l}$, are conserved. Gravitational collapse calculations of the iron core of massive stars indicate that, at the onset of trapping, the electron lepton number is $Y_e \simeq 0.4$; since no muons are present at this stage we can impose also $Y_\mu = 0$. For neutrino free matter we just set $\mu_{\nu_l} = 0$ in Eq.(1) and neglect the above constraints on lepton numbers.

We assume a constant value of temperature throughout the core and perform some calculations for a value of temperature ranging from 0 to 50 MeV, with and without neutrinos. The EOS employed is that of the BHF approach at finite temperature for the hadron phase and that of the MIT bag model for QM. Many more details on the nuclear matter EOS employed together with plots for the chemical composition and pressure at increasing density and temperature can be found in [5].

The condition of isothermality adopted for the core cannot be extended to the outer part of the star. Dynamical calculations suggest that the temperature drops rapidly to zero at the surface of the star; this is due to the fast cooling of the outer part of the PNS where the stellar matter is transparent to neutrinos. Moreover, in the early stage, the outer part of a PNS is characterized by a high value of the entropy per baryon ranging from 6 to 10 in units of Boltzmann's constant [1].

For a low enough core temperature ($T \leq 10$ MeV) in [5] a temperature profile in the shape of a step function was assumed, joining the hadronic EoS (BHF) with the BPS [6] plus FMT [7] EoS for the cold crust. When the core temperature T_{core} is greater than 10 MeV, we consider an isentropic envelope in the range of baryon density from 10^{-6} fm^{-3} to 0.01 fm^{-3} based on the EoS of LS [8] with the incompressibility modulus of symmetric nuclear matter $K = 220$ MeV (LS220). Within this EoS, fixing the entropy per baryon s to the above values $(6, 8, 10)$ and imposing beta-equilibrium (neutrino-free regime) we get temperature profiles which rise quikly from 0 (at 10^{-6} fm^{-3}) to values of temperature typical of the hot interior of a PNS (respectively $T_{core} = 30, 40, 50$ MeV) at a baryon density of about 0.01 fm^{-3}. This explicitly suggests a natural correspondance between the entropy of the envelope s_{env} and T_{core} [9].

Since the energy of neutrinos, emerging from the interior, possesses some spreading [10] and their transport properties vary quite a lot during the PNS evolution [11], we do not think that the location of the neutrino-sphere, affected by the same spreading, is a good criterion to fix the matching density between core and envelope. All this instead indicates the possibility to have a blurred region inside the star where we are free to choose the matching density. To build up our model of PNS, a fine tuning of s_{env} is performed

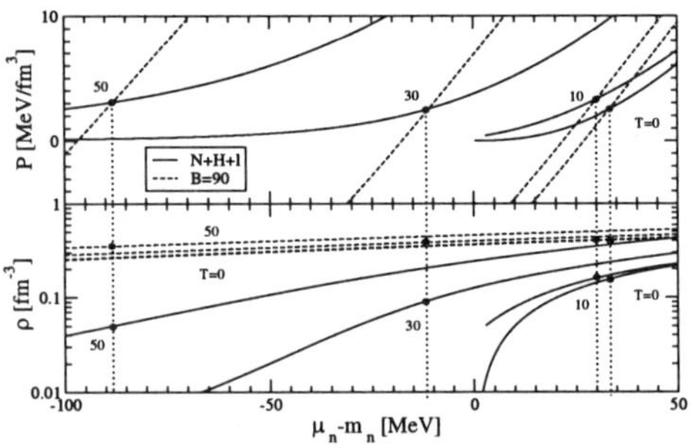

Fig. 1. Baryon density (lower panel) and pressure (upper panel) as a function of baryon chemical potential of beta-stable baryonic matter (solid curves) and quark matter (dashed curves) for the neutrino-free case at different temperatures $T = 0, 10, 30, 50$ MeV. The vertical dotted lines indicate the positions of the phase transitions. A bag constant $B = 90$ MeV/fm^3 is used for QM.

in order to have an exact matching between core and envelope of all the thermodynamic quantities (energy density, free energy density, and temperature). In this sense we can consider this static description of PNS with only one free parameter: the temperature of the core. Once T_{core} is chosen, s_{env} is fixed by matching conditions (see [9] for more details and results for pure baryonic PNS).

3. Hadron-quark phase transition

We review briefly the description of the bulk properties of uniform QM at finite temperature, deconfined from the beta-stable hadronic matter discussed in the previous section, by using the MIT bag model [12]. In its simplest form, the quarks are considered to be free inside a bag and the thermodynamic properties are derived from the Fermi gas model, where the quark $q = u, d, s$ baryon density and the energy density, are given by $\rho_q = \frac{g}{3}\int\frac{d^3k}{(2\pi)^3}\left[f_q^+(k) - f_q^-(k)\right]$ and $\epsilon_Q = g\sum_q\int\frac{d^3k}{(2\pi)^3}\left[f_q^+(k) + f_q^-(k)\right]E_q(k)+B$, where $g = 6$ is the quark degeneracy, $E_q(k) = \sqrt{m_q^2 + k^2}$, B is the bag constant and $f_q^\pm(k)$ are the Fermi distribution functions for the quarks and anti-quarks. We have used massless u and d quarks, and $m_s = 150$ MeV. It has been found [13,14] that within the MIT bag model (without color superconductivity) with a density-independent bag constant B, the maximum mass of a NS cannot exceed a value of about 1.6 solar masses. Indeed, the maximum mass increases as the value of B decreases, but too small values of B are incompatible with a hadron-quark transition density $\rho >$ 2–3 ρ_0 in nearly symmetric nuclear matter, as demanded by heavy-ion collision phenomenology. In order to overcome these restrictions of the model, one can introduce a density-dependent bag parameter $B(\rho)$, and this approach was followed in Ref. [14]. This allows one to lower the value of B at large density, providing a stiffer QM EOS and increasing the value of the maximum mass, while at the same time still fulfilling the condition of no phase transition below $\rho \approx 3\rho_0$ in symmetric matter. In the following we present results based on the MIT model using both a constant value of the bag parameter, $B = 90$ MeV/fm^3, and a gaussian parametrization for the density dependence, $B(\rho) = B_\infty + (B_0 - B_\infty) \exp\left[-\beta\left(\frac{\rho}{\rho_0}\right)^2\right]$ with $B_\infty = 50$ MeV/fm^3, $B_0 = 400$ MeV/fm^3, and $\beta = 0.17$, see Ref. [14]. The introduction of a density-dependent bag has to be taken into account properly for the computation of various thermodynamical quantities; in particular the quark chemical potentials μ_q and the pressure p are modified as

$$\mu_q \to \mu_q + \frac{dB(\rho)}{d\rho}\,, \quad p \to p + \rho\frac{dB(\rho)}{d\rho}\,. \tag{2}$$

Nevertheless, due to a cancelation of the second term in (2), occurring in relations (1) for the beta-equilibrium, the composition at a given total baryon density remains unaffected by this term (and is in fact independent of B). At this stage of investigation, we disregard possible dependencies of the bag parameter on temperature and individual quark densities. For a more extensive discussion of this topic, the reader is referred to Refs. [14].
The individual quark chemical potentials are fixed by Eq. (1) with $b_q = 1/3$, which implies: $\mu_d = \mu_s = \mu_u + \mu_l - \mu_{\nu_l}$. The charge neutrality condition and the total baryon number conservation together with the constraints on the lepton number Y_l conservation allow us to determine the composition $\rho_q(\rho)$ and then the pressure of the QM phase. In both phases the contribution of leptons is that of a Fermi gas. In the range of temperature considered here ($0 \div 50$ MeV) thermal effects are rather weak, the presence of neutrinos instead influences quite strongly the composition: In this case the relative fraction of u quarks increases substantially from 33% to about 42%, compensating the charge of the electrons that are present at an average percentage of 25% throughout the considered range of baryon density, whereas d and s quark fractions are slightly lowered (see [15]). We now consider the hadron-quark phase transition in beta-stable matter at finite temperature. In the present work we adopt the simple Maxwell construction for the phase transition from the plot of pressure versus chemical potential. The more general Glendenning (Gibbs) construction [16] is still affected by many theoretical uncertainties and in any case influences very little the fi-

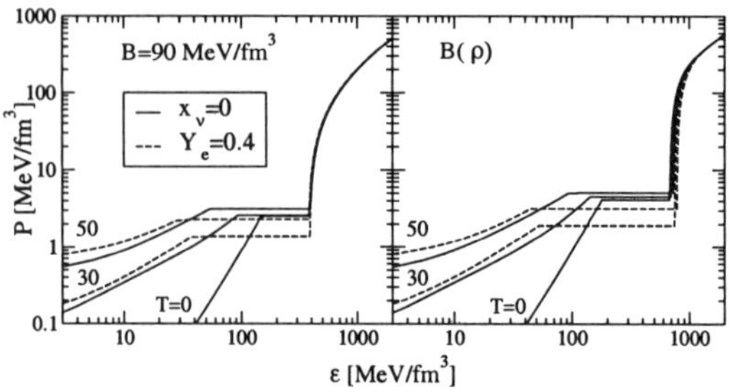

Fig. 2. Pressure as a function of energy density for beta-stable matter with (dashed curves) and without (solid curves) neutrino trapping at different temperatures $T = 0$, 30, and 50 MeV with a bag constant $B = 90$ MeV/fm^3 (left panel) or a density-dependent bag parameter (right panel).

nal mass-radius relations of massive (proto)neutron stars [14]. We therefore display in Fig. 1 the pressure p (upper panels) and baryon density ρ (lower panels) as functions of the baryon chemical potential μ_n for both baryonic and QM phases at temperatures $T = 0, 10, 30, 50$ MeV. The crossing points of the baryon and quark pressure curves (marked with a dot) represent the transitions between baryon and QM phases. The projections of these points (dotted lines) on the baryon and quark density curves in the lower panels indicate the corresponding transition densities from low-density baryonic matter, ρ_H, to high-density QM, ρ_Q.

The main aspects of the EoS for such stellar matter are displayed in Fig. 2 and are well summarized as follows. The transition density ρ_H is rather low, of the order of the nuclear matter saturation density. The phase transition density jump $\rho_Q - \rho_H$ is large, several times ρ_H, and the model with density-dependent bag parameter predicts larger transition densities ρ_H and larger jumps $\rho_Q - \rho_H$ than those with bag constant $B = 90$ MeV/fm^3. The plateaus in the Maxwell construction are thus wider for the former case. Thermal effects and neutrino trapping shift ρ_H to lower values of subnuclear densities and increase the density jump $\rho_Q - \rho_H$. For the cold case the presence of neutrinos even inhibits completely the phase transition [15].

4. Structure and stability

The stable configurations of a PNS can be obtained from the hydrostatic equilibrium equation of Tolman, Oppenheimer, and Volkov [17] for the pressure P and the gravitational mass m, once the EoS $P(\epsilon)$ is specified, being ϵ the total energy density. We schematize the entire evolution of a PNS as

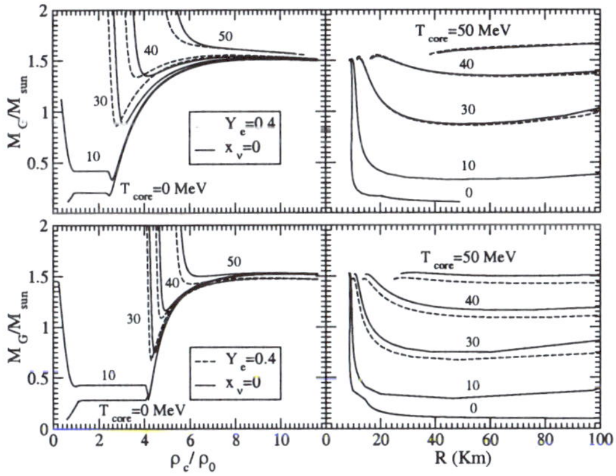

Fig. 3. (Proto)Neutron star mass-central density (left panel) and mass-radius (right panel) relations for different core temperatures $T = 0, 10, 30, 40, 50$ MeV and neutrino-free (solid curves) or neutrino-trapped (dashed curves) matter. A bag constant $B = 90$ MeV/fm^3 is used for QM. Same for lower panels but with a density-dependent bag parameter.

divided in two main stages. In the first, representing the early stage, the PNS is in a hot ($T_{core} = 30 \div 40$ MeV) stable configuration with a neutrino-trapped core and a high-entropy envelope ($s_{env} \simeq 6 \div 8$). The second stage represents the end of the short-term cooling where the neutrino-free core possesses a low temperature ($T_{core} = 10$ MeV) and the outer part can be considered as a cold crust (BPS+FMT).

The results are plotted in Fig. 3, where we display the gravitational mass M_G (in units of the solar mass $M_\odot$) as a function of the radius R (right panels) and the central baryon density ρ_c (left panels), for QM EOS with $B = 90$ MeV/fm^3 and $B(\rho)$, respectively. Due to the use of the Maxwell construction, the curves are not continuous [16]. PNS in our approach are thus practically hybrid stars and the heaviest ones have only a thin outer layer of baryonic matter. For completeness we display the complete set of results at core temperatures $T = 0, 10, 30, 40, 50$ MeV with and without neutrino trapping, although only the curves with high temperatures and neutrino trapping and low temperatures without trapping are the physically relevant ones. We observe in any case a surprising insensitivity of the results to the presence of neutrinos, in particular for the $B = 90$ MeV/fm^3 case, which can be traced back to the fact that the QM EOS $p(\epsilon)$ in Fig. 2 is practically insensitive to the neutrino fraction (see [15] for more details). On the other hand, the temperature dependence of the curves is quite pronounced for intermediate and low-mass stars, showing a strong increase of the minimum mass with temperature, whereas the maximum mass remains practically constant under all possible circumstances. Above core temperatures of about 40–50 MeV all stellar configurations become unstable. Concerning the dependence on the QM EOS, we observe again only a slight variation of the maximum PNS masses between 1.55 $M_\odot$ for $B = 90$ MeV/fm^3 and 1.48 $M_\odot$ for $B(\rho)$. Clearer differences exist for the radii, which for the same mass and temperature are larger for the $B = 90$ MeV/fm^3 model.

In conclusion, in this article we have extended a previous work on baryonic PNS [5] to the case of hybrid PNS. We combined the most recent microscopic baryonic EOS in the BHF approach involving nuclear three-body forces and hyperons with two versions of a generalized MIT bag model for QM. The EoS employed for both phases are checked by phenomenological constraints.

We modelled the entropy and temperature profile of a PNS in a simplified way taking as much as possible care about results coming from dynamical calculations [9]. This approach allows us to study the stability of a PNS varying both the temperature of the core and the central density.

References

1. A. Burrows and J.M. Lattimer, Astrophys. J. **307**, 178 (1986); J.A. Pons *et al.*, Astrophys. J. **513**, 780 (1999); Sumiyoshi *et al.*, Astrophys. J. **629**, 922 (2005).
2. M. Prakash *et al.*, Phys. Rep. **280**, 1 (1997).
3. K. Strobel, Ch. Schaab, and M.K. Weigel, Astr. Astroph. **350**, 497 (1999).
4. D. Gondek, P. Haensel, and J. Zdunik, Astr. Astroph. **325**, 217 (1997).
5. O.E. Nicotra *et al.*, Astr. Astroph. **451**, 213 (2006).
6. G. Baym, C. Pethick, and D. Sutherland, Astrophys. J. **170**, 299 (1971).
7. R. Feynman, F. Metropolis, and E. Teller, Phys. Rev. **75**, 1561 (1949).
8. J.M. Lattimer and F.D. Swesty, Nucl. Phys. **A535**, 331 (1991).
9. O.E. Nicotra, arXiv:nucl-th/0607055.
10. H.A. Bethe, Rev. Mod. Phys. **62**, 801 (1990).
11. H.J. Janka, Astropart. Phys. **3**, 377 (1995).
12. A. Chodos *et al.*, Phys. Rev. D **9**, 3471 (1974).
13. M. Alford and S. Reddy, Phys. Rev. D **67**, 074024 (2003).
14. G.F. Burgio *et al.*, Phys. Rev. C **66**, 025802 (2002); M. Baldo *et al.*, Phys. Lett. **B562**, 153 (2003); C. Maieron *et al.*, Phys. Rev. D **70**, 043010 (2004).
15. O.E. Nicotra *et al.*, Phys. Rev. D **74**, 123001 (2006); arXiv:astro-ph/0608021.
16. N.K. Glendenning, Phys. Rev. D **46**, 1274 (1992).
17. S.L. Shapiro and S.A. Teukolsky, 1983, Black Holes, White Dwarfs and Neutron Stars, ed. John Wiley & Sons, New York; Ya.B. Zeldovich and I.D. Novikov, 1971, Relativistic astrophysics, vol. I, University of Chicago Press, Chicago.

PROCESSES OF BURNING AND CONVECTION IN COMPACT OBJECTS

I. PARENTI[1] and A. DRAGO[2]
Dip. di Fisica, Univ. di Ferrara
and INFN Sez. di Ferrara
I-44100 Ferrara, Italy
[1] *E-mail: parenti@fe.infn.it*
[2] *E-mail: drago@fe.infn.it*

A. LAVAGNO
Dip. di Fisica, Politecnico di Torino
and INFN Sez. di Torino
I-10129 Torino, Italy
E-mail: andrea.lavagno@polito.it

We study the hydrodynamical transition from an hadronic star into a quark or a hybrid star. We discuss the possible mode of burning, using a fully relativistic formalism and realistic Equations of State. We take into account the possibility that quarks form a diquark condensate. We find that the conversion process always corresponds to a deflagration and never to a detonation. Hydrodynamical instabilities can develop on the front but the increase in the conversion's velocity is not sufficient to transform the deflagration into a detonation. Concerning convection, it does not always develop. Instead the process of conversion from ungapped quark matter to gapped quarks always allows the formation of a convective layer.

1. Introduction

The central density of compact stellar objects may reach values up to ten times nuclear matter saturation density and therefore it is common opinion that a phase transition to deconfined Quark Matter (QM) may take place at least in the central region. Recent studies on the QCD phase diagram at finite densities and temperatures have revealed the existence of several possible types of quark phases. Many theoretical works have investigated the possible formation of a diquark condensate in the quark phase, at densities reachable in the core of a compact star [1–4] and the formation of this condensate can deeply modify the structure of the star [5–7].

It has been proposed several times that the transition from a star containing only hadrons to a star composed, at least in part, of deconfined quarks can release enough energy to power a GRB [8–11]. A crucial question concerns the way in which the conversion takes place, either via a detonation or a deflagration. It has been shown in the past that the mechanical wave associated to a detonation would expel a relatively large amount of baryon from the star surface [12]. In the case of a detonation the region in which the electron-photon plasma forms (via neutrino-antineutrino annihilation near the surface of the compact star and also via direct photon and electron-positron pair production if a quark star is formed) would be contaminated by the baryonic load and it would be impossible to accelerate the plasma up to the enormous Lorentz factors needed to explain the GRBs.
We have studied this process from the hydrodinamical point of view to understand in which way a neutron star trasforms to a more compact object.

2. Fluidodynamics equations of the conversion process

The starting point of our analysis is given by the fluidodynamics equations describing the conservation of the energy-momentum tensor and of the baryon flux density across the conversion front. Following [13], the continuity equations for the energy-momentum tensor, in the frame in which the front is at rest, read:

$$(e_h + p_h)v_h\gamma_h^2 = (e_q + p_q)v_q\gamma_q^2 \,, \tag{1}$$

$$(e_h + p_h)v_h^2\gamma_h^2 + p_h = (e_q + p_q)v_q^2\gamma_q^2 + p_q \,, \tag{2}$$

while the continuity of the baryon flux density reads:

$$\rho_B^h v_h \gamma_h = \rho_B^q v_q \gamma_q \,. \tag{3}$$

In the previous equations e is the energy density, p is the pressure, ρ_B is the baryon density (all in the rest frame); v is the velocity of a given phase (in the front frame) and finally γ is the Lorentz factor. We use the labels h and q to indicate the hadron and the quark phase, respectively.
It is possible to classify the various conversion mechanisms by comparing the velocities of the phases (in the front frame) with the corresponding sound velocities v_s. The conditions are:

$$v_h > v_{sh} \; v_q < v_{sq} \text{ strong detonation} \tag{4}$$

$$v_h > v_{sh} \; v_q > v_{sq} \text{ weak detonation} \tag{5}$$

$$v_h < v_{sh} \; v_q < v_{sq} \text{ weak deflagration} \tag{6}$$

$$v_h < v_{sh} \; v_q > v_{sq} \text{ strong deflagration.} \tag{7}$$

In order to understand if the process is exothermic we have compared the energy per baryon of each phase in the hadronic matter rest frame. From thermodinamics first principle we obtain:

$$\Delta\left(\frac{E}{A}\right)(T,\rho_B^h) \equiv \frac{e_h(u_h,\rho_B^h,T_h)}{\rho_B^h(u_h)} - \frac{e_q(u_q,\rho_B^q,T)}{\rho_B^q(u_q)} = c_V^q(T-T_h) \quad (8)$$

Here the energy density $e(u,\rho_B,T)$ and the baryon density $\rho_B(u)$ are computed at finite velocity u and they read [14]:

$$e(u) = \left(e + p\,u^2\right)\gamma(u)^2 \quad (9)$$

$$\rho_B(u) = \rho_B\,\gamma(u)^2 \quad (10)$$

where e and ρ_B are the rest frame quantities. An exothermic process corresponds to a positive $\Delta(E/A)$.

3. Deflagration or detonation

We are interested in answering two questions, namely if a detonation is possible and, if this is not the case, which type of deflagration is obtained. We have solved Eqs. (1,2,3) using various Equations of State (EoSs). For the hadronic phase we have used a relativistic mean field theory corresponding to the GM3 parametrization [15]. For the quark phase we have used two different structures: a Normal Quark phase (NQ) described by the MIT bag model and a phase containing a diquark condensate. We have also investigated the possibility of forming a mixed phase of quarks and hadrons. We have analyzed many scenarios [16], discussing both the possibility of a not β-stable phase and of a β-stable phase immediately after the front, the presence of hyperons, the effect of temperature, etc. In all cases the conditions for detonation are not fulfilled.

We have also considered the possibility that the formation of diquark condensate takes place not immediately after the deconfinement transition but it is delayed. It is therefore interesting to discuss also the transition from NQ to gapped QM, which in our case we assume to be in Color Flavor Locked (CFL) phase. Also in this case no detonation is obtained and we are always in the regime of strong deflagration.

Finally, it is important to remark that the velocity of sound in the center of a compact star is typically of the order of $(0.5 \div 0.8)c$ and the velocity of the deflagrative front is marginally lower, $(0.4 \div 0.7)c$.

A remark is in order. Concerning the *actual* velocity of the conversion front, it can be determined by solving the fluidodynamics equations only if the conversion is a detonation. If it is a deflagration, other physical scales determine the velocity (significantly reducing its value), as e.g. heat diffusion or the production and diffusion of strangeness.

The main result obtained in this Section is that a detonative regime is never directly reached after imposing the continuity conditions on the front. On the other hand two problems still need to be discussed: the estimate of the velocity of the deflagrative front taking into account the effect of hydrodynamical instabilities which in principle can increase the conversion velocity transforming a deflagration into a detonation and the possibility that convection develops. These two points are discussed in the next Sections.

4. Hydrodynamical instabilities

In the previous section we have shown that the conversion process always takes place as a deflagration. In this case it is extremely difficult to estimate the velocity of the conversion front. The velocity is governed by the slowest among the processes which need to take place for the combustion to continue. In the seminal work of [17] it was shown that, in the absence of hyperons, the conversion velocity crucially depends on the rapidity by which strangeness is produced in the quark sector and diffuses into the hadronic sector. The final expression for the velocity, assuming a stable laminar front (as if it was a slow combustion), reads:

$$v_{\rm sc} = \frac{2}{\sqrt{g}} \left(\frac{\mu}{T}\right) \,\mathrm{m\,s^{-1}} \tag{11}$$

where $g \simeq 2(1-a_0)/a_0^4$, a_0 is the down-strange asymmetry parameter, μ is the chemical potential of quarks and T is the temperature.

As already remark by [18] the conversion velocity can be significantly increased taking into account hydrodynamical instabilities. Indeed in the previous section we have shown that the conversion is always a strong deflagration and not a slow combustion. Therefore the conversion front is unstable and wrinkles can form.

There are at least two types of hydrodynamical instabilities discussed in literature, the Landau-Darrieus (LD) and the Rayleigh-Taylor (RT) (for an introduction to these problems see [19]). Both these instabilities can develop when $\Delta e \equiv e_1 - e_2 > 0$. The LD instability is the one which characterizes the strong deflagration regime and the amplification of the wrinkles on the conversion front is directly due to the conservation of the energy-momentum flux, as imposed by Eqs. (1,2,3). That instability can develop independently on the presence of gravity. At the contrary, the RT instability develops only

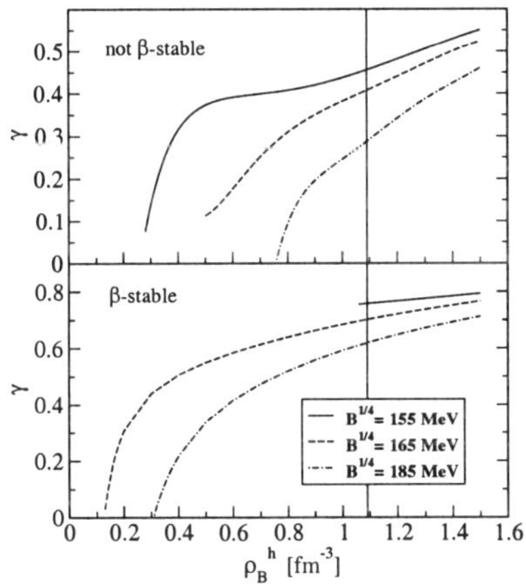

Fig. 1. The γ-factor entering the fractal dimension of the conversion front, as a function of the baryonic density of the hadronic phase. The vertical line corresponds to the central density of the most massive stable configuration of a nucleonic star obtained using GM3 model (ρ_h^{max}=1.09 fm^{-3}).

if a gravitational field is present and if the direction of the density gradient is opposite to the direction of the gravitational force.

Due to RT and LD instabilities the area of the conversion front increases. The conversion velocity also increases since all exchanges between the burned and the unburned zone are now more efficient. A way of estimating the effective velocity $v_{\rm eff}$ is through the introduction of the fractal dimension of the surface [20–22]. The larger is the excess of the fractal dimension respect to the dimension of a spherical front, the huger is the increase of the front velocity respect to the laminar case velocity ($v_{\rm sc}$). In the absence of new dimensional scales between the minimal dimension $l_{\rm min}$ and the maximal dimension $l_{\rm max}$ of the wrinkle, $v_{\rm eff}$ is given by:

$$v_{\rm eff} = v_{\rm sc} \left(\frac{l_{\rm max}}{l_{\rm min}}\right)^{D-2} . \tag{12}$$

Here D is the fractal dimension of the surface of the front and it can be estimated as [21,22]:

$$D = 2 + D_0 \gamma^2 , \tag{13}$$

where $D_0 \sim 0.6$ and $\gamma = 1 - e_2/e_1$.

In this analysis a crucial role is played by neutrino trapping which does not allow the system to reach β-equilibrium on the same timescale of the conversion process. Taking into account this delay of the weak processes, then γ.45 at all densities. The effect of neutrino trapping is displayed in Fig. 1. Our numerical analysis shows that, although the effective velocity can be significantly enhanced respect to the laminar velocity, it is unlikely that $v_{\rm eff}$ exceeds the speed of sound and therefore the process remains a deflagration.

5. Convection

Convective instability can also instaure, because in a regime of strong deflagration the energy density of the newly formed phase is smaller than the energy density of the old phase. The condition for convection to develop reads:

$$e(P_B, S_B, Y_e^B) < e(P, S, Y_e)\,, \tag{14}$$

where S is the entropy and Y_e is electron fraction and the suffix B indicates the same quantities for the blob. It is called *mixing length* the distance traveled by the blob before being so modified by the surrounding medium that condition (14) is no more satisfied. There are various types of convection. In order for the so-called "quasi-Ledoux" convection to develop, the inequality in Eq. (14) has to be satisfied with $P_B = P$, in every point of the convection layer, whose size is actually defined through Eq. (14) itself (for a recent review of hydrodynamical problems see e.g. [23]). More explicitly, as soon as the quarks' drop enters the hadronic phase, the pressure of the blob starts equilibrating with the pressure of the surrounding material. Quasi-Ledoux convection develops only if the energy density of the new phase remains smaller than that of the old phase *after* the pressure has equilibrated. When the quasi-Ledoux convection can take place, the typical size of the convective layer (that corresponds to the maximum value of the mixing length) is of the order of a few km. In Fig. 2 (where $B^{1/4} = 155$ MeV) we show the results of the analysis of the quasi-Ledoux convection using realistic EoSs. We also show examples of trajectories in the pressure vs energy-density plane of a drop of quarks after its formation.
It is important to remark that the existence of a convective layer during the conversion of unpaired QM into gapped QM is independent of the model parameters. In this second transition convection always develops.

6. Conclusions

Summarizing, the results of our analysis are the followings:

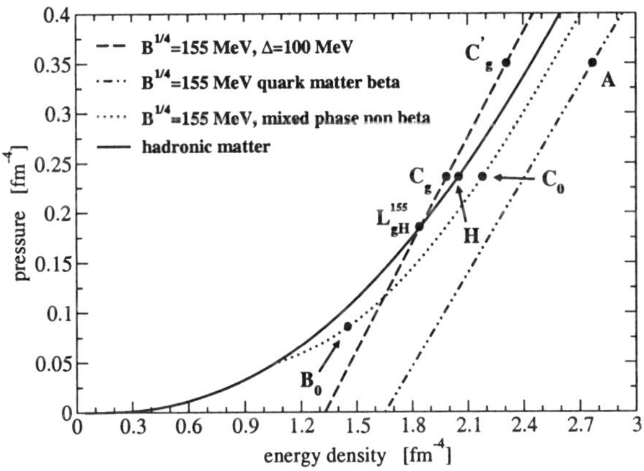

Fig. 2. Scheme for convection: H represents the drop of hadronic matter just before deconfinement, B_0 represents the drop of newly formed QM, C stays for the drop of QM after pressure equilibration and L indicates the end point of the convective layer. Finally A represents a drop of ungapped quark matter before its transition to CFL phase. Here $B^{1/4} = 155$ MeV and hyperons are not included.

- the conversion always takes place as a strong deflagration and never as a detonation
- fluidodynamical instabilities are present and they significantly increase the conversion velocity but, in realistic cases, the conversion process does not transform from a deflagration to a detonation
- convection can develop in specific cases, in particular it takes place if hyperons are present or if diquark condensate does form.

Our results about the not-detonative mode of the transition allow the formation of a GRB via the conversion of a neutron star in a more compact object.
Moreover, as discussed in [24] the conversion process can take place in two steps, with a first transition from hadrons to ungapped quarks and a second transition in which a CFL phase is produced. In order to associate an emission peak with each of the two transitions, the conversion process must be rapid enough to deposit in a few seconds (or less) a huge energy inside the star. Neutrinos will then transport the energy to the exterior on a time scale of order $(10 \div 20)$ s. Clearly, the result of our calculation provides these large velocities, since the conversion process occurs on a time scale of $(0.1 \div 1)$ s for the first transition in the case of a laminar front and it is much more rapid if the hydrodynamical instabilities are taken into

account. The second transition lasts only some 10^{-3} s due to the formation of a convective layer. If the two processes takes place one after the other it is even possible that the formation of diquark condensate accelerates the conversion process by developing a convective layer inside the hadronic phase. It is interesting to remark that this scenario is compatible with an analysis of the time-structure of the light curves of GRBs [25].

References

1. M. Alford, J. Berges, and K. Rajagopal, Nucl. Phys. **B558**, 219 (1999).
2. M. Alford, K. Rajagopal, and F. Wilczek, Nucl. Phys. **B537**, 443 (1999).
3. M. Alford, C. Kouvaris, and K. Rajagopal, Phys. Rev. D **71**, 054009 (2005).
4. S. B. Ruster, V. Werth, M. Buballa, I. A. Shovkovy, and D. H. Rischke, Phys. Rev. D **73**, 034025 (2006).
5. M. Alford and S. Reddy, Phys. Rev. D **67**, 074024 (2003).
6. M. Baldo *et al.*, Phys. Lett. **B562**, 153 (2003).
7. A. Drago, A. Lavagno, and G. Pagliara, Phys. Rev. D **69**, 057505 (2005).
8. K.S. Cheng and Z.G. Dai, Phys. Rev. Lett. **77**, 1210 (1996).
9. X.Y. Wang *et al.*, Astron. Astrophys. **357**, 543 (2000).
10. Z. Berezhiani *et al.*, Astrophys. J. **586**, 1250 (2003).
11. B. Paczynski and P. Haensel, Mon. Not. Roy. Astron. Soc. **362**, L4 (2005).
12. C.L. Fryer and S.E. Woosley, Astrophys. J. **501**, 780 (1998).
13. L. D. Landau and E. M Lifshitz, *Fluid Mechanics* (Pergamon Press, 1987).
14. R. Tolman, *Relativity Thermodynamics and Cosmology* (Oxford University Press, 1934).
15. N. K. Glendenning and S. A. Moszkowski, Phys. Rev. Lett. **67**, 2414 (1991).
16. A. Drago, A. Lavagno, and I. Parenti, astro-ph/0512652.
17. A. Olinto, Phys. Lett. **B192**, 71 (1987).
18. J. E. Horvath and O. G. Benvenuto, Phys. Lett. **B213**, 516 (1988).
19. Ya. B. Zeldovich, G. I. Barenblatt, V. B. Librovich, and G. M. Makhviladze, *The Mathematical Theory of Combustion and Explosions* (Plenum, New York, 1985).
20. S. E. Woosley, in *Supernovae*, edited by A. G. Petschek (Springer-Verlag, New York, 1990), p.182.
21. S. Iv. Blinnikov, P. V. Sasorov, and S. E. Woosley, Space Science Rev. **74**, 299 (1995).
22. S. Iv. Blinnikov and P. V. Sasorov, Phys. Rev. E **53**, 4827 (1995).
23. J. R. Wilson and G. J. Mathews, *Relativistic Numerical Hydrodynamics* (Cambridge University Press) (2003).
24. A. Drago, A. Lavagno, and G. Pagliara, Nucl. Phys. **A774**, 823 (2006).
25. A. Drago and G. Pagliara, astro-ph/0512602.

Nuclear Physics with Electroweak Probes

NUCLEAR PHYSICS WITH ELECTROWEAK PROBES

C. GIUSTI

Dipartimento di Fisica Nucleare e Teorica, Università degli Studi, and INFN, Sezione di Pavia, via Bassi 6 I-27100 Pavia, Italy

The research activitities carried out in Italy during the last two years in the field of theoretical nuclear physics with electroweak probes are reviewed. Different models for electron-nucleus and neutrino-nucleus scattering are compared. The results obtained for electromagnetic reactions on few-nucleon systems and on complex nuclei are discussed. The recent developments in the study of electron- and photon-induced reactions with one and two-nucleon emission are presented.

1. Introduction

Several decades of theoretical and experimental investigation in the field of nuclear physics with electromagnetic probes have yielded a wealth of information on nuclear structure and interaction mechanisms [1]. An important contribution has been given by the Italian theory groups that since many years have been working in the field, within many international collaborations and in close connection with the experimental activities.

In spite of so many years of investigation and of the great progress achieved, some interesting aspects are still unclear and there is yet much to be learned. In the last few years theories have improved remarkably but some data are old, incomplete or not accurate enough to disentangle interesting effects. The need of a more complete experimental program to investigate nuclei with electromagnetic probes was already pointed out by G. Co' in the previous report two years ago [2].

In recent years almost all the Italian groups working in electron scattering have applied their models to ν-nucleus scattering. Although the two situations are different, the extension to neutrino scattering of the electron scattering formalism is straightforward. Moreover, electron scattering is the best available guide to determine the prediction power of a nuclear model.

The observation of neutrino oscillations and the proposal and realization of new experiments, aimed at determining neutrino properties with high accuracy, renewed interest in neutrino scattering on complex nuclei. In fact, neutrinos are elusive particles. They are chargeless, almost massless, and only weakly interacting. Their presence can only be inferred detecting the particles they create when colliding or interacting with matter. Nuclei are often used as neutrino detectors providing relatively large cross sections. The interest in ν-nucleus scattering extends to different fields, such as astrophysics, cosmology, particle and nuclear physics. In hadronic and nuclear physics neutrinos can excite nuclear modes unaccessible in electron scattering, can give information on the hadronic weak current and on the strange nucleon form factors [3]. Thus, neutrino physics is of great interest and involves many different phenomena. The interpretation of data, however, requires a detailed knowledge of the ν-nucleus interaction as well as reliable cross section calculations where nuclear effects are properly taken into account.

The different models used to treat electron- and ν-nucleus scattering are discussed in Sec. 2. The treatment of the two-body weak axial current is presented in Sec. 3. Scaling and superscaling in lepton-nucleus scattering are considered in Sec. 4. Electromagnetic reactions in few-body and complex nuclei are reviewed in Sects. 5 and 6, respectively.

2. Electron-Nucleus and Neutrino-Nucleus Scattering

Lepton-nucleus scattering is usually described in the one-boson exchange approximation. The exchanged boson is the photon in the case of the electromagnetic interaction and the Z^0 or $W^{\pm}$ in the case of the weak interaction. In electron scattering the invariant amplitude is given by the sum of the one-photon and the one-Z^0 boson exchange term. The first term is parity conserving and the second-one is Parity Violating (PV). PV Electron Scattering (PVES) requires a polarized incident electron and is interesting to study the strange nucleon form factors [3]. In ν ($\bar{\nu}$)-nucleus scattering the boson exchanged is the Z^0 for neutral-current (NC) scattering, i.e. (ν, ν') $[(\bar{\nu}, \bar{\nu}')]$, and the W^+ (W^-) for charged-current (CC) scattering, where a charged lepton is obtained in the final state, i.e. (ν, l^-) $[(\bar{\nu}, l^+)]$.

Different processes can thus be considered. In any case, in the one-boson exchange approximation, the cross section is given in the form

$$\mathrm{d}\sigma = K L^{\mu\nu} \, W_{\mu\nu}, \tag{1}$$

where K is a kinematical factor, the lepton tensor $L^{\mu\nu}$ depends only on the lepton kinematics, and the nuclear response is contained in the hadron tensor $W^{\mu\nu}$, whose components are given by products of the matrix elements of the nuclear current J^μ between the initial and final nuclear states, i.e.,

$$W_{\mu\nu} = \sum_f \langle \Psi_f \mid J^\mu(\boldsymbol{q}) \mid \Psi_i \rangle \langle \Psi_i \mid J^{\nu\dagger}(\boldsymbol{q}) \mid \Psi_f \rangle \, \delta(E_i + \omega - E_f), \quad (2)$$

where ω and $\boldsymbol{q}$ are the energy and momentum transfer, respectively. Similar models are used to calculate $W^{\mu\nu}$ in electron- and ν-nucleus scattering.

In a schematic representation of the nuclear response to the electroweak probe, different regions can be identified. At low energy transfer, below the threshold for the emission of a nucleon from the target, the response is dominated by discrete states, that can be treated in large model spaces. Above the continuum threshold there are giant resonance levels, collective excitations that can be described within the Random Phase Approximation (RPA). Then, a large broad peak occurs at about $\omega = q^2/2m_{\rm N}$. Its position corresponds to the elastic peak in electron scattering by a free nucleon. In the region of the quasielastic (QE) peak the response is dominated by the single-particle (s.p.) dynamics and by one-nucleon knockout processes, where the interaction occurs on a quasifree nucleon which is emitted from the nucleus with a direct one-step mechanism. At higher energies mesons and nucleon resonances can be produced. For instance, at an excitation energy of $\sim$ 300 MeV the Δ-peak corresponds to the first nucleon excitation.

Schematically, the response is the same for electron and neutrino scattering. However, the cross sections calculated for the (e,e$'$), (ν,ν'), (ν,e$^-$), and ($\bar{\nu}$,e$^+$) reactions on O^{16}, in the same kinematic conditions within the continuum RPA, have a quite different behavior [2,4]. The difference is due to the different current in the electromagnetic and weak interactions: the neutrino cross sections are dominated by the axial vector term that does not contribute to electron scattering. As a consequence, it is necessary to be careful in relying on the fact that a model able to give a good description of electron scattering will give a good description of neutrino scattering. In spite of this warning, a model able to describe electron scattering data can be considered as a good basis to treat also ν-nucleus scattering.

2.1. *Random Phase Approximation*

Electron-nucleus and ν-nucleus cross sections calculated within the RPA are compared in Refs. [4,5]. The RPA is an effective theory, aiming to describe the excitations of many-body systems, that has been widely and successfully applied in nuclear physics over a wide range of excitation energies.

The RPA describes the nuclear excited states as a linear combination of particle-hole (ph) and hole-particle (hp) excitations. The combination coefficients for each state are obtained solving the secular RPA equations that contain s.p. energies and wave functions as input. They are generated by a Woods-Saxon potential whose parameters have been fixed to reproduce the energies of the levels close to the Fermi surface and the rms radii [4,5]. The other input of the theory is the effective interaction, V^{eff}, that is an effective interaction in the medium. Calculations are compared in Ref. [4,5] with various zero-range and finite-range interactions, in order to point out the sensitivity of the RPA results to the choice of V^{eff}.

RPA calculations are necessary to produce giant resonances and collective low-lying states. The results, however, strongly depend on the effective interaction used. Interactions equivalent from the spectroscopic point of view can produce different excited states. This indicates the limits of the RPA. In the evaluation of some observables, the use of effective interactions cannot substitute the explicit treatment of degrees of freedom beyond 1p-1h excitations. The use of quenching factors reduces the spreading of the results obtained with different interactions in the electron excitation and increases the spreading in the neutrino excitation. This is further evidence that electrons and neutrinos excite the same states in a different way. The uncertainty on the cross section is large for neutrinos of 20-40 MeV. These uncertainties have heavy consequences on the cross sections of low energy neutrinos. When the neutrino energy is above 50 MeV, the results are rather independent of the interaction and the inclusion of many-particle many-hole excitations reduces the RPA cross sections by a factor of $\simeq 10 - 15\%$.

In the QE region RPA effects are rather small when a finite-range V^{eff} is used [4,5]. A zero-range interaction overestimates RPA effects. However, many-body effects beyond RPA are important, as it is shown in Fig. 1, where QE ^{16}O(e,e$'$) and ^{12}C(e,e$'$) cross sections calculated within the mean-field (MF) model, i.e., setting $V^{\text{eff}} = 0$, are compared to data. In the QE region these many-body effects, which in the RPA language are described as many-particle many-hole excitations, are usually called Final-State Interactions (FSI). Different treatments of FSI are available in the literature. In Fig. 1 [4,5] FSI are folded with a Lorentz function whose parameters are fixed by hadron scattering data [6] and give a redistribution of the strength that is essential to reproduce (e,e$'$) data.

2.2. *Nuclear Effects and FSI in the Quasielastic Region*

In the QE region the nuclear response is dominated by one-nucleon knockout processes. A lot of theoretical and experimental work has been done to study the QE inclusive (e,e$'$) and exclusive (e,e$'$p) reactions [1,8]. This

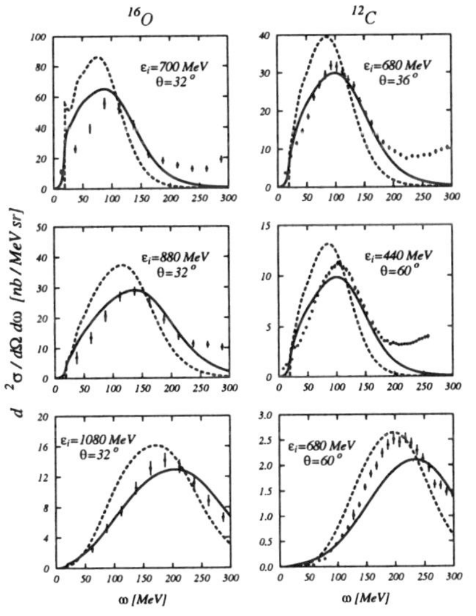

Fig. 1. $^{16}O(e,e')$ and $^{12}C(e,e')$ cross sections in the QE region. The MF results are shown by the dashed lines. The inclusion of FSI produces the full lines. Data from Ref. [7]. (From Ref. [5]).

work can be helpful to treat ν-nucleus scattering.

In the QE $\nu(\bar{\nu})$-nucleus NC and CC scattering we assume that one nucleon is emitted

$$\nu(\bar{\nu}) + \mathrm{A} \rightarrow \nu'(\bar{\nu}') + \mathrm{N} + (\mathrm{A}-1) \qquad \mathrm{NC} \qquad (3)$$

$$\nu(\bar{\nu}) + \mathrm{A} \rightarrow l^-(l^+) + \mathrm{p(n)} + (\mathrm{A}-1) \qquad \mathrm{CC} \qquad (4)$$

In the NC scattering only the emitted nucleon can be detected. Thus, the cross section must be integrated over the energy and angle of the final lepton. Also the state of the residual (A-1)-nucleus is not determined and the cross section is summed over all the available states of the residual nucleus. The same situation occurs for the CC reaction if only the outgoing nucleon is detected. The cross sections are therefore semi-inclusive in the hadronic sector and inclusive in the leptonic one and can be treated as an (e,e′p) reaction where only the outgoing proton is detected. The exclusive CC process where, as in the case of (e,e′p), the charged final lepton is detected in coincidence with the emitted nucleon can be considered as well. The inclusive CC scattering where only the charged lepton is detected can be treated with the same nuclear models used for the inclusive (e,e′) scattering.

QE ν-nucleus scattering has been treated in Refs. [9–11] using the same relativistic models that were developed for the inclusive (e,e′) and the exclusive (e,e′p) reactions. These models include nuclear effects and FSI.

In the first order perturbation theory and in the Impulse Approximation (IA), the transition amplitude of the NC and CC processes where the outgoing nucleon is detected is described as the sum of terms similar to those appearing in the Relativistic Distorted Wave IA (RDWIA) for the (e,e′p) knockout reaction. The amplitudes for the transition to a specific state n of the residual nucleus are obtained in a one-body representation

$$\langle n; \chi^{(-)}_{\boldsymbol{p}_{\rm N}} \mid J^{\mu}(\boldsymbol{q}) \mid \Psi_i \rangle = \langle \chi^{(-)}_{\boldsymbol{p}_{\rm N}} \mid j^{\mu}(\boldsymbol{q}) \mid \varphi_n \rangle \tag{5}$$

and contain three ingredients: the one-body nuclear weak current j^{μ}, the one-nucleon overlap $\varphi_n = \langle n|\Psi_i\rangle$, that is a s.p. bound state wave function whose normalization gives the spectroscopic factor, and the s.p. scattering wave function $\chi^{(-)}$ for the outgoing nucleon, that is eigenfunction of a complex optical potential describing the FSI between the outgoing nucleon and the residual nucleus.

Bound and scattering states are calculated with the same phenomenological ingredients used for the (e,e′p) calculations. A pure SM description is assumed for the states n, i.e., n is a one-hole state in the SM and a sum over all the occupied SM states is carried out. In these calculations FSI are described by a complex optical potential whose imaginary part gives an absorption that reduces the cross sections by $\sim$ 50%. The imaginary part accounts for the flux lost in a particular channel towards other channels. This approach is conceptually correct for an exclusive reaction, where only one channel contributes, but it would be conceptually wrong for an inclusive reaction, where all the channels contribute and the total flux must be conserved. For the semi-inclusive process where an emitted nucleon is detected, some of the reaction channels which are responsible for the imaginary part of the optical potential are not included in the experimental cross section and, from this point of view, it is correct to include the absorptive imaginary part of the potential.

In the inclusive scattering FSI can be treated in the Green's Function Approach (GFA), that was firstly applied to the QE (e,e′) scattering in a nonrelativistic [12] and in a relativistic [13] framework, and then adapted to the CC scattering [14] and to the PVES [15].

Recently, the GFA has been reformulated [16], within a nonrelativistic framework, including antisymmetrization and nuclear correlations, that were neglected in previous applications. Correlations are included by means of realistic one-body density matrices. Their numerical effects on the (e,e′) reaction are, however, small, within $\sim 5\%$ when only Short-Range Correlations (SRC) are included and within 10% when tensor correlations are added. These effects are in substantial agreement with those obtained in Ref. [17].

In the GFA the components of the nuclear response are written in terms of the s.p. optical model Green's function. This is the result of suitable approximations, such as the assumption of a one-body current and subtler approximations related to the IA. The explicit calculation of the Green's function can be avoided by its spectral representation, which is based on a biorthogonal expansion in terms of a non Hermitian optical potential $\mathcal{H}$ and of its Hermitian conjugate $\mathcal{H}^\dagger$. In practice, the calculation requires matrix elements of the same type as the RDWIA ones in Eq. (5), but involves eigenfunctions of both $\mathcal{H}$ and $\mathcal{H}^\dagger$, where the different sign of the imaginary part gives in one case an absorption and in the other case a gain of flux. Thus, the total flux is conserved and the imaginary part is responsible for the redistribution of the strength among different channels. This approach guarantees a consistent treatment of FSI in the exclusive and in the inclusive scattering.

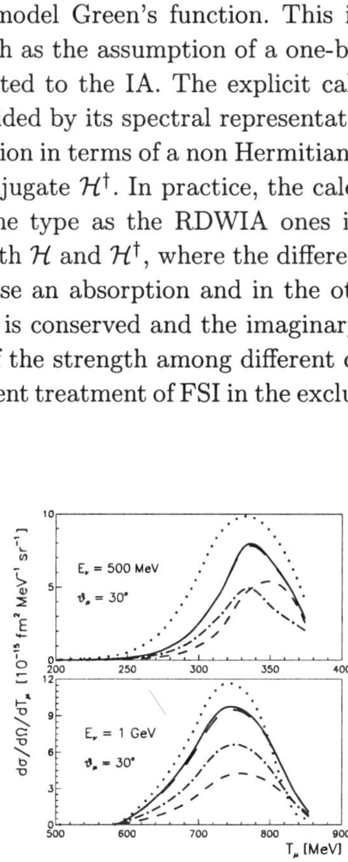

Fig. 2. The cross sections of the $^{16}\mathrm{O}(\nu_\mu, \mu^-)$ reaction for $E_\nu = 500$ and 1000 MeV at $\theta_\mu = 30^\circ$ as a function of the muon kinetic energy T_μ. Results for GFA (solid) RPWIA (dotted), rROP (long-dashed) are compared. The dot-dashed lines give the contribution of the integrated exclusive reactions with one-nucleon emission. Short dashed lines give the cross sections of the $^{16}\mathrm{O}(\bar{\nu}_\mu, \mu^+)$ reaction calculated within the GFA. (From Ref. [9]).

An example of the role of FSI is displayed in Fig. 2, where the $^{16}\mathrm{O}(\nu_\mu, \mu^-)$ cross sections calculated within the GFA are compared with the results of the Relativistic Plane Wave IA (RPWIA), where the plane wave approximation is assumed for the outgoing nucleon and FSI are neglected. The results obtained when only the real part of the Relativistic Optical Potential (rROP) is retained and the imaginary part is neglected are also shown in the figure. This approximation conserves the flux, but is inconsistent with the exclusive process. Although the use of a complex optical potential is conceptually important from a theoretical point of view, the very small differences given by the GFA and rROP results mean that the conservation of the flux is the most important condition in the present situation. The partial contribution given by the sum of all the integrated exclusive one-nucleon knockout reactions, also shown in the figure, is much smaller than the complete result. The difference is due to the spurious loss of flux produced by the absorptive imaginary part of the optical potential.

For the analysis of data a precise knowledge of ν-nucleus cross sections is needed, where theoretical uncertainties are reduced as much as possible. To this purpose, it is important to check the differences and the consistencies of the different models and the validity of the approximations used.

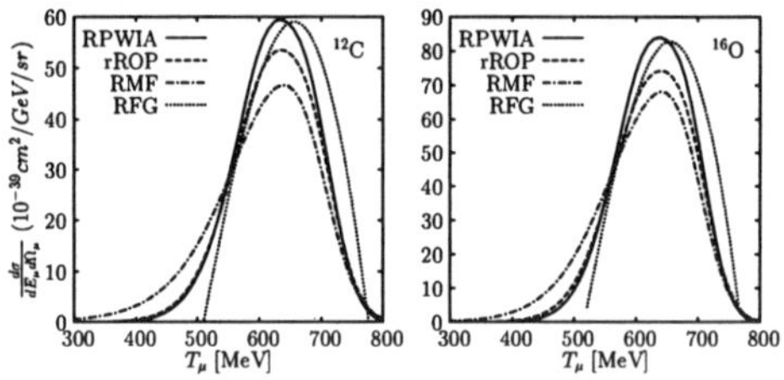

Fig. 3. The cross sections of the $^{12}\mathrm{C}(\nu_\mu, \mu^-)$ and $^{16}\mathrm{O}(\nu_\mu, \mu^-)$ reactions for $E_\nu = 1000$ MeV at $\theta_\mu = 45^o$ as a function of T_μ. The results of RPWIA (solid), rROP (dashed), RMF (dot-dashed), RFG (dotted) are compared. (From Ref. [18]).

Different treatments of FSI are considered in Fig. 3 [18]. The RPWIA and rROP cross sections are compared with those of a Relativistic Mean Field (RMF) approach, where the distorted waves are calculated with the same potential used for the initial bound states. The results of the Relativistic Fermi Gas (RFG) are also displayed in the figure.

CC and NC cross sections calculated in the RPWIA and in the nonrelativistic PWIA are compared in Fig. 4. The differences are small at E_ν=500 Mev and somewhat larger at 1 GeV.

Relativistic and nonrelativistic models have been developed and ap-

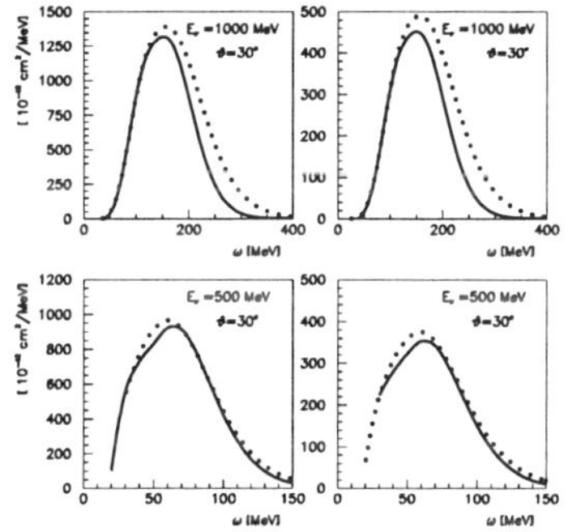

Fig. 4. The cross sections of the $^{16}O(\nu_e, e^-)$ (left panel) and $^{16}O(\nu, \nu')$ (right panel) reactions for $E_\nu = 500$ and 1000 MeV at $\theta = 30^o$. Solid lines: RWPIA (A. Meucci) dotted lines non relativistic PWIA (G. Co').

plied in nuclear physics with electroweak probes. Relativistic effects have been widely investigated. They increase with the energy and for energies of about 1 GeV a fully relativistic model should be used. In order to account for these effects, relativistic corrections in the kinematics and in the current operators are often included in nonrelativistic models. Even with these corrections, however, a nonrelativistic approach cannot reproduce all the relativistic aspects in the dynamics of a relativistic one. Moreover, calculations are generally carried out in the two cases with different theoretical ingredients. Relativistic and nonrelativistic models have thus to be considered as different and alternative approaches. All the available models make use of approximations and have merits and shortcomings. Only a relativistic approach can fully account for relativistic effects. At present, however, nonrelativistic models may allow to include specific nuclear effects, e.g., due to correlations and two-body currents, in a more consistent and clear framework.

Nuclear correlations and FSI in electron and neutrino scattering off ^{16}O are considered in Refs. [19–21]. The approach is based on the nonrelativistic nuclear many-body theory and on the IA. In the IA the scattering process off a nuclear target reduces to the incoherent sum of elementary processes involving only one nucleon. For each term in the sum the cross section is factorized into the product of the elementary lepton-nucleon cross section and the nuclear spectral function, describing the momentum and energy distribution of nucleons in the target. The correlated spectral function of ^{16}O

is obtained with a local density approximation in which nuclear matter results for a wide range of density values are combined with the experimental information from the ^{16}O(e,e′p) knockout reaction. FSI are treated within a Correlated Glauber Approximation (CGA), which rests on the premises that: i) the struck nucleon moves along a straight trajectory with constant velocity (eikonal approximation), ii) the spectator nucleons are seen by the struck particle as a collection of fixed scattering centers (frozen approximation). Under these assumptions, the propagator of the struck nucleon after the electroweak interaction is factorized in terms of the free space propagator and of a part that is related to the nuclear transparency measured in (e,e′p). The cross section is written in the convolution form in terms of the IA cross section and of a folding function including FSI.

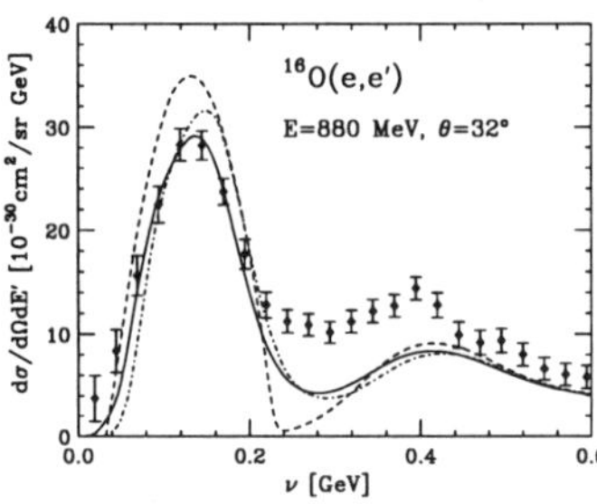

Fig. 5. Cross section of the ^{16}O(e,e′) reaction at beam energy 880 MeV and electron scattering angle 32°. Dot-dashed line: IA, solid line: full calculation including FSI, dashed line: FG. Data from Ref. [7]. (From Ref. [19]).

A numerical example is presented in Fig. 5 for the ^{16}O(e,e′) reaction. The model gives a good description of data in the region of the QE peak. FSI produce a shift and a redistribution of the strenght leading to a quenching of the peak and to an enhancement of the tail. The FG model overestimates the data. The failure of the calculations to reproduce the data in the Δ-peak region is likely to be mostly due [22] to the poor knowledge of the neutron structure function at low Q^2. The ability to yield quantitative predictions over a wide range of energies is critical to the analysis of neutrino experiments, in which the energy of the incident neutrino is not known, and must be reconstructed from the kinematics of the outgoing lepton.

3. Two-Body Weak Axial Current

Two-body Weak Axial Exchange Currents (WAECs) are considered in Refs. [23,24]. The issue of the pion pair axial current is addressed in Ref. [23], showing how the interplay of the chiral invariance and the double-counting problem restricts uniquely its form. The complete treatment of the WAECs, including the heavy meson exchange contributions, that is, besides π-, ρ-ω-a_1-exchange, is the main result of Ref. [24].

The Partial Conservation of the Axial Current (PCAC) reads

$$q_\mu < \Psi_f | j^a_{5\mu}(q) | \Psi_i >= i f_\pi m^2_\pi \Delta^\pi_F(q^2) < \Psi_f | m^a_\pi(q) | \Psi_i > , \qquad (6)$$

where the weak axial current $j^a_{5\mu}(q)$ is the sum of the one- and two–nucleon components

$$j^a_{5\mu}(q) = \sum_{i=1}^{A} j^a_{5\mu}(1, i, q_i) + \sum_{i<j}^{A} j^a_{5\mu}(2, ij, q), \qquad (7)$$

$m^a_\pi(q)$ is the pion source (the pion production/absorption amplitude). The wave functions $|\Psi_{i,f}>$ describe the initial or final nuclear states, which are eigenfunctions of the Schrödinger equation with the nuclear Hamiltonian $H = T + V$, where V is the potential describing the interaction between nucleon pairs. From Eqs. (6) and (7), taking for simplicity $A = 2$, the following set of equations are obtained in the operator form for the one- and two-nucleon components of the axial current

$$\vec{q}_i \cdot \vec{j}^a_5(1, \vec{q}_i) = [T_i \, , \, \rho^a_5(1, \vec{q}_i)] + i f_\pi m^2_\pi \Delta^\pi_F(q^2) m^a_\pi(1, \vec{q}_i) \, , \quad i = 1, 2 \, , \qquad (8)$$

$$\begin{aligned} \vec{q} \cdot \vec{j}^a_5(2, \vec{q}) &= [T_1 + T_2 \, , \, \rho^a_5(2, \vec{q})] + ([V \, , \, \rho^a_5(1, \vec{q})] + (1 \leftrightarrow 2) \\ &\quad + i f_\pi m^2_\pi \Delta^\pi_F(q^2) m^a_\pi(2, \vec{q}). \end{aligned} \qquad (9)$$

If the WAECs are constructed in order to satisfy these conditions, the matrix element of the total current between the solutions of the nuclear equation of motion satisfies the PCAC.

The WAECs are constructed from an effective Lagrangian possessing the chiral symmetry and respecting the vector dominance model. The exchange amplitudes of range B (B=π, ρ, ω, a_1) are derived as Feynman tree graphs and satisfy the PCAC. The currents are constructed from the amplitudes, in analogy with the electromagnetic Meson Exchange Currents (MEC) [25], as the difference between the relativistic amplitudes and the first Born iteration of the weak axial one–nucleon current contribution to the two-nucleon scattering amplitude, satisfying the Lippmann-Schwinger equation.

For practical calculations a nonrelativistic reduction of the currents is performed. The nuclear wave functions are generated by the same One-Boson Exchange Potential (OBEP) of Eq. (9) and the same couplings and strong form factors as in the potential are applied in the WAECs. Consistent calculations are carried out employing the realistic OBE potentials OBEPQG [26], Nijmegen 93 and Nijmegen I [27].

The WAECs have been used to calculate ν- and $\bar{\nu}$-deuteron disintegration cross sections at the typical solar neutrino energies. The results indicate that the main two-body effect comes from the Δ excitation and that the heavy-meson exchange contributions are of the same order of magnitude as the π-exchange one. The uncertainty of standard nuclear physics calculations has been reduced from 5-10% to about 3%.

4. Scaling and Superscaling in Lepton-Nucleus Scattering

The analyses of ongoing and future neutrino experiments require reliable predictions of ν-nucleus cross sections. Any nuclear model should first be tested in comparison with electron scattering data. Sophisticated models have been developed to describe electron-nucleus scattering. In spite of all these efforts, however, the uncertainty due to the treatment of nuclear effects in different models is still high when compared with the required precision.

The analogies between ν-nucleus and electron-nucleus scattering suggest an alternative approach to extract model independent ν-nucleus cross sections from experimental electron-nucleus cross sections [28]. Instead of using a specific model, one can exploit the scaling properties of (e,e$'$) data and i) extract a scaling function from (e,e$'$) data, ii) invert the procedure to predict CC ν-nucleus cross sections.

This scaling approach [18,28–31] relies on the superscaling properties of the electron scattering data [32]. At sufficiently high momentum transfer a scaling function is derived dividing the experimental (e,e$'$) cross sections by an appropriate single-nucleon cross section. This is basically the idea of the IA. If this scaling function depends only upon one kinematical variable, the scaling variable, one has scaling of first kind. If the scaling function is roughly the same for all nuclei, one has scaling of second kind. When both kinds of scaling are fulfilled, one says that superscaling occurs.

In the QE region the scaling variable is [28]

$$\psi_{\mathrm{QE}} = \pm\sqrt{1/(2T_F)\left(q\sqrt{1+1/\tau} - \omega - 1\right)}, \tag{10}$$

where T_F is the Fermi kinetic energy, $4m_N^2\tau = q^2 - \omega^2$ and the $-$ $(+)$ sign corresponds to energy transfers lower (higher) than the QE-peak ($\psi = 0$).

An extensive analysis of electron scattering data [32] has shown that scaling of first kind is fulfilled at the left of the QEP and broken at its right, whereas scaling of second kind is well satisfied at the left of the peak and not so badly violated at its right. As a consequence, a scaling function f^{QE} can be extracted from the data.

The superscaling analysis has been extended to the first resonance peak [28]. The contribution of the Δ has been (approximately) isolated in the data by subtracting the QE scaling contribution from the total experimental cross sections. Then, the scaling function has been studied as a function of a new scaling variable

$$\psi_\Delta = \pm\sqrt{1/(2T_F)\left(q\sqrt{\rho + 1/\tau} - \omega\rho - 1\right)}\,, \tag{11}$$

where $\rho = 1 + (m_\Delta^2 - m_N^2)/(4\tau m_N^2)$. The results show that also in this region superscaling is satisfied and a second superscaling function, f^Δ, can be extracted from the data to account for the nuclear dynamics. Clearly this approach can work only at $\psi_\Delta < 0$, since at $\psi_\Delta > 0$ other resonances and the tail of the deep-inelastic scattering start contributing.

The two scaling functions can firstly be tested in comparison with (e,e′) data and can then be used to predict ν-nucleus cross sections.

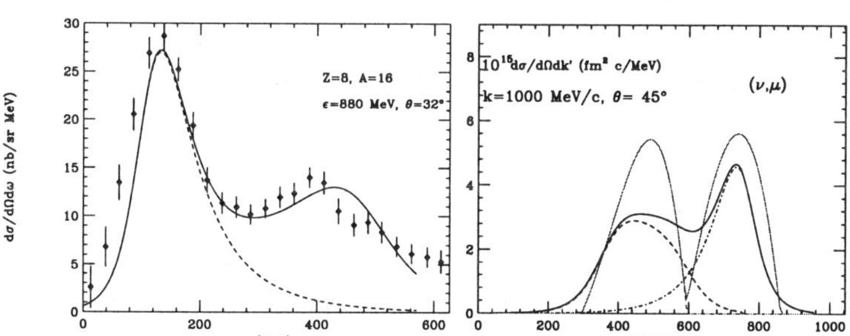

Fig. 6. Left panel: cross section of the ^{16}O(e,e′) reaction. The solid line is obtained using f^{QE} and f^Δ, the dashed line using only f^{QE}. Data from Ref. [7]. Right panel: cross section of the $^{12}\mathrm{C}(\nu,\mu^-)$ reaction as a function of the muon momentum. Solid line: superscaling prediction; heavier line: RFG. The separate QE and Δ contribution are shown (dotted lines). (From Ref. [28]).

In the left panel of Fig. 6 an example of the (e,e′) cross section reconstructed by multiplying the empirical superscaling functions by the appropriate single-nucleon functions is shown in comparison with data. In this as well as in many other cases it turns out that typical deviations are 10% or less [28], thus confirming that the scaling approach offers a reliable description of the nuclear dynamics.

A numerical prediction for CC ν-scattering is displayed in the right panel of Fig. 6. The result obtained using the empirical scaling functions $f^{\rm QE}$ and f^{Δ} is compared with the result of the RFG model. The RFG cross section differs significantly from the scaling prediction, which lies somewhat lower and extends over a wider range in k'.

In the case of NC reactions, the kinematics are different from the ones of CC and (e,e′) processes since the detected final state is the outgoing nucleon and the neutrino kinematic variables are integrated over. This implies an integration region in the residual nucleus variables which is different in the two cases and it is not obvious that the superscaling procedure, based on the analogy with inclusive electron scattering, is still valid. The scaling method is based on a factorization assumption which has to be tested numerically. The outcome is that the procedure can be applied also to NC reactions [30].

The properties of the empirical scaling functions should be accounted for by microscopic calculations. In particular, the asymmetric shape of $f^{\rm QE}$ should be explained. In Ref. [18] the scaling properties of different models (RPWIA, rROP, RMF) are verified. Superscaling is fulfilled to high accuracy in the QE region by the three descriptions of FSI considered. Then, the associated scaling functions are compared with the experimental scaling function. Only the RMF model is able to reproduce the asymmetric shape of the experimental function. This result deserves further investigation.

More results on the scaling approach can be found in [33].

5. Electromagnetic Reactions on Few-Nucleons Systems

Electromagnetic reactions on few-body nuclei are investigated in Refs. [34–36]. The theoretical study of the electromagnetic structure of few-body nuclei requires the knowledge of the nuclear wave functions and the electromagnetic transition operators. For the low-energy observables considered in Ref. [34] and for processes involving two and three nucleons, accurate bound and scattering states are calculated using the pair-correlated Hyperspherical Harmonics (HH) method [37] from the Argonne v_{18} (AV18) two-nucleon [38] and Urbana IX (UIX) [39] or Tucson-Melbourne (TM) [40] three-nucleon interactions. For two- and three-nucleon interactions the nuclear electromagnetic current operator includes, in addition to the one-body terms, also two- and three-body terms. Different models for conserved two-

and three-body currents are constructed using either meson exchange (ME) mechanisms or minimal substitution (MS) in the momentum dependence of the interactions. The connection between these two schemes is elucidated [34].

The electromagnetic current operator must satisfy the Current Conservation Relation (CCR)

$$\mathbf{q}\cdot\mathbf{j}(\mathbf{q}) = [H, \rho(\mathbf{q})] \ , \tag{12}$$

where the Hamiltonian H contains two- and three-body interactions, v_{ij} and V_{ijk}, respectively. To lowest order in $1/m$, Eq. (12) separates into

$$\mathbf{q}\cdot\mathbf{j}_i(\mathbf{q}) = \left[\frac{\mathbf{p}_i^2}{2m}, \rho_i(\mathbf{q})\right] \ , \quad \mathbf{q}\cdot\mathbf{j}_{ij}(\mathbf{q}) = [v_{ij}, \rho_i(\mathbf{q}) + \rho_j(\mathbf{q})] \ , \tag{13}$$

and similarly for the three-body current. The one-body current satisfies the CCR. It is rather difficult to construct conserved two- and three-body currents because H includes momentum and isospin dependent terms that do not commute with ρ.

The two-body current can be separated into model-independent (MI) and model-dependent (MD) parts. The MD current is purely transverse and is unconstrained by the CCR. The MI current has longitudinal and transverse components and is constrained by the CCR. The longitudinal part is constructed so as to satisfy the CCR. The MI currents from the momentum-independent terms of AV18 have been constructed following the ME scheme and satisfy the CCR. The currents from the momentum-dependent terms of the interaction obtained in the ME scheme do not strictly satisfy the CCR. If these currents are obtained in the MS scheme they satisfy the CCR. Both the ME and MS schemes can be generalized to calculate the three-body currents induced by a three-nucleon interaction.

Several electronuclear observables have been calculated [34] to test the model of the current operator. For the A=3 nuclear systems, cross sections and polarization observables in the energy range 0–20 MeV are compared with data and with earlier results [42] where the current operator retains only two-body terms, all of them obtained within the ME scheme, and the CCR is only approximately satisfied.

The differences with respect to the previous results are generally small, but for some of the polarization parameters measured in pd radiative capture, specifically the tensor polarizations T_{20} and T_{21}, where the exactly conserved currents resolve the discrepancies between theory and data obtained in Ref. [42]. An example is shown in Fig. 7.

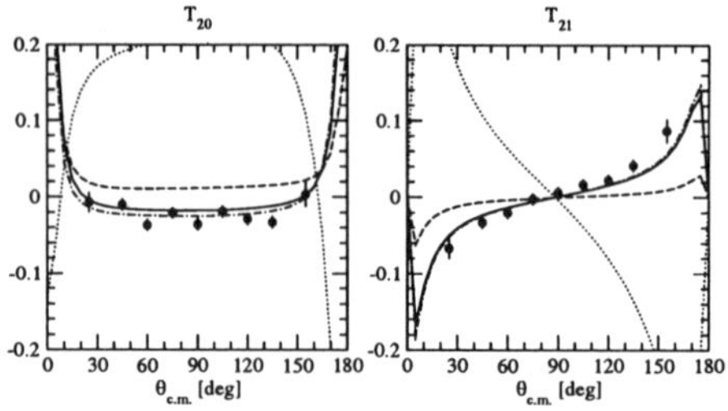

Fig. 7. Deuteron tensor polarization observables T_{20} and T_{21} for pd radiative capture at $E_{c.m.}$= 2 MeV. The dotted curves include only the one-body current, the dashed curves are the results of Ref. [42], the dot-dashed curves are obtained in the long-wavelength-approximation (LWA), applying the Siegert theorem. The solid curves are the results of Ref. [34]. Data from Ref. [41]. (From Ref. [34]).

An overall nice description has been reached for all the observables. Also, some small three-body currents effects are noticeable, which is an indication of the fact that if a Hamiltonian model with two- and three-nucleon interactions is used, then the model for the nuclear current operator should include the corresponding two- and three-body contributions [34] .

A comparison with the ^{3}He(e,e′p)d [43] and ^{4}He($\vec{e}$,e′$\vec{p}$)^{3}H [44] JLab data is presented in Refs. [35] and [36], respectively. Accurate bound state wave functions are still calculated within the HH method. The electromagnetic current operator includes one- and two-body terms. Since at the high energies of the JLab experiments a theory of the interaction and few-nucleon systems is not available, different treatments of FSI are used. In Ref. [35], a Glauber approximation is used, where the profile operator in the Glauber expansion is derived from a NN scattering amplitude, which retains its full spin and isospin dependence, and is consistent with phaseshift analyses of NN scattering data. In Ref. [36], a nonrelativistic optical potential is used including charge-exchange terms. In both cases a fair agreement with data is found. Of particular interest is the agreement obtained [36] for the ratio of transverse to longitudinal polarization transfers in the ^{4}He($\vec{e}$,e′$\vec{p}$)^{3}H reaction. In the elastic process $\vec{e}$p $\rightarrow$ e$\vec{p}$, the ratio is proportional to that of the electric to magnetic form factors of the proton, and a measurement in a nucleus by QE proton knockout can shed light, in principle, on the question of whether these form factors are modified in medium. Thus, the agreement found when free-nucleon electromagnetic form factors are used in the current operator [36] challenges the current interpretation of data in terms of medium-modified form factors.

6. Electromagnetic Reactions on Complex Nuclei

Complementary polarization measurements suited to study nucleon properties in the nuclear medium are proposed in Ref. [45], where the general formalism of the $\vec{A}(\vec{e},e'\vec{p})B$ reaction is presented within the RPWIA. The simultaneous polarization of the target and the ejected proton provides information which is not contained in the $\vec{A}(\vec{e},e'p)B$ and $A(\vec{e},e'\vec{p})B$ reactions. The polarization transfer mechanism in which the electron interacts with the initial nucleon carrying the target polarization, making the proton exit with a fractional polarization in a different direction, is referred to as "skewed polarization". Although difficult to measure, these new observables would provide information on nucleon properties complementing the results for the ratio of transverse to longitudinal polarization transfers.

Proton emission induced by polarized photons of energies above the giant resonance region and below the pion production threshold is studied in Ref. [46]. With respect to (e,e′p), a different kinematics is explored in the (γ,p) reaction. In fact, for a real photon the energy and momentum transfer are constrained by the condition $\omega = |\boldsymbol{q}|$, and only the high-momentum components of the nuclear wave function are probed. Moreover, the validity of the direct knockout mechanism, which is clearly stated for (e,e′p), is questionable for (γ,p), where important contributions are due to two-nucleon processes, such as those involving MEC [47,48] . The sensitivity of various polarization observables in the ($\vec{\gamma}$,p) reaction to FSI, MEC, and SRC is investigated [46] using the same model [48] applied to calculate (γ,p) cross sections.

The sensitivity to FSI, MEC, and NN correlations, in particular SRC, are the same issues considered in the theoretical studies of electron- and photon-induced two-nucleon knockout [49,50]. Since a long time these reactions have been devised as a preferential tool to investigate SRC [1]. In fact, direct insight into SRC can be obtained from the situation where the electromagnetic probe hits, through a one-body current, either nucleon of a correlated pair and both nucleons are then ejected from the nucleus. This process is entirely due to correlations. Additional complications have, however, to be taken into account, such as competing mechanisms, like contributions of two-body MEC and Δ isobar excitations, as well as the FSI between the two outgoing nucleons and the residual nucleus.

The calculated cross sections are sensitive to the different ingredients of the model and to their treatment. The role and relevance of competing reaction mechanisms and of different contributions is different in different reactions and kinematics. It is thus possible, in principle, with the help of theoretical predictions, to envisage appropriate situations where specific effects can be disentangled and separately investigated. Data from NIKHEF [51] and MAMI [52] for the exclusive ^{16}O(e,e$'$pp)^{14}C reaction confirmed the validity of the direct knockout mechanism and gave clear evidence of SRC for the transition to the ground state of ^{14}C. This important result, that was obtained from a close collaboration between experimentalists and theorists, means that further studies on these reactions would make it possible to disentagle SRC in experimental cross sections. From the experimental side, however, during the last few years only the results of a first measurement of the ^{16}O(e,e$'$pn)^{14}N reaction have been published [53]. From the theoretical side, the recent studies [49,50] focussed on specific aspects of the theoretical models, such as a consistent treatment of different types of correlations in the two-nucleon wave function, the competing contribution of correlations and two-body currents, the uncertainties in the treatment of the Δ-current, the effects of FSI, whose consistent treatment requires in general a genuine three-body approach for the mutual interaction of the two nucleons and the residual nucleus.

The outcome of this work is that electromagnetic two-nucleon knockout contains a wealth of information on correlations and on the behavior of the Δ-current in a nucleus. The uncertainties in the treatment of the theoretical ingredients and the large number of parameters involved in the models make it difficult to extract clear and unambiguous information from one ideal kinematics. Data are therefore needed for electron- and photon-induced pp and pn emission and in various kinematics which mutually supplement each other. The choice of suitable conditions for the experiments and the interpretation of data require close collaboration between experimentalists and theorists.

I dedicate this report to the memory of Adelchi Fabrocini, who gave significant contributions to the study of correlations in electromagnetic reactions and with whom, during the years, I had so many fruitful and pleasant conversations. I thank Franco Pacati for his valuable help and Giampaolo Co' for useful discussions.

References

1. S. Boffi, C. Giusti, F.D. Pacati, and M. Radici, *Electromagnetic Response of Atomic Nuclei*, Oxford Studies in Nuclear Physics (Clarendon Press, Oxford, 1996); S. Boffi, C. Giusti, and F.D. Pacati, Phys. Rep. **226**, 1 (1993).
2. G. Co', in *Theoretical Nuclear Physics in Italy*, Proceedings of the 10th Conference on Problems in Theoretical Nuclear Physics, Cortona, 2004, edited

by S. Boffi, A. Covello, M. Di Toro, A. Fabrocini, G. Pisent, S. Rosati (World Scientific, Singapore, 2005), p. 333.
3. A. Meucci, C. Giusti, and F.D. Pacati, these proceedings.
4. A. Botrugno and G. Co', Nucl. Phys. **A761**, 200 (2005); Eur. Phys. J. **A24**, s1, 109 (2005).
5. G. Co', Nucl. Phys. **B159** (Proc. Suppl.), 192 (2006); Acta Phys. Polon. **B37**, 2235 (2006).
6. G. Co', K.F. Quader, R.D. Smith, and J. Wambach, Nucl. Phys. **A485**, 61 (1988); C. Mahaux and N. Ngô, Nucl. Phys. **A378**, 205 (1982).
7. M. Anghinolfi *et al.*, Nucl. Phys. **A602**, 405 (1996).
8. O. Benhar, D. Day, and I. Sick, nucl-ex/0603029.
9. A. Meucci, C. Giusti, and F.D. Pacati, nucl-th/0501047.
10. A. Meucci, C. Giusti, and F.D. Pacati, Nucl. Phys. **A773**, 250 (2006); Acta Phys. Polon. **B37**, 2279 (2006).
11. C. Giusti, A. Meucci, and F.D. Pacati, nucl-th/0607037.
12. F. Capuzzi, C. Giusti, and F.D. Pacati, Nucl. Phys. **A524**, 681 (1991).
13. A. Meucci, F. Capuzzi, C. Giusti, and F.D. Pacati, Phys. Rev. C **67**, 054601 (2003).
14. A. Meucci, C. Giusti, and F.D. Pacati, Nucl. Phys. **A739**, 277 (2004).
15. A. Meucci, C. Giusti, and F.D. Pacati, Nucl. Phys. **A756**, 359 (2005).
16. F. Capuzzi, C. Giusti, F.D. Pacati, and D.N. Kadrev, Annals Phys. **317**, 492 (2005).
17. G. Co' and A.M. Lallena, Annals Phys. **287**, 101 (2001).
18. J.A. Caballero *et al.*, Phys. Rev. Lett. **95**, 252502 (2005).
19. O. Benhar, Nucl. Phys. **B139** (Proc. Suppl.), 15 (2005).
20. O. Benhar and N. Farina, Nucl. Phys. **B139** (Proc. Suppl.), 230 (2005); O. Benhar, N. Farina, H. Nakamura, M. Sakuda, and R. Seki, Phys. Rev. D **72**, 053005 (2005); Nucl. Phys. **B155** (Proc. Suppl.), 155 (2006).
21. O. Benhar, Nucl. Phys. **B159** (Proc. Suppl.), 168 (2006); Acta Phys. Polon. **B37**, 2243 (2006).
22. O. Benhar and D. Meloni, hep-ph/0604071; hep-ph/0610403.
23. B. Mosconi, P. Ricci, and E. Truhlík, Eur. Phys. J. **A25**, 283 (2005).
24. B. Mosconi, P. Ricci, and E. Truhlík, Eur. Phys. J. **A27**, s1, 67 (2006); Nucl. Phys. **A772**, 81 (2006).
25. J. Adam, Jr., E. Truhlík, and D. Adamová, Nucl. Phys. **A494**, 556 (1989).
26. R. Machleidt, *Advances in Nuclear Physics*, **19**, 189 (1989); P. Obersteiner, W. Plessas, and E. Truhlík, *in Proceedings of the XIII International Conference on Particles and Nuclei, Perugia 1993* edited by A. Pascolini, (World Scientific, Singapore, 1994), p. 430.
27. V.G.J. Stoks, R.A.M. Klomp, C.P.F. Terheggen, and J.J. de Swart, Phys. Rev. C **49**, 2950 (1994).
28. J.E. Amaro, M.B. Barbaro, J.A. Caballero, A. Molinari, and I. Sick, Phys. Rev. C **71**, 015501 (2005).
29. J.E. Amaro, M.B. Barbaro, J.A. Caballero, T.W. Donnelly, and C. Maieron, Phys. Rev. C **71**, 065501 (2005).
30. J. E. Amaro, M. B. Barbaro, J.A. Caballero, and T. W. Donnelly, Phys.

Rev. C **73**, 035503 (2006).
31. M. B. Barbaro, J. E. Amaro, J.A. Caballero, T. W. Donnelly, A. Molinari, and I. Sick, Nucl. Phys. **B155** (Proc. Suppl.), 257 (2006); M.B.Barbaro, nucl-th/0602011; M. B. Barbaro, J. E. Amaro, J.A. Caballero, and T.W. Donnelly, nucl-th/0609057; A.N. Antonov *et al.*, Phys. Rev. C **74**, 054603 (2006).
32. W.M. Alberico, A. Molinari, T.W. Donnelly, E.L. Kronenberg, and J.W. Van Orden, Phys. Rev. C **38**, 1801 (1988); D.B. Day, J.S. MacCarthy, T.W. Donnelly, and I. Sick, Ann. Rev. Nucl. Part. Sci. **40**, 357 (1990); T.W. Donnelly and I. Sick, Phys. Rev. C **60**, 065502 (1999); Phys. Rev. Lett. **82**, 3212 (1999); C. Maieron, T. W. Donnelly, and I. Sick, Phys. Rev. C **65**, 025502 (2002).
33. M. Martini, these proceedings.
34. L.E. Marcucci. M. Viviani. R. Schiavilla, A. Kievsky, and S. Rosati, Eur. Phys. J. **A24**, s1, 95 (2005); Phys. Rev. C **72**, 014001 (2005).
35. R. Schiavilla, O. Benhar, A. Kievsky, L.E. Marcucci, and M. Viviani, Phys. Rev. C **72**, 064003 (2005).
36. R. Schiavilla, O. Benhar, A. Kievsky, L.E. Marcucci, and M. Viviani, Phys. Rev. Lett. **94**, 072303 (2005).
37. A. Kievsky, M. Viviani, and S. Rosati, Nucl. Phys. **A551**, 241 (1993); Nucl. Phys. **A577**, 511 (1994); A. Kievsky, S. Rosati, and M. Viviani, Phys. Rev. C **64**, 024002 (2001).
38. R.B. Wiringa, V.G.J. Stoks, and R. Schiavilla, Phys. Rev. C **51**, 38 (1995).
39. B.S. Pudliner, V.R. Pandharipande, J. Carlson, and R.B. Wiringa, Phys. Rev. Lett. **74**, 4396 (1995).
40. S.A. Coon, *et al.*, Nucl. Phys. **A317**, 242 (1979).
41. M.K. Smith and L.D. Knutson, Phys. Rev. Lett. **82**, 4591 (1999).
42. M. Viviani *et al.*, Phys. Rev. C **61**, 064001 (2000).
43. M.M. Rvachev *et al.*, Phys. Rev. Lett. **94**, 192302 (2005).
44. S. Strauch *et al.*, Phys. Rev. Lett. **91**, 052301 (2003).
45. J.E. Amaro, M.B. Barbaro, and J.A. Caballero, Annals Phys. **319**, 123 (2005).
46. M. Anguiano, G. Co', and A.M. Lallena, Phys. Rev. C **74**, 044603 (2006).
47. C. Giusti and F.D. Pacati, Phys. Rev. C **66**, 034610 (2002).
48. M. Anguiano, G. Co', A.M. Lallena, and S.R. Mokhtar, Annals Phys **296**, 235 (2002).
49. M. Anguiano, G. Co', and A.M. Lallena, J. Phys. **G29**, 1119 (2003); Nucl. Phys. **A744**, 168 (2004).
50. C. Barbieri, C. Giusti, F.D. Pacati, and W.H. Dickhoff, Phys. Rev. C **70**, 014606 (2004); M. Schwamb, S. Boffi, C. Giusti, and F.D. Pacati, Eur. Phys. J. **A20**, 233 (2004); M. Schwamb, C. Giusti, F.D. Pacati, and S. Boffi, Eur. Phys. J. **A26**, 209 (2005); C. Giusti, BGNS Transactions **10** n.2, 136, (2005).
51. C.J.G. Onderwater *et al.*, Phys. Rev. Lett. **78** 4893 (1997); **81**, 2213 (1998); R. Starink *et al.*, Phys. Lett. **B474**, 33 (2000).
52. G. Rosner, Prog. Part. Nucl. Phys. **44**, 99 (2000).
53. D.G. Middleton *et al.*, Eur. Phys. J. **A29**, 261 (2006).

^{4}HE PHOTODISINTEGRATION WITH A REALISTIC NUCLEAR FORCE

S. BACCA

Gesellschaft für Schwerionenforschung,
Planckstr. 1, 64291 Darmstadt, Germany
E-mail: s.bacca@gsi.de

D. GAZIT and N. BARNEA

Racah Institute of Physics,
Hebrew University, 91904, Jerusalem, Israel

W. LEIDEMANN and G. ORLANDINI

Dipartimento di Fisica,
Università di Trento and INFN
(Gruppo Collegato di Trento),
via Sommarive 14, I-38050 Povo, Italy

The photodisintegration cross section of ^{4}He is calculated using the realistic nucleon-nucleon potential Argonne V18 (AV18) and the three-body force Urbana IX (UIX). Final State interaction is taken into account via the Lorentz Integral Transform method. A comparison with other potential models and with the available experimental data is discussed.

Keywords: Few-body systems; Photodisintegration; Realistic interaction; Three-body force.

1. Introduction

One of the main challenges in nuclear physics is to give a microscopic description of nuclear properties, using a realistic nuclear interaction. For a given force one has to find observables sensitive to the different parts of the potential to test the interaction model. If the calculation is based on an exact solution of the quantum mechanical problem, as it is in the framework of few-particle physics, then the comparison of theoretical prediction with experimental data can give valuable information about the nuclear dynamics. In this work we report on our recent calculation of the total ^{4}He photodisintegration with a realistic two- and three-body force [1]. Photodisintegration cross sections shed light on the implicit degrees of freedom of the interaction, as meson exchange mechanisms, and show impact of three-body forces. The theoretical treatment of the ^{4}He(γ) reactions is very challenging, since the four-nucleon continuum problem and the different channels have to be considered. We circumvent the difficulty in the calculation of the scattering states using the Lorentz Integral Transform (LIT) method [2], where the problem is reduced to the solution of a bound-state-like equation. The photodisintegration of the alpha particle has recently attracted much attention also from the experimental point of view [3,4]. Unfortunately, a fairly large disagreement is found, and a discriminant test of realistic interaction models on this observable is therefore limited.

Our work is organized as follows. In the first section we give a theoretical overview. Then we present our results and at the end we draw some conclusions.

2. Theoretical Overview

The total photoabsorption cross section, at low photon energy ω, is given by

$$\sigma(\omega) = 4\pi^2 \omega \alpha R(\omega)\,, \tag{1}$$

where α is the electromagnetic coupling constant and $R(\omega)$ is the inclusive dipole response function, generally defined as:

$$R(\omega) = \frac{1}{3(2J_0+1)} \sum_{M_0} \sum\!\!\!\!\!\!\!\int_f \left|\langle \Psi_f |\, \mathbf{D}\, | \Psi_0 \rangle\right|^2 \delta(E_f - E_0 - \omega)\,. \tag{2}$$

Here, J_0 and M_0 indicate the total angular momentum and the corresponding projection of the nucleus in its initial ground state, while $|\Psi_{0/f}\rangle$ and $E_{0/f}$ denote wave function and energies of the ground and final states, respectively. The dipole operator is

$$\mathbf{D} = \frac{1}{2} \sum_i \mathbf{r}_i \tau_i^3\,, \tag{3}$$

where $\mathbf{r}_i$ and τ_i^3 are the position in the center of mass reference frame and the third component of the isospin of the i-th particle, respectively. The dipole approximation has been proven to be very good at low photon energy for the deuteron [5] and for the triton case [6]. In this way, the major part of the meson exchange currents (MEC) is implicitly taken into account, via the Siegert theorem [7].

In the LIT method [2] one obtains $R(\omega)$ after the inversion of an integral transform with a Lorentzian kernel

$$L(\sigma_R, \sigma_I) = \int d\omega \frac{R(\omega)}{(\omega - \sigma_R)^2 + \sigma_I^2} = \langle \widetilde{\Psi} | \widetilde{\Psi} \rangle \,. \tag{4}$$

The state $\widetilde{\Psi}$ is the unique solution of the inhomogeneous "Schrödinger-like" equation

$$(H - E_0 - \sigma_R + i\sigma_I)|\widetilde{\Psi}\rangle = \mathbf{D}|\Psi_0\rangle. \tag{5}$$

Because of the presence of an imaginary part σ_I in Eq. (5) and the fact that the right-hand side of this same equation is localized, one has an asymptotic boundary condition similar to a bound state. Thus, one can apply bound-state techniques for its solution. We expand $|\Psi_0\rangle$ and $|\widetilde{\Psi}\rangle$ in terms of the four-body symmetrized hyperspherical harmonics (HH) [8]. The expansion is performed up to maximal value K_m of the HH grand-angular momentum quantum number. We improve the convergence of the HH expansion using the effective interaction hyperspherical harmonics (EIHH) approach [9], where the bare potential is replaced by an effective potential constructed via the Lee-Suzuki method [10]. When convergence is reached, the same results as with the bare potential are obtained (see Ref. [9]). We use the EIHH approach only in case of the nucleon-nucleon (NN) interaction, while in case of the three-body force its bare form is kept. The LIT of Eq. (4) is calculated using the Lanczos techniques as described in [11].

3. Result

We perform our calculation using as NN interaction the AV18 potential [12] and as three-body force the UIX [13]. With the local AV18 we obtain ground state energies $E_0 = -24.27(2)$ and $-7.62(2)$ MeV for ^{4}He and ^{3}H, respectively, which are higher than the experimental value. With realistic non-local NN interactions, as the Nijmegen potentials (I and II) and the CD-Bonn potential, the discrepancy with respect to experiment is reduced. In Fig. 1 we show ground state energies of ^{4}He versus ^{3}H for different potential models: exact few-body calculation predict for different phase-shift equivalent interactions results which lay on the so-called Tjon line. When the three-body force UIX is added to AV18, we get $E_0 = -28.42(2)$ and $-8.468(1)$ MeV for ^{4}He and ^{3}H [14], respectively, in agreement with experiment. It is now interesting to investigate whether this realistic interaction model can predict photoabsorption reactions as well.

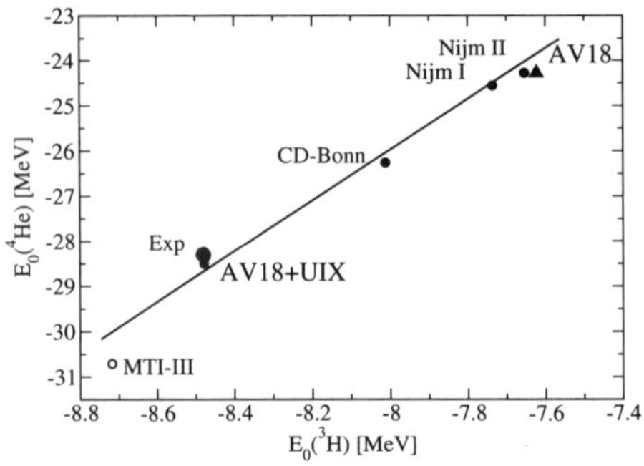

Fig. 1. Tjon line: ground state energies of ^{4}He *versus* ^{3}H for different interaction models.

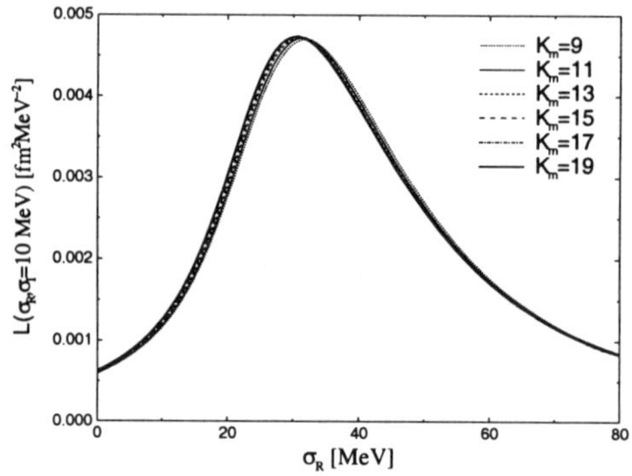

Fig. 2. Convergence of the LIT for different K_m as a function of the parameter σ_R (for fixed $\sigma_I = 10$ MeV).

In Fig. 2, we show the transforms $L(\sigma_R, \sigma_I = 10 \text{ MeV})$ in case of ^{4}He as a function of the parameter σ_R. The convergence in terms of the HH expansion is studied as a function of the K_m. One readily notes that the convergence of the transform is rather good for high σ_R, but it is weaker for small σ_R. The case of $K_m = 19$ correspond to the largest Hilbert space we can use, where we have about 400000 basis states. In order to get a deeper insight into the convergence pattern we show the relative errors of the LITs, defined as

$$\Delta_{K_m} = \frac{L_{K_m} - L_{19}}{L_{19}}. \tag{6}$$

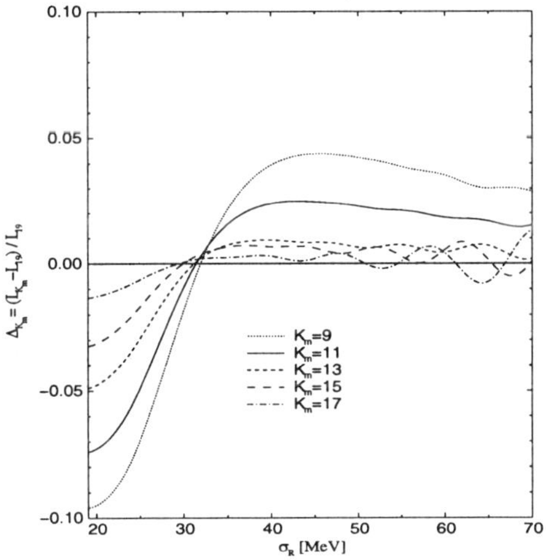

Fig. 3. Convergence of Δ_{K_m} as a function of the parameter σ_R.

In Fig. 3 the quantity Δ_{K_m} *versus* σ_R is presented. The regularity of the convergence pattern as a function of K_m for small σ_R allows us to perform a stable Padé extrapolation of the LIT. After the inversion of the extrapolated transform, making use of Eq. (1) we obtain the photoabsorption cross section prediction for AV18+UIX.

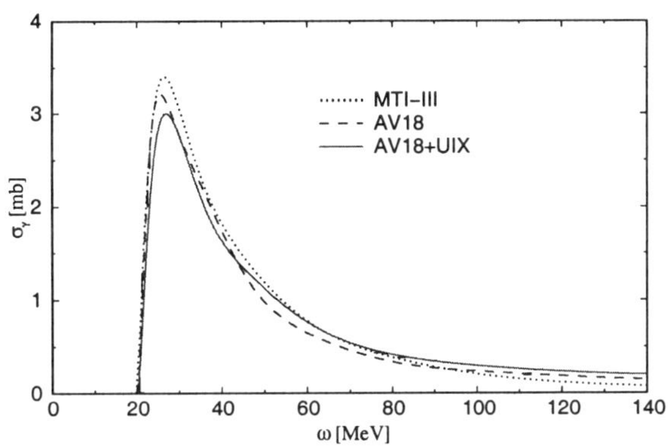

Fig. 4. Photoabsorption cross section of ^{4}He with different potential models.

At this point it is interesting to compare our result with σ_γ obtained with the AV18 potential alone and the semirealistic Maliflet-Tjon [15] from

[16,17], as done in Fig. 4. Due to the effect of the three-body force we observe a reduction of the peak height by about 6% and a shift of the peak position by about 1 MeV towards higher energy. Larger differences are found above 50 MeV with an enhancement of the cross section by *e.g.* 18%, 25%, and 35% at $\omega = 60$, 100, and 140 MeV, respectively. Here, we would like to mention that three-body effect at higher energies could change slightly when considering other multipoles beyond the dipole. The semirealistic potential gives for ^{4}He a rather "realistic" result in the giant resonance region with an overestimation of the peak by about 10-15% and quite a correct result for the peak position; however, at higher energy σ_γ is strongly underestimated, by a factor of three at pion threshold.

It is interesting to compare the ^{4}He situation with the $\sigma_\gamma(^3\text{H}/^3\text{He})$ [16, 18]. The peak height reduction due to the three-body force is smaller for ^{4}He. For ^{3}H/^{3}He the size of the reduction is similar to the decrease in E_0 (10%), whereas for ^{4}He the differences introduced by the UIX potential in the continuum cannot be interpreted as a simple binding effect, since E_0 is reduced by 17%. Furthermore, the increase of $\sigma_\gamma(^4\text{He})$ due to the three-body force is about two times larger than for the ^{3}H/^{3}He.

Table 1. Photonuclear sum rules.

Interaction	m_{-2} 10^{-2} [mb MeV $^{-1}$]	m_{-1} [mb]	m_0 10^2 [mb]
AV18	7.681	2.696	1.383
AV18+UIX	6.473	2.410	1.462

Using σ_γ one can calculate important photonuclear sum rule, defined as

$$m_n(\bar{\omega}) = \int_{\omega_{tr}}^{\bar{\omega}} d\omega\, \omega^n \sigma_\gamma(\omega), \tag{7}$$

from direct integration of the cross section or from the Lanczos coefficients, as explained in [19]. In Table 2 we present the result for the polarizability, m_{-2}, the bremsstrahlung, m_{-1} (proportional to total dipole strength), and the Thomas-Reich Kuhn, m_0, sum rules. One can see that the three-body force reduces the polarizability and the total dipole strength of ^{4}He, while it enhances the TRK sum rule, which contains information about meson exchange mechanism.

In Fig. 5 we compare our low-energy results to experimental data. Close to threshold several data were taken in different experiments, which unfortunately show fairly large discrepancies. In [21] the peak cross section is de-

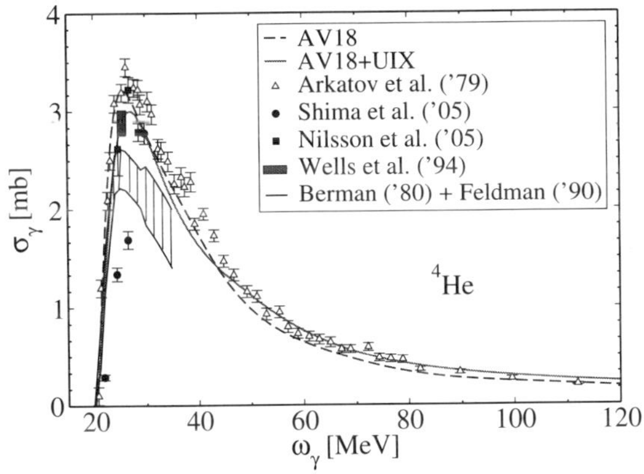

Fig. 5. ^{4}He total photoabsorption cross section: theoretical predictions in comparison with available experimental data: triangles from Arkatov *et al.* [20], circles from Shima *et al.* [4], squares from Nilsson *et al.* from [3], box from Wells *et al.* [21] and finally the shaded area is the sum of Berman *et al.* [22] and Feldman *et al.* [23] data.

termined from Compton scattering via dispersion relations, the dashed area corresponds to the sum of cross sections for (γ, n) from [22] and $(\gamma, p)^3$H from [23]. The data from the recent (γ, n) experiment [3] are included only up to about the three-body break-up threshold, where one can rather safely assume that $\sigma_\gamma \simeq 2\sigma(\gamma, n)$, whereas in [4] all open channels are considered. Only data from Arkatov *et al.* [20] cover a larger energy range. The theoretical prediction of AV18+UIX agrees quite well with the low-energy data of [22,23]. In the peak region, however, the situation is very unclear. There is a rather good agreement between the theoretical σ_γ and the data of [3] and [21]. It is evident that the experimental situation is rather unsatisfactory and further improvement is urgently needed.

4. Conclusion

We have performed the first *ab initio* calculation of the ^{4}He total photoabsorption cross section with the AV18 and the UIX potential. The full final state interaction is consistently taken into account via the LIT method. Our results show a rather pronounced giant dipole peak. The three-body force produces a reduction of the peak height and a strong enhancement of the tail of the cross section, up to $20 - 30\%$ depending on the photon energy range. Unfortunately experiments do not lead to a unique picture. Thus a stringent test of a realistic two- and three-body interaction on this observable is limited. Further experimental investigations are mandatory.

5. Acknowledgments

Numerical calculations were partially performed on at CINECA (Bologna). This work was supported by the Israel Science Foundation (Grant No. 202/02) and by the Italian Ministry of Research (COFIN03).

References

1. D. Gazit, S. Bacca, N. Barnea, W. Leidemann, and G. Orlandini, Phys. Rev. Lett. **96**, 112301 (2006).
2. V. D. Efros, W. Leidemann, and G. Orlandini, Phys. Lett. **B338**, 130 (1994).
3. B. Nilsson *et al.*, Phys. Lett. **B626**, 65 (2005).
4. T. Shima *et al.*, Phys. Rev. C **72**, 044004 (2005).
5. H. Arenhövel and M. Sanzone, Few-Body. Syst. Suppl. **3**, 1 (1991).
6. J. Golak *et al.*, Nucl. Phys. **A707**, 365 (1992).
7. J. M. Eisenberg and Greiner, *Excitation mechanisms of the nucleus* North-Holland Publishing Company, Amsterdam, (1970).
8. N. Barnea and A. Novoselsky, Ann. Phys (N.Y.) **256**, 192 (1997); Phys. Rev. A **57**, 48 (1998).
9. N. Barnea, W. Leidemann, and G. Orlandini, Phys. Rev. C **61**, 054001 (2000); Nucl. Phys. **A693**, 565 (2001).
10. K. Suzuki and S. Y. Lee, Prog. Theor. Phys. **64**, 2091 (1980).
11. M. A. Marchisio, N. Barnea, W. Leidemann, and G. Orlandini, Few-Body Syst. **33**, 259 (2003).
12. R. B. Wiringa, V. G. J. Stoks, and R. Schiavilla, Phys. Rev. C **51**, 38 (1995).
13. B. S. Pudliner *et al.*, Phys. Rev. C **56**, 1720 (1997).
14. N. Barnea, V. D. Efros, W. Leidemann, and G. Orlandini, Few-Body Syst. **35**, 155 (2004).
15. R. A. Malfliet and J. A. Tjon, Nucl. Phys. **A127**, 161 (1969).
16. V. D. Efros, W. Leidemann, G. Orlandini, and E. L. Tomusiak, Phys. Lett. **B484**, 223 (2000).
17. N. Barnea, V. D. Efros, W. Leidemann, and G. Orlandini, Phys. Rev. C **63**, 057002 (2001).
18. J. Golak *et al.*, Nucl. Phys. **A707**, 365 (2002).
19. D. Gazit, N. Barnea, S. Bacca, W. Leidemann, and G. Orlandini, to appear on Phys. Rev. C. (2006).
20. Y. M. Arkatov *et al.*, Yad. Konst. **4**, 55 (1979).
21. D. Wells *et al.*, Phys. Rev. C **46**, 449 (1992).
22. B. Berman *et al.*, Phys. Rev. C **22**, 2273 (1980).
23. G. Feldman *et al.*, Phys. Rev. C **42**, R1167 (1990).

TESTING SUPERSCALING PREDICTIONS IN ELECTROWEAK EXCITATIONS OF NUCLEI

M. MARTINI and G. CO'

Dipartimento di Fisica, Università di Lecce
and Istituto Nazionale di Fisica Nucleare sez. di Lecce, I-73100 Lecce, Italy

M. ANGUIANO and A. M. LALLENA

Departamento de Física Atómica, Molecular y Nuclear,
Universidad de Granada, E-18071 Granada, Spain

Superscaling analysis of electroweak nuclear response functions is done for momentum transfer values from 300 to 700 MeV/c. Some effects, absent in the Relativistic Fermi Gas model, where the superscaling holds by construction, are considered. From the responses calculated for the ^{12}C, ^{16}O and ^{40}Ca nuclei, we have extracted a theoretical universal superscaling function similar to that obtained from the experimental responses. Theoretical and empirical universal scaling functions have been used to calculate electron and neutrino cross sections. These cross sections have been compared with those obtained with a complete calculation and, for the electron scattering case, with the experimental data.

1. Introduction

The properties of the Relativistic Fermi Gas (RFG) model of the nucleus [1] have inspired the idea of superscaling. In the RFG model, the responses of the system to an external perturbation are related to a universal function of a properly defined scaling variable which depends upon the energy and the momentum transferred to the system. The adjective universal means that the scaling function is independent on the momentum transfer, this is called scaling of first kind, and it is also independent on the number of nucleons, and this is indicated as scaling of second kind. The scaling function can be defined in such a way to result independent also on the specific type of external one-body operator. This feature is usually called scaling of zeroth-kind [2–4]. One has superscaling when the three kinds of scaling are verified. This happens in the RFG model.

The theoretical hypothesis of superscaling can be empirically tested by extracting response functions from the experimental cross sections and by studying their scaling behaviors. Inclusive electron scattering data in the quasi-elastic region have been analyzed in this way [2,5]. The main result of these studies is that the longitudinal responses show superscaling behavior. The situation for the transverse responses is much more complicated.

The presence of superscaling features in the data is relevant not only by itself, but also because this property can be used to make predictions. In effect, from a specific set of longitudinal response data [6], an empirical scaling function has been extracted [2], and has been used to obtain neutrino-nucleus cross sections in the quasi-elastic region [3].

We observe that the empirical scaling function is quite different from that predicted by the RFG model. This indicates the presence of physics effects not included in the RFG model, but still conserving the scaling properties. We have investigated the superscaling behavior of some of these effects. They are: the finite size of the system, its collective excitations, the Meson Exchange Currents (MEC) and the Final State Interactions (FSI). The inclusion of these effects produce scaling functions rather similar to the empirical one. Our theoretical universal scaling functions, $f_{\mathrm{U}}^{\mathrm{th}}$, and the empirical one $f_{\mathrm{U}}^{\mathrm{ex}}$, have been used to predict electron and neutrino cross sections.

2. Superscaling beyond RFG model

The definitions of the scaling variables and functions, have been presented in a number of papers [1–4] therefore we do not repeat them here. The basic quantities calculated in our work are the electromagnetic, and the weak, nuclear response functions. We have studied their scaling properties by direct numerical comparison (for a detailed analysis see Ref. [7]).

We present in Fig. 1 the experimental longitudinal and transverse scaling function data for the ^{12}C, ^{40}Ca and ^{56}Fe nuclei given in Ref. [6] for three values of the momentum transfer. We observe that the f_L functions scale better than the f_T ones. The f_T scaling functions of ^{12}C, especially for the lower q values, are remarkably different from those of ^{40}Ca and ^{56}Fe.

The observation of the figure, indicates that the scaling of first kind, independence on the momentum transfer, and of zeroth kind, independence on the external probe, are not so well fulfilled by the experimental functions. These observations are in agreement with those of Refs. [2,5].

To quantify the quality of the scaling between a set of M scaling functions, each of them known on a grid of K values of the scaling variable Ψ,

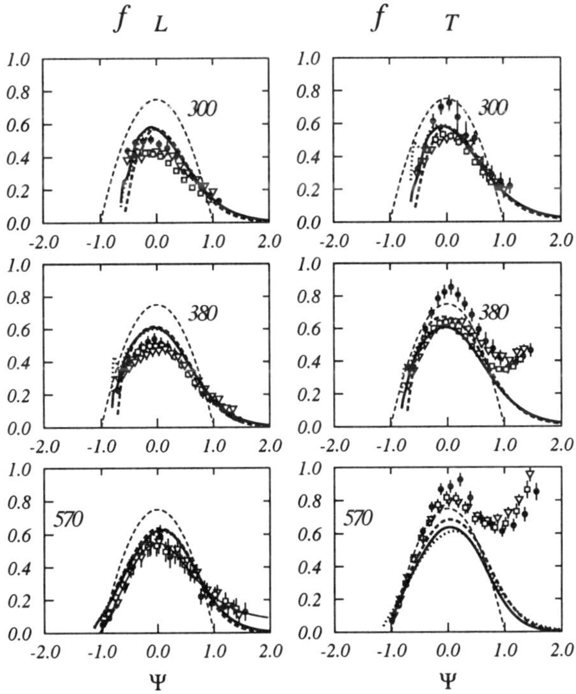

Fig. 1. Empirical longitudinal, f_L, and transverse, f_T, scaling functions obtained from the experimental electromagnetic responses of Ref. [6]. The numbers in the panels indicate the values of the momentum transfer in MeV/c. The full circles refer to ^{12}C, the white squares to ^{40}Ca , and the white triangles to ^{56}Fe. The thin black line in the f_L panel at 570 MeV/c, is the empirical scaling function obtained from a fit to the data. The thick lines show the results of our calculations when all the effects beyond the RFG model have been considered. The full lines have been calculated for ^{12}C, the dotted lines for ^{16}O, and the dashed lines for ^{40}Ca . The dashed thin lines show the RFG scaling functions.

we define the two indexes:

$$\mathcal{D} = \max_{i=1,\ldots,K} \left\{ \max_{\alpha=1,\ldots,M} [f_\alpha(\Psi_i)] - \min_{\alpha=1,\ldots,M} [f_\alpha(\Psi_i)] \right\} , \tag{1}$$

and

$$\mathcal{R} = \frac{1}{K f^{\max}} \sum_{i=1,\ldots,K} \left\{ \max_{\alpha=1,\ldots,M} [f_\alpha(\Psi_i)] - \min_{\alpha=1,\ldots,M} [f_\alpha(\Psi_i)] \right\} \tag{2}$$

where $f^{\max}$ is the largest value of the f_α.

The two indexes give complementary information. The $\mathcal{D}$ index is related to a local property of the functions: the maximum distance between the various curves. Since the value of this index could be misleading if the responses have sharp resonances, we have also used the $\mathcal{R}$ index which is instead sensitive to global properties of the differences between the functions. Since we know that the functions we want to compare are roughly bell shaped, we have inserted the factor $1/f^{\max}$ to weight more the region of the maxima of the functions than that of the tails.

Table 1. Values of the $\mathcal{D}$ and $\mathcal{R}$ indexes, for the experimental scaling functions of Fig. 1.

	f_{L}	
q [MeV/c]	$\mathcal{D}$	$\mathcal{R}$
300	0.107 ± 0.002	0.152 ± 0.013
380	0.079 ± 0.003	0.075 ± 0.009
570	$\mathbf{0.101 \pm 0.009}$	$\mathbf{0.079 \pm 0.017}$
	f_{T}	
300	0.223 ± 0.004	0.165 ± 0.017
380	0.235 ± 0.005	0.155 ± 0.014
570	0.169 ± 0.003	0.082 ± 0.007

In Tab. 1 we give the values of the indexes calculated by comparing the experimental scaling functions of the various nuclei at fixed value of the momentum transfer. We consider that the scaling between a set of functions is fulfilled when $\mathcal{R} < 0.096$ and $\mathcal{D} < 0.11$. These values have been obtained by adding the uncertainty to the values of $\mathcal{R}$ and $\mathcal{D}$ for f_L at 570 MeV/c. From a best fit of this last set of data we extracted an empirical universal scaling function [7] represented by the thin full line in the lowest left panel of Fig. 1. This curve is rather similar to the universal empirical function given in Ref. [2].

Let's consider now the scaling of the theoretical functions. The thin dashed lines of Fig. 1 show the RFG scaling functions. The thick lines show the results of our calculations when various effects beyond the RFG are introduced, *i.e.*: nuclear finite size, collective excitations, final state interactions, and, in the case of the f_T functions, meson-exchange currents.

We have studied the effects of the nuclear finite size, by calculating scaling functions within a continuum shell model. At q=700 MeV/c, these scaling functions are very similar to those of the RFG model. At lower values of the momentum transfer, the shell model scaling functions show sharp peaks, produced by the shell structure, not present in the RFG model. We found that shell model scaling functions fulfill the scaling of first kind, the most likely violated, down to 400 MeV/c.

We have estimated the effects of the collective excitations by doing continuum RPA calculations with two different residual interactions [8]. The RPA effects become smaller the larger is the value of the momentum transfer. At $q > 600$ MeV/c, the RPA effects are negligible if calculated with a finite-range interaction. Collective excitations breaks scaling properties, but we found that scaling of first kind is satisfied down to about 500 MeV/c.

The presence of the MEC violates the scaling of the transverse responses. We included the MEC by using the model of Ref. [9]. In our calculations only one-pion exchange diagrams are considered, including those with virtual excitation of the Δ. In our model MEC effects start to be relevant for $q \sim$ 600 MeV/c. We found that MEC do not destroy scaling in the kinematic range of our interest.

The main modification of the shell model scaling functions, are produced by the FSI, we have considered by using the model developed in Ref. [8]. We obtained scaling functions very different from those predicted by the RFG model, and rather similar to the empirical ones. In any case, the FSI do not heavily break the scaling properties. We found that the scaling of first kind is conserved down to q=450 MeV/c.

The same type of scaling analysis applied to (ν_e, e^-) reaction leads to very similar results [7].

3. Superscaling predictions

To investigate the prediction power of the superscaling hypothesis, we compared responses, and cross sections, calculated by using RPA, FSI and eventually MEC, with those obtained by using $f_{\rm U}^{\rm th}$ and $f_{\rm U}^{\rm exp}$.

We show in Fig. 2 double differential electron scattering cross sections calculated with complete model (full) and those obtained with $f_{\rm U}^{\rm th}$ (dashed lines) and $f_{\rm U}^{\rm exp}$ (dotted lines). These results are compared with the data of Refs. [10–12].

The excellent agreement between the results of the full calculations and those obtained by using $f_{\rm U}^{\rm th}$, indicates the validity of the scaling approach in this kinematic region where the q values are larger than 500 MeV/c. The differences with the cross sections obtained by using the empirical scaling functions, reflect the differences between the various scaling functions shown in Fig. 1. The disagreement with the experimental data is probably due to the fact that our models do not consider the excitation of the real Δ resonance, and the pion production mechanism.

The situation for the double differential cross sections is well controlled,

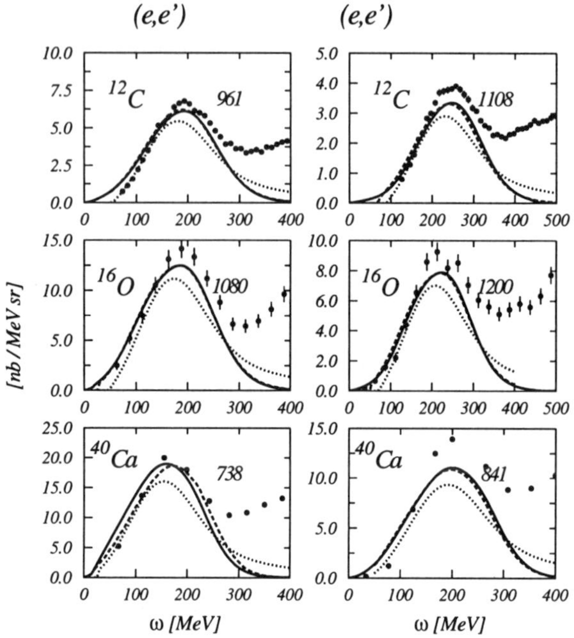

Fig. 2. Inclusive electron scattering cross sections. The numbers in the panels indicate, in MeV, the energy of the the incoming electron. The ^{12}C data [10] have been measured at a scattering angle of θ=37.5^o, the ^{16}O data [11] at θ=32.0^o and the ^{40}Ca data [12] at θ=45.5^o. The full lines show the results of our complete calculations. The cross sections obtained by using $f_{\rm U}^{\rm th}$ are shown by the dashed lines, and those obtained with $f_{\rm U}^{\rm ex}$ by the dotted lines.

since all the kinematic variables, beam energy, scattering angle, energy of the detected lepton, are precisely defined, and consequently also energy and momentum transferred to the target nucleus. This situation changes for the total cross sections which are of major interest for the neutrino physics. The total cross sections are only function of the energy of the incoming lepton, therefore they consider all the scattering angles and of the possible values of the energy and momentum transferred to the nucleus, with the only limitation of the global energy, and momentum, conservations. This means that, in the total cross sections, kinematic situations where the scaling is valid and also where it is not valid are both present.

We show in the first three panels of Fig. 3 various differential charge-exchange cross sections obtained for 300 MeV neutrinos on ^{16}O target. In the panel (a) we show the double differential cross sections calculated for a

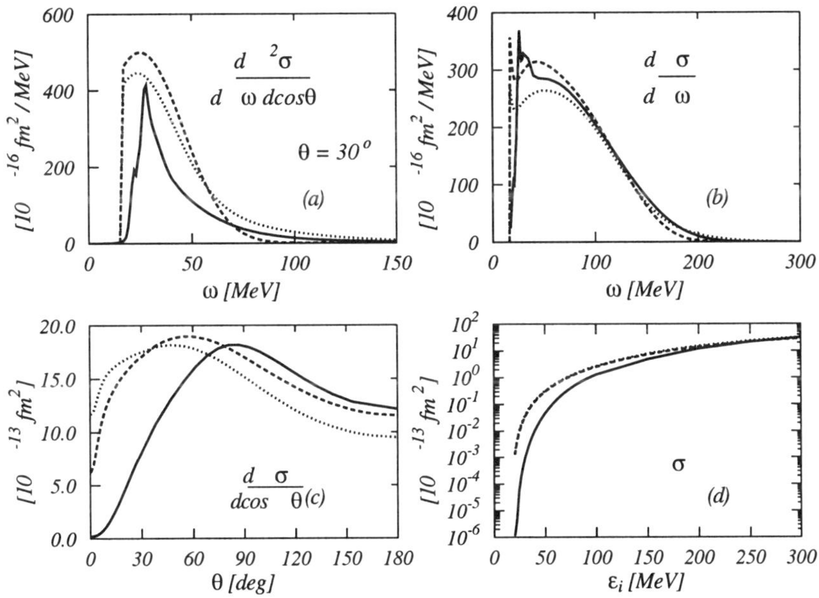

Fig. 3. Neutrino charge exchange cross sections on ^{16}O . In all the panels the full lines show the result of our complete calculation, the dashed (dotted) lines the result obtained with our universal (empirical) scaling function. The results shown in panels (a), (b) and (c) have been obtained for neutrino energy of 300 MeV. Panel (a): double differential cross sections calculated for the scattering angle of 30° as a function of the nuclear excitation energy. Panel (b): cross sections integrated on the scattering angle, always as a function of the nuclear excitation energy. Panel (c): cross sections integrated on the nuclear excitation energy, as a function of the scattering angle. Panel (d): total cross sections, as a function of the neutrino energy.

scattering angle of 30°, as a function of the nuclear excitation energy. The values of the momentum transfer vary from about 150 to 200 MeV/c. This is not the quasi-elastic regime where the scaling is supposed to hold, and this explains the large differences between the various cross sections.

The cross sections integrated on the scattering angle are shown as a function of the nuclear excitation energy in the panel (b) of the figure, while the cross sections integrated on the excitation energy as a function of the scattering angle are shown in the panel (c). The first three panels of the figure illustrate in different manner the same physics issue. The calculation with the scaling functions fails in reproducing the results of the full calculation in the region of low energy and momentum transfer, where surface and collective effects are important. This is shown in panel (b) by the bad agreement between the three curves in the lower energy region, and in panel (c) at low values of the scattering angle, where the q valued are minimal.

Total charge-exchange neutrino cross sections are shown in panel (d) as a function of the neutrino energy ϵ_i. The scaling predictions for neutrino energies up to 200 MeV are unreliable. These total cross sections are dominated by the giant resonances, and more generally by collective nuclear excitation. We have seen that these effects strongly violate the scaling. At ϵ_i =200 MeV the cross section obtained with our universal function is still about 20% larger than those obtained with the full calculation. This difference becomes smaller with increasing energy and is about the 7% at ϵ_i = 300 MeV. This is an indication that the relative weight of the non scaling kinematic regions becomes smaller with the increasing neutrino energy.

References

1. W. M. Alberico, A. Molinari, T. W. Donnelly, E. L. Kronenberg, and J. W. V. Orden, Phys. Rev. C **38**, 1801 (1988).
2. C. Maieron, T. W. Donnelly, and I. Sick, Phys. Rev. C **65**, 025502 (2002).
3. J. E. Amaro, M. B. Barbaro, J. A. Caballero, T. W. Donnelly, A. Molinari, and I. Sick, Phys. Rev. C **71**, 015501 (2005).
4. J. E. Amaro, M. B. Barbaro, J. A. Caballero, T. W. Donnelly, and C. Maieron, Phys. Rev. C **71**, 065501 (2005).
5. T. W. Donnelly and I. Sick, Phys. Rev. C **60**, 065502 (1999).
6. J. Jourdan, Nucl. Phys. **A603**, 117 (1996).
7. M. Martini, G. Co', M. Anguiano, and A. M. Lallena, to be published in Phys. Rev. C.
8. A. Botrugno and G. Co', Nucl. Phys. **A761**, 203 (2005).
9. M. Anguiano, G. Co', A. M. Lallena, and S. R. Mokhtar, Ann. Phys. (N.Y.) **296**, 235 (2002).
10. R. M. Sealock *et al.*, Phys. Rev. Lett. **62**, 1350 (1989).
11. M. Anghinolfi *et al.*, Nucl. Phys. **A602**, 405 (1996).
12. C. Williamson *et al.*, Phys. Rev. C **56**, 3152 (1997).

STRANGE QUARK EFFECTS IN ELECTRON AND NEUTRINO-NUCLEUS QUASI-ELASTIC SCATTERING

A. MEUCCI, C. GIUSTI, and F. D. PACATI

Dipartimento di Fisica Nucleare e Teorica, Università di Pavia, and Istituto Nazionale di Fisica Nucleare, Sezione di Pavia, I-27100 Pavia, Italy

The role of the sea quarks to ground state nucleon properties with electroweak probes is discussed. A relativistic Green's function approach to parity violating electron scattering and a distorted-wave impulse-approximation applied to charged- and neutral-current neutrino-nucleus quasi-elastic scattering are presented in view of the possible determination of the strangeness content of the nucleon.

1. Introduction

The nucleon is a bound state of three valence quarks. However, a sea of virtual $q\bar{q}$ pairs and gluons surrounds each valence quark and play an important role at distance scales of the bound state, where the QCD coupling constant is large and the effects of the color field cannot be calculated accurately. One simple way to probe the effects of the sea is to investigate whether strange quarks contribute to the static properties of the nucleon. The first evidence that the strange axial form factor $g_{\mathrm{A}}^{\mathrm{s}} = G_{\mathrm{A}}^{\mathrm{s}}(Q^2 = 0)$ is different from zero and large was found at CERN by the EMC collaboration [1] in a measurement of deep inelastic scattering of polarized muons on polarized protons. In order to study the role of the strange quark to the spin structure of the nucleon various reactions have been proposed. Here we are interested in parity-violating (PV) electron scattering and neutrino-nucleus scattering. These two kinds of reactions can give us complementary information about the contributions of the sea quarks to the properties of the nucleon. While PV electron scattering is essentially sensitive to the electric and magnetic strangeness of the nucleon, neutrino-induced reactions are primarily sensitive to the axial-vector form factor of the nucleon.

A number of PV electron scattering measurements have been carried out in recent years. They are sensitive to the strangeness contribution by measuring the helicity-dependent PV asymmetry

$$A_{\rm PV} = \frac{{\rm d}\sigma_+ - {\rm d}\sigma_-}{{\rm d}\sigma_+ + {\rm d}\sigma_-}\,, \tag{1}$$

where ${\rm d}\sigma_{+(-)}$ is the cross section for incident right(left)-handed electrons from unpolarized targets (usually protons). $A_{\rm PV}$ arises from the interference between electromagnetic and weak processes and depends on the electric (magnetic) form factors $G^{\gamma}_{\rm E(M)}$, their weak counterparts $G^{Z}_{\rm E(M)}$, and the axial form factor as seen in electron scattering $G^{e}_{\rm A}$. It has been noted that the contribution of radiative correction must be calculated in order to allow a precise extraction of the strange axial form factor $G^{s}_{\rm A}$ from a PV measurement of $G^{e}_{\rm A}$. This usually prevents from a final determination of $G^{s}_{\rm A}$ from this data. The SAMPLE [2] results at backward angle on proton and deuteron targets reported results for $G^{\rm s}_{\rm M}$ and $G^{e}_{\rm A}$ at $Q^2 \simeq 0.1$ (GeV/c)^2. The HAPPEX [3] , A4 [4] , and G0 [5] results at forward angles provided a linear combination of $G^{\rm s}_{\rm E}$ and $G^{\rm s}_{\rm M}$ over the range $0.1 \leq Q^2 \leq 1$ (GeV/c)^2, where the contribution from the axial term is usually suppressed by kinematical conditions. Three independent measurements are needed to extract $G^{\rm s}_{\rm E}$, $G^{\rm s}_{\rm M}$, and $G^{e}_{\rm A}$ separately. The PV asymmetry from a spinless, isoscalar target, such as ^{4}He, depends only on the electric form factors [6] and represents an interesting via to avoid the problem of the axial term.

Neutrino reactions are a well-established alternative to PV electron scattering and give us complementary information about the contributions of the sea quarks to the properties of the nucleon. A measurement of $\nu(\bar{\nu})$-proton elastic scattering at Brookhaven National Laboratory (BNL) [7] suggested a non-zero value for the strange axial-vector form factor of the nucleon. However, it has been shown in Ref. [8] that the BNL data cannot provide us decisive conclusions about the strange form factors when also strange vector form factors are taken into account. The FINeSSE [9] experiment at Fermi National Laboratory aims at performing a detailed investigation of the strangeness contribution to the proton spin via measurements of the ratio of neutral-current to the charged-current $\nu(\bar{\nu})N$ processes. When combined with the existing data on PV scattering, a determination of the strange form factors in the range $0.25 \leq Q^2 \leq 0.75$ would have to be possible with an uncertainty at each point of $\simeq \pm 0.02$ [10] . Since a significant part of the event will be from scattering on ^{12}C, nuclear structure effects have to be clearly understood in order to give a reliable interpretation of the data.

2. PV asymmetry in inclusive electron scattering on nuclei

The helicity asymmetry for the scattering of a polarized electron on a target nucleus through an angle ϑ can be written from Eq. (1) as the ratio between the PV and the parity-conserving (PC) cross section, i.e.,

$$A = A_0 \frac{v_L R_L^{AV} + v_T R_T^{AV} + v_T' R_T^{VA}}{v_L R_L + v_T R_T} \; , \tag{2}$$

where $A_0 \simeq 1.799 \times 10^{-4}\, Q^2\, (\mathrm{GeV}/c)^{-2}$ is a scale factor. The coefficients v are derived from the lepton tensor components and are taken from Ref. [11]. The response functions R are given in terms of the components of the hadron tensor and contain the interference between the electromagnetic and the weak neutral part of the current operator [11] . The single-particle electromagnetic part of the current is

$$j^\mu = F_1 \gamma^\mu + i \frac{\kappa}{2M} F_2 \sigma^{\mu\nu} q_\nu \; . \tag{3}$$

The single-particle current operator related to the weak neutral current is

$$j^\mu = F_1^V \gamma^\mu + i \frac{\kappa}{2M} F_2^V \sigma^{\mu\nu} q_\nu - G_A \gamma^\mu \gamma^5 \; . \tag{4}$$

The vector form factors F_i^V can be expressed in terms of the corresponding electromagnetic form factors for protons (F_i^p) and neutrons (F_i^n), plus a possible isoscalar strange-quark contribution (F_i^s), i.e.,

$$F_i^{V;\, p(n)} = \pm \left\{ F_i^p - F_i^n \right\}/2 - 2 \sin^2 \theta_W F_i^{p(n)} - F_i^s/2 \; , \tag{5}$$

where $+(-)$ stands for proton (neutron) knockout and θ_W is the Weinberg angle ($\sin^2 \theta_W \simeq 0.2313$). The strange vector form factors are taken as

$$F_1^s(Q^2) = \frac{(\rho^s + \mu^s)\tau}{(1+\tau)(1 + Q^2/M_V^2)^2}, \quad F_2^s(Q^2) = \frac{(\mu^s - \tau\rho^s)}{(1+\tau)(1 + Q^2/M_V^2)^2}, \tag{6}$$

where $\tau = Q^2/(4M_p^2)$ and $M_V = 0.843$ GeV. The quantities μ^s and ρ^s are related to the strange magnetic moment and radius of the nucleus. The axial form factor is expressed as

$$G_A(Q^2) = \frac{1}{2} \left(\pm g_A - g_A^s \right) G \; , \tag{7}$$

where $g_A \simeq 1.26$, g_A^s describes possible strange-quark contributions, and

$$G = (1 + Q^2/M_A^2)^{-2} \; . \tag{8}$$

The axial mass has been taken from Ref. [12] as $M_A = (1.026 \pm 0.021)$ GeV.

The inclusive PV electron scattering may be treated using the same relativistic approach which was already applied to the inclusive PC electron scattering [13] and to the inclusive quasi-elastic $\nu(\bar{\nu})$-nucleus scattering [14] . The components of the nuclear response are written in terms of the single-particle optical-model Green's function, that is based on a bi-orthogonal expansion in terms of the eigenfunctions of the non-Hermitian optical potential and of its Hermitian conjugate. As it is discussed in Refs. [13,14], and [11], the flux is preserved and final state interactions (FSI) are treated in the inclusive reaction consistently with the exclusive one.

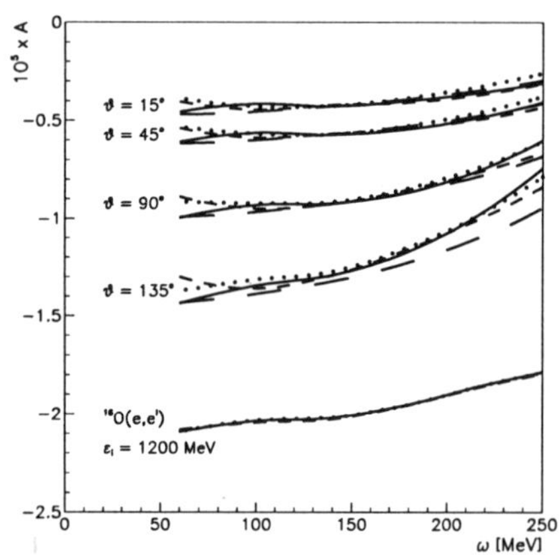

Fig. 1. PV asymmetry for ^{12}C at q = 400 MeV/c and ^{16}O at ε_i = 1200 MeV and $\vartheta = 32^\circ$ with different bound states and optical potentials as explained in the text. The results are rescaled by the factor 10^5.

In order to evaluate the uncertainties of the model, we compare in Fig. 1 the asymmetries for ^{12}C at $q = 400$ MeV and ^{16}O at $\varepsilon_i = 1200$ MeV and $\vartheta = 32^\circ$ calculated with different bound states and optical potentials. The full lines give the results with the NL-SH bound states [15] and the EDAD1 optical potentials [16] . The dotted lines give the same calculations, but with the NL3 bound states of Ref. [17] and the EDAD1 optical potentials. The dashed lines are calculated with the NL-SH bound states and the energy-dependent and A-independent EDAI-12C or EDAI-16O optical potentials [16] . The differences are very small everywhere, but for ^{12}C at low

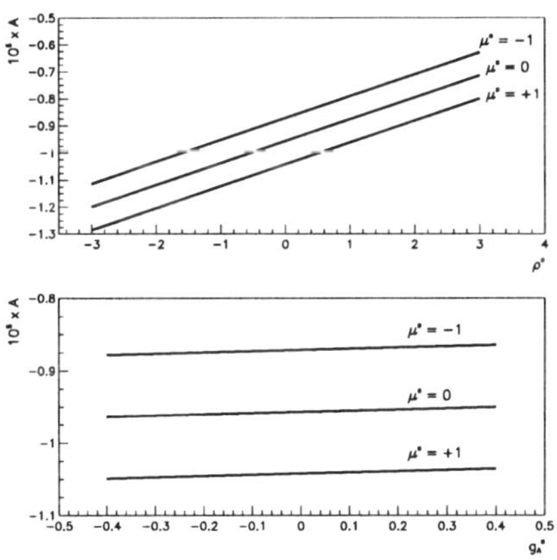

Fig. 2. PV asymmetry for ^{12}C at q = 500 MeV/c, ω = 120 MeV, and $\vartheta = 30^o$ as a function of ρ^s and μ^s (upper panel) and as a function of g_A^s and μ^s (lower panel).

and high energy transfers, where they are in any case less than 10%. The long dashed lines show the results for the RPWIA. Only small differences are found when they are compared with the Green's function results. This means that the ratio that gives the asymmetry cancels most of the effect of the FSI.

The sensitivity of PV electron scattering to the effect of strange-quark contribution to the vector and axial-vector form factors, is shown in Fig. 2 for ^{12}C at q = 500 MeV/c, ω = 120 MeV, and $\vartheta = 30^o$ as a function of the strangeness parameters, ρ^s, μ^s, and g_A^s. The range of their values is chosen according to Refs. [3] and [18]. The asymmetry reduces in absolute value up to $\simeq$40% as ρ^s varies in the range $-3 \leq \rho^s \leq +3$, whereas it changes up to $\simeq$15% for $-1 \leq \mu^s \leq +1$. We note that, according to HAPPEX [3] results, ρ^s and μ^s might have opposite sign, thus leading to a partial cancellation of the effects. The sensitivity to g_A^s is very weak, as can be seen in the lower panel of Fig. 2.

3. The quasi-elastic neutrino-nucleus scattering

The $\nu(\bar{\nu})$-nucleus cross section for the semi-inclusive process may be written as a contraction between the lepton and the hadron tensor. The lepton tensor is defined in a similar way as in electromagnetic knockout and it separates into a symmetrical and an anti-symmetrical component which are written as in Refs. [14,19], and [20]. The hadron tensor is given in its general form by suitable bilinear products of the transition matrix elements of the nuclear weak-current operator. Assuming that the final states are given by the product of a discrete (or continuum) state of the residual nucleus and a scattering state of the emitted nucleon and using the impulse approximation, the transition amplitude reduces to the sum of terms similar to those appearing in the electron scattering case [19,21] . The single-particle current operator related to the weak current is

$$j^\mu = \left[F_1^{\rm V}\gamma^\mu + i\frac{\kappa}{2M}F_2^{\rm V}\sigma^{\mu\nu}q_\nu - G_{\rm A}\gamma^\mu\gamma^5 + F_{\rm P}q^\mu\gamma^5\right]\mathcal{O}_\tau \ , \qquad (9)$$

where $\mathcal{O}_\tau = \tau^\pm$ are the isospin operators for charged-current (CC) reactions, while $\mathcal{O}_\tau = 1$ for neutral-current (NC) scattering. The induced pseudoscalar form factor $F_{\rm P}$ contributes only to CC scattering but its effects are almost negligible. For NC reactions, the weak isovector form factor, $F_1^{\rm V}$ and $F_2^{\rm V}$, and the axial form factor are expressed as in Eqs. (5) and (7), whereas for CC scattering they are

$$F_i^{\rm V} = F_i^{\rm p} - F_i^{\rm n} \ , \quad G_{\rm A} = g_{\rm A} G \ , \qquad (10)$$

where $g_{\rm A} \simeq 1.26$ and G is defined in Eq. (8). The single differential cross section for the quasi-elastic $\nu(\bar{\nu})$-nucleus scattering with respect to the outgoing nucleon kinetic energy $\rm T_N$ is obtained after performing an integration over the solid angle of the final nucleon and over the energy and angle of the final lepton. We use in our calculations a relativistic optical potential with a real and an imaginary part which produces an absorption of flux. This is correct for an exclusive reaction, but would be incorrect for an inclusive one. Here we consider situations where an emitted nucleon is detected and treat the quasi-elastic neutrino scattering as a process where the cross section is obtained from the sum of all the integrated exclusive one-nucleon knockout channels. Some of the reaction channels which are responsible for the imaginary part of the optical potential, like multi-step processes, fragmentation of the nucleus, absorption, etc. are not included in the experimental cross section as an emitted proton is always detected. The outgoing proton can be re-emitted after re-scattering in a detected channel, thus simulating the kinematics of a quasi-elastic reaction, only in few cases. The relevance of this contributions depends on kinematics and should not be too large in the situations considered here.

Since an absolute cross section measurement is a very hard experimental task due to difficulties in the determination of the neutrino flux in Ref. [22] was suggested to measure the ratio of proton to neutron (p/n) yield as an

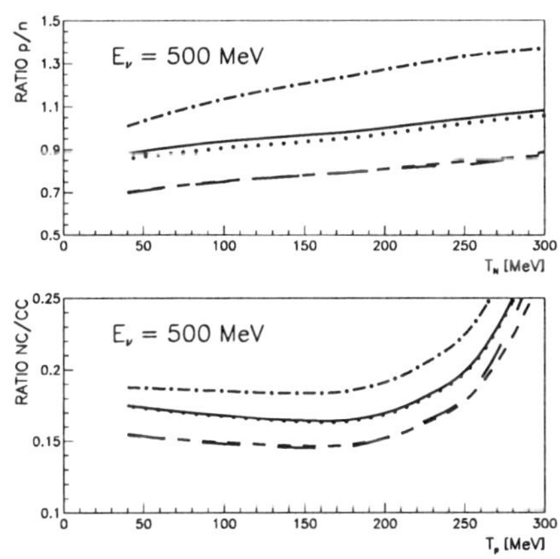

Fig. 3. Upper panel: ratio of proton-to-neutron NC cross sections of the ν scattering on ^{12}C . Lower panel: ratio of neutral-to-charged current cross sections of the ν scattering on ^{12}C. Dashed lines are the results with no strangeness contribution, solid lines with $g_A^s = -0.10$, dot-dashed lines with $g_A^s = -0.10$ and $\mu^s = -0.50$, dotted lines with $g_A^s = -0.10$ and $\rho^s = +2$. Long dashed lines are the RPWIA results without strangeness contribution.

alternative way to separate the effects of the strange-quark contribution. This ratio is very sensitive to the strange-quark contribution as the axial-vector strangeness g_A^s interferes with the isovector contribution g_A with one sign in the numerator and with the opposite sign in the denominator (see Eq. (7)) and it is expected to be less sensitive to distortion effects than the cross sections themselves. In the upper panel of Fig. 3 the p/n ratio of the quasi-elastic ν scattering on ^{12}C is displayed as a function of T_N. The RPWIA results are shown in the figure and they are almost coincident with RDWIA ones. The p/n ratio for an incident neutrino is enhanced by a factor $\simeq$ 20-30% when g_A^s is included and by $\simeq$ 50% when both g_A^s and μ^s are included. A minor effect is produced by ρ^s, which gives only a slight reduction of the p/n ratio. Precise measurements of the p/n ratio appear however problematic due to the difficulties associated with neutron detection. This is the reason why the most attractive quantity to extract experimental information about the strangeness content seems the ratio of the neutral-to-charged (NC/CC) cross sections. In fact, although sensitive to the strange-quark effects only in the numerator, the NC/CC ratio is simply related to the number of events with an outgoing proton and a missing mass with respect to the events with an outgoing proton in coincidence with a muon. Our RDWIA results for the NC/CC ratio of the quasi-elastic ν scattering on ^{12}C are presented in the lower panel of Fig. 3 as a function of the kinetic energy of the outgoing proton. The fact that the CC cross section goes to zero more rapidly than the corresponding NC one (because of the muon mass) causes the enhancement of the ratio at large values of T_p. The simultaneous inclusion of g_A^s and μ^s gives an enhancement that is about a factor of 2 larger than the one corresponding to the case with only g_A^s included. The effect of ρ^s is very small.

References

1. J. Ashman *et al.*, [European Muon Collaboration], Nucl. Phys. **B328**, 1 (1989).
2. D.T. Spayde *et al.*, Phys. Lett. **B583**, 79 (2004).
3. K. Aniol *et al.*, [HAPPEX Collaboration], Phys. Rev. Lett. **82**, 1096 (1999); Phys. Rev. C **69**, 065501 (2004); Phys. Lett. **B635**, 275 (2006).
4. F.E. Maas *et al.*, [A4 Collaboration], Phys. Rev. Lett. **94**, 152001 (2005).
5. D.S. Armstrong *et al.*, [G0 Collaboration], Phys. Rev. Lett. **95**, 092001 (2005).
6. A. Acha *et al.*, [HAPPEX Collaboration], nucl-ex/0609002.
7. L.A. Ahrens *et al.*, Phys. Rev. D **35**, 785 (1987).
8. G.T. Garvey *et al.*, Phys. Rev. C **48**, 761 (1993).
9. S. Brice *et al.*, hep-ex/0402007. See also `http://www-finesse.fnal.gov/index.html` and `http://www-boone.fnal.gov/`.
10. S.F. Pate, Phys. Rev. Lett. **92**, 082002 (2004); S.F. Pate, G. MaLachlan, D. McKee, and V. Papavassiliou, hep-ex/0512032.
11. A. Meucci, C. Giusti, and F.D. Pacati, Nucl. Phys. **A756**, 359 (2006).
12. V. Bernard, L. Elouadrhiri, and Ulf-G. Meissner, J. Phys. G **28**, R1 (2002).
13. A. Meucci, F. Capuzzi, C. Giusti, and F.D. Pacati, Phys. Rev. C **67**, 054601 (2003).
14. A. Meucci, C. Giusti, and F.D. Pacati, Nucl. Phys. **A739**, 277 (2004).
15. M.M. Sharma, M.A. Nagarajan, and P. Ring, Phys. Lett. **B312**, 377 (1993).
16. E.D. Cooper *et al.*, Phys. Rev. C **47**, 297 (1993).
17. G.A. Lalazissis, J. König, and P. Ring, Phys. Rev. C **55**, 540 (1997).
18. D.H. Beck and R.D. McKeown, Ann. Rev. Nucl. Part. Sci. **51**, 189 (2001).
19. S. Boffi *et al.*, *Electromagnetic Response of Atomic Nuclei*, Oxford Studies in Nuclear Physics, Vol. 20 (Clarendon, Oxford, 1996); S. Boffi *et al.*, Phys. Rep. **226**, 1 (1993).
20. A. Meucci, C. Giusti, and F.D. Pacati, Nucl. Phys. **A773**, 250 (2006).
21. A. Meucci, C. Giusti, and F.D. Pacati, Phys. Rev. C **64**, 014604 (2001); Phys. Rev. C **64**, 064615 (2001).
22. G.T. Garvey *et al.*, Phys. Lett. **B289**, 249 (1992); G.T. Garvey *et al.*, Phys. Rev. C **48**, 1919 (1993); C.J. Horowitz *et al.*, Phys. Rev. C **48**, 3078 (1993); W.M. Alberico *et al.*, Nucl. Phys. **A623**, 471 (1997); W.M. Alberico *et al.*, Phys. Lett. **B438**, 9 (1998).

Structure of Hadrons and Hadronic Matter

HADRON STRUCTURE AND HADRONIC MATTER

M. M. GIANNINI

Dipartimento di Fisica, Università degli Studi di Genova,
Istituto Nazionale di Fisica Nucleare, Sezione di Genova
Genova, I-16146 Italy
E-mail: giannini@ge.infn.it

Some of the most relevant and recent results of the italian theoretical groups active in hadronic physics are briefly reviewed.

Keywords: 12.39.-x; 13.40.Gp; 12.38.-t; 13.60.-r.

1. Introduction

The studies of hadron properties can be considered to follow roughly two main approaches. The first one, which can be called microscopic or systematic, starts from the internal hadron dynamics, that is from the quark degrees of freedom and their interactions. The widely accepted framework is of course provided by Quantum ChromoDynamics (QCD), which however can be presently treated only in the perturbative regime (pQCD). A big effort is devoted to the analysis of hadron structure in the Lattice QCD (LQCD): there are now many important results and more are expected in the future, when the computer capabilities will allow to reach the required precision of the calculations in order to extract the hadron properties in a systematic way. This will presumably take still some time and in the meanwhile one can rely on models, eventually based on QCD or LQCD. The second approach, which can be denoted as "phenomenological", uses some parametrization of single hadron properties within a theoretical framework, based on general aspects of quarks and gluons dynamics. The two approaches are not sharply separated, but on the contrary they can be strongly correlated, they are introduced mainly for simplifying the discussion.

Many models have been built and applied to the description of hadron properties. An important class is provided by Constituent Quark Models (CQM), in which quarks are considered as effective internal degrees of freedom and can acquire a mass and even, in certain approaches, a finite size. The idea of quark as constituent particles of hadrons has been introduced very soon [1]; quark confinement was not yet known and quarks were considered to have a very large mass, thereby supporting a non relativistic approach. The modern version of the non relativistic CQM has been formulated by Isgur and Karl [2] (IK), using a h.o. confinement potential and a hyperfine interaction, which is spin dependent and inspired by the One-Gluon-Exchange mechanism (OGE). The IK model has been subsequently reformulated using a relativistic kinetic energy operator and a three-body linear confinement, inspired by a flux tube approach [3] (CI). Further models have been recently proposed. In the algebraic model [4] an underlying $U(7)$ symmetry is assumed and group theoretical tecniques are used in order to describe the nucleon excitation spectrum and other properties, such as elastic form factors and helicity amplitudes. The hypercentral CQM (hCQM) is characterized by a linear confinement and a coulomb-like term [5]; the hypercentral interaction depends on the coordinates of the three quarks and can include three-body effects. The main feature of the chiral CQM [6] is the quark interaction derived from the exchange of Golstone Bosons (GBE), like pions and kaons. In a recent CQM [7], instanton interactions are introduced in order to describe the baryon properties.

Besides CQM, there are various models formulated in a consistent way within a chiral theory. Many papers have been written along this line of thought and the interested reader is referred to the abundant literature (see e.g. ref. [8]). A completely different approach is provided by instanton liquid models [9].

In the following, the recent work done by the italian theoretical groups in the field of hadron physics will be briefly reviewed.

2. Properties of the nucleon

The properties to be discussed are mainly the excitation spectrum of the nucleon and its electromagnetic form factors, that is the elastic form factors and the amplitudes for the excitation of the resonances. A particular attention will be devoted to the time-like form factors of the nucleon, also in view of an interest testified by plans of new experiments to be performed in various laboratories, including the Dafne facility.

2.1. *The spectrum*

The nucleon has a rich spectrum, which however becomes very hard to be studied experimentally in the higher energy part because of the large widths of the states and the increasing overlapping of the various resonances. The CQM studies usually are limited to the states up to $2GeV$. In any case, the description of the spectrum is the first task of a model builder: it serves to determine a quark interaction to be used for the calculation of other physical quantitites. The form factors (elastic and inelastic) are particularly suited for a test of the models, since they use explicitly the baryon wave functions determined by the description of the spectrum.

Since the beginning, the modern CQM have made use of some considerations based on the fundamental theory. According to De Rújula, Georgi and Glashow [10], LQCD indicates that the quark interaction contains a long range spin-independent confinement, to which a short range spin dependent term is added. This means that one can start in principle by considering a spin independent quark interaction and that the corresponding three-quark states can be classified according to the group $SU(6)$. In fact, in the experimental baryon spectrum the states can be easily grouped into $SU(6)$ configurations, as it is shown in Fig. 1. The states within the multiplets are not degenerate, however the multiplets are clearly separated and the splittings can be ascribed to some spin (and/or flavour) dependent interaction, which violates the $SU(6)$ symmetry and acts as a perturbation.

In connection with the spectrum, we shall illustrate the features and the results of three CQM, the Isgur-Karl in the relativized version [2,3] (CI), the Goldstone-Boson-Exchange model [6] (GBE) and the hypercentral one [5] (hCQM). The CI and the GBE have been used by the Rome and Pavia groups, respectively, while the hCQM has been built and used by the Genoa group.

CI [3]. The confinement is provided by a three-body term corresponding to a $Y-$shaped configuration. The multiplet splittings are mainly given by an interaction, which is inspired by the One-Gluon-Exchange mechanism and as such it contains a spin-spin term and a tensor force. The three-body equation, with relativistic kinetic energy, is solved by means of a variational approach in a large h.o. basis.

GBE [6]. The confinement interaction, in the more recent version of the model, is given by a linear two-body term. Consistently with the idea that at low energies pseudoscalar mesons are the relevant degrees of freedom as Goldstone Bosons, an explicit quark-quark potential due to meson exchange is introduced. The splittings within multiplets are then provided by the spin

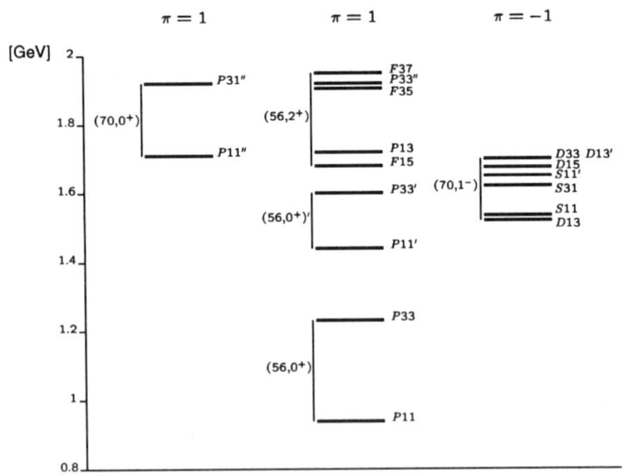

Fig. 1. The three- and four-star non strange resonances [11] grouped into $SU(6)$ multiplets; the states are labeled by (d, L^P), where d is the dimension of the $SU(6)$ representation, L is the total orbital angular momentum of the three quark states and P the corresponding space parity.

and isospin dependence of the pseudoscalar meson exchange.

hCQM [5]. The quark potential is assumed to be hypercentral, that is to depend only on the hyperradius x, defined as $x = \sqrt{\vec{\rho}^2 + \vec{\lambda}^2}$, where $\vec{\rho}$ and $\vec{\lambda}$ are the Jacobi coordinates describing the quark internal motion. The hyperradius x assumes the meaning of a collective variable, describing the size of the baryon state. The explicit form of the potential is given by [5]

$$V_{hCQM} = -\frac{\tau}{x} + \alpha x \tag{1}$$

A potential containing a coulomb-like and a linear confinement term has been used since long time in the description of the meson sector (Cornell potential). Such structure has been recently supported by LQCD calculations for static quarks [12–14]. In this respect the hCQM potential can be considered as the hypercentral approximation of a quark-quark interaction of the Cornell type. Thanks to the x-dependence, the hCQM interaction may also include many body contributions, corresponding for instance to the already mentioned Y-shaped string configuration. The multiplet splittings are provided by a hyperfine interaction of the standard form [2]. The fit of the spectrum (see Fig. 2) leads to the values $\tau = 4.59, \alpha = 1.61(fm)^{-2}$ [5], which are kept fixed in the subsequent applications of the model to various quantities of interest. It should be mentioned that the model has been improved by including an isospin dependent term in the potential [15] and also by considering a relativistic kinetic energy [16].

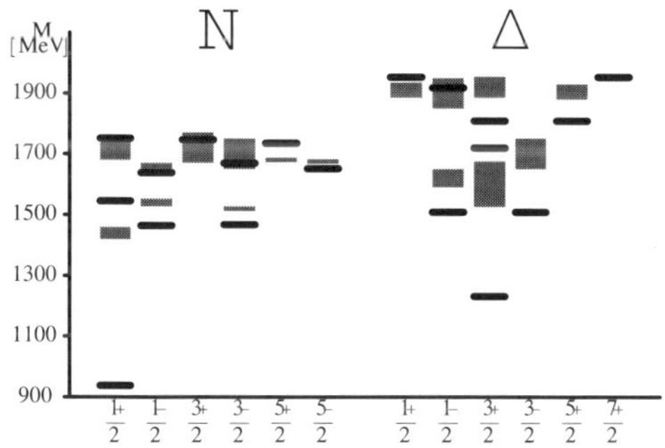

Fig. 2. The spectrum obtained with the hCQM [5]; the fitted parameters are $\alpha = 1.61(fm)^{-2}$, $\tau = 4.59$ and the strength of the hyperfine interaction, which is determined in order to reproduce the $\Delta - N$ mass difference.

All the CQMs have the problem of the so called missing resonances, i.e. states predicted by the models in the higher part of the spectrum but not seen experimentally. Recently a quark-diquark model has been contructed with no missing resonances [17]. Encoding Wilczek's idea of two types of diquarks (scalar and vector) one can obtain a good description of both the baryon spectrum and the charge form factor of the proton.

Practically all the mentioned CQMs are able to describe the baryon spectrum in a more or less satisfactory way. If an explicit isospin dependence is introduced in the quark interaction [4,6,15], also the Roper position can be correctly reproduced. A more stringent test of the models is provided by the comparison of the theoretical predictions with other types of hadron observables. The form factors, both elastic and inelastic, are very important in this respect.

2.2. *Form factors*

2.2.1. *The elastic form factors*

The study of the elastic form factors of the nucleon has always been considered an important source of information on the hadron internal structure. Recently, the interest on these quantities has been renewed by the polarization experiments performed at Jlab [18], which showed an unexpected decrease of the ratio G_E^p/G_M^p between the magnetic and electric proton form factors.

From the theoretical point of view, the analysis of these new data has shown that relativity is an important issue. Relativity can be included in the dynamical formulation of quark models by means of a Bethe-Salpeter approach, as done by the Bonn group [7]. An alternative way is the relativistic hamiltonian formulation introduced by Dirac [19], which can be performed according to three forms [20], light front, point form, instant form, which differ in the realization of the Poincaré group and in the number of interaction independent generators.

Some results of the Rome group [21] are presented in Fig. 3; the calculations are performed using the CI model and light front wave functions. Similarly in Fig. 3 the calculation of the Graz-Pavia group [23] are reported: the curves are obtained in the GBE model using a Point Form Spectator Approximation (PFSA) (more details will be given in the talk by T. Melde [27]).

The hCQM has also been applied to the calculation of the nucleon form factors [28], but it was necessary to include some relativistic corrections, such as Lorentz boost applied to the three quark states. In this way it has been shown [29] that relativistic effects produce a significant decrease in the ratio between the proton form factors. Therefore, the model has been relativized [16], introducing a relativistic kinetic energy and including the dynamical formulation in a point form approach. The resulting potential parameters are not much different from the non relativistic ones and, keeping them fixed, the elastic nucleon form factors are calculated [25]. The predicted values are very near to the experimental ones; however, in order to reach a good description of data (see Fig. 3), one has to introduce quark form factors.

2.2.2. *Time-like form factors*

The information on the time-like form factors is of course of great importance for the understanding of the nucleon structure and the present knowledge is poor (for a review, see e.g. [30]). A further motivation for an extensive study of the time-like form factors is provided by some apparent inconsistencies displayed by the present data. For instance, the dispersion relation technique leads to the prediction that at very high momentum transfer q^2 the proton magnetic form factor should be of the same order of magnitude in the time- and space-like region, i.e. $G_M^p(q^2 < 0) \approx G_M^p(q^2 > 0)$, while from the data there seems to be a factor $2 : G_M^p(q^2 < 0) \approx 2G_M^p(q^2 > 0)$. The Fenice data on the neutron [31] indicate that $|G_M^p| \leq |G_M^n|$, while the expectation is that they should scale according to the quark charges, that

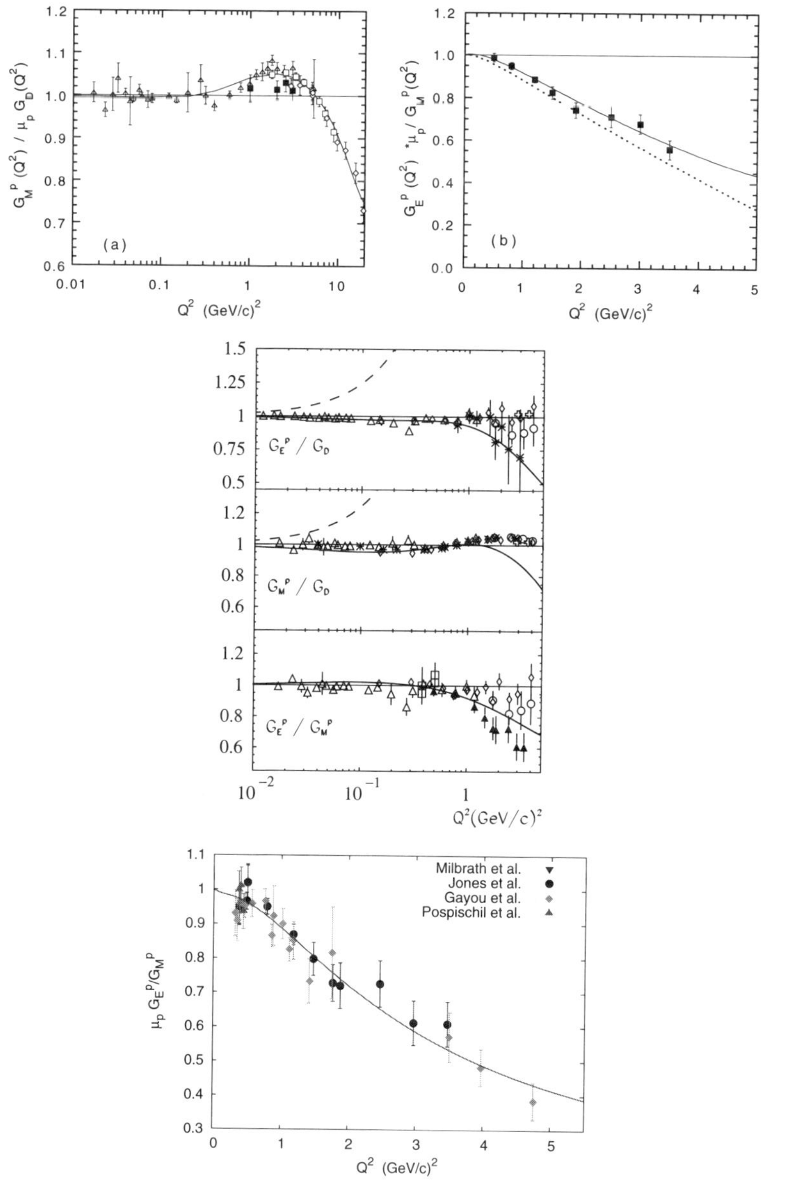

Fig. 3. Various calculations of the nucleon elastic form factors. Upper part: the proton magnetic form factor, divided by the dipole form (left) and the ratio $G_E^p/\mu_p G_M^p$ (right), evaluated by the Rome group [21]. The dotted (full) curve is obtained without (with) quark form factors. Data in the left figure are as quoted in ref. [22], in the right figure from ref. [18]. Middle part: the proton electric and magnetic form factor, divided by the dipole form and their ratio, evaluated by the Graz-Pavia group [23] (Data are from [24]). Lower part: the proton electric and magnetic form factor ratio, evaluated by the Genoa group [25] . Data are from [18,26].

is $|G_M^p|/|G_M^n| \approx (e_d^2/e_u^2)^2 = 1/4$. The ratio between the two proton form factors decreases, according to the recent Jlab data, in contrast to what is observed at Babar [32] for $q^2 > 0$. The present data show an interesting structure in the neighborhood of the physical theshold $q^2 = 4m_p^2$, however data are obtained by integrating over angles, in view of the low available statistics, and assuming perfect equality between the electric and magnetic form factors, which is true only at the physical theshold and in the remaining regions the time-like electric form factor is presently unknown. The time-like form factors are complex functions of the momentum transfer and in order to get information on their phases, which are completely unknown, one should perform polarization experiments. A systematic knowledge of the form factors for $q^2 > 0$ would allow a stringent test of the available theoretical models. Measurements are planned in various experiments, both in the reaction in which a nucleon-antinucleon pair is produced in the channel e^+e^- (Dafne) or in the inverse process (PANDA). There has been recently a great interest, also from the theoretical italian community [33], in the proposed upgrading of the Dafne facility up to $2.5GeV^2$. There are in any case poblems in extracting the data on the time-like form factors from the observed cross section and a careful study is therefore necessary [34]. In the unpolarized process the cross section can be written as the sum of a non-Born term, due to two-photon exchange, and a Born one, depending on the ratio $|G_E/G_M|$; from this process information on the ratio and on the two-γ terms can be extracted. In a similar way one can get information on phases studying the same process where however the produced nucleon is polarized [34].

2.2.3. *The electromagnetic transitions*

A very important test of the models is provided by the helicity amplitudes fro the electromagnetic excitation of the nucleon resonances. The transverse amplitudes, $A_{1/2}$ and $A_{3/2}$, are defined as the matrix elements of the transverse electromagnetic interaction, $H_{e.m.}^t$, between the nucleon, N, and the resonance, B, states:

$$\begin{aligned} A_{1/2} &= \langle B, J', J'_z = \tfrac{1}{2} \, |H_{em}| N, J \;=\; \tfrac{1}{2}, J_z = -\tfrac{1}{2} \,\rangle \\ A_{3/2} &= \langle B, J', J'_z = \tfrac{3}{2} \, |H_{em}| N, J \;=\; \tfrac{1}{2}, J_z = \tfrac{1}{2} \,\rangle \end{aligned} \qquad (2)$$

The building of a CQM implies the knowledge of model wave functions and therefore the above matrix elements can be calculated. The results of the IK model were soon available [35], however they were obtained adjusting the proton size to a value of about $0.5 fm$ [36]. More recently, results from the CI in the light front approach have been obtained by the Rome group [21]. A systematic study of the helicity amplitudes for the negative parity resonances has been performed within the hCQM [37]; this study is now extended to all resonances with an observable electromagnetic transition strength and to the longitudinal excitation amplitudes as well [38]. In the case of the negative parity non strange resonances, the experimental behaviour is fairly well reproduced by the parameter-free calculations made with the hCQM [37], specially for the S_{11}, whose trend has been predicted in advance with respect to the recent Jlab data [39]. In the medium Q^2 region the theoretical amplitudes are not damped, at variance with the h.o. calculations. This softer dependence is due, in the hCQM calculation, to the presence of the Coulomb-like term in the interaction. At low Q^2, in particular for the $A_{3/2}$ amplitudes, there is a lack of strength, which is however typical of all CQMs. Such discrepancy can be attributed to the fact that all CQMs miss some fundamental mechanism, such as the quark-antiquark pair production. A confirmation of this statement is provided by a recent analysis performed by a collaboration of the Genoa and Mainz groups [40]. In Fig. 4 [40] the transition amplitudes for the Δ excitation, calculated with the hCQM, are reported in comparison with a global fit performed by the Mainz group [41]. The meson cloud contributions, evaluated by means of a dynamical model [42], are also reported. Their importance decreases with increasing Q^2, going rapidly to zero, as expected; however, for low Q^2, they are able to fill most of the gap between data and the hCQM results. This feature is quite general, since it happens systematically also for the excitation of higher resonances, such as $P_{11}(1440), S_{11}(1535), D_{13}(1520), F_{15}(1680)$ [40].

To conclude, it should be noted that the proton radius calculated in the hCQM is about $0.5 fm$, just the value previously fitted [35,36] to the $D13$ transition strength. In the outer region, that is for low Q^2, there is a lack of strength, which can be at least partially accounted for by pion cloud contributions. The emerging picture is therefore that of a quark core with a dimension of about $0.5 fm$ plus a meson (or sea-quark) cloud.

2.3. *Interlude*

The Trento group has devoted much attention to the interpaly between models and LQCD [43]. The reason stems from some features of LQCD calculation. In fact, many observables of interest (time-like form factors, GPDs) cannot be related to quantities calculable on the lattice. Moreover,

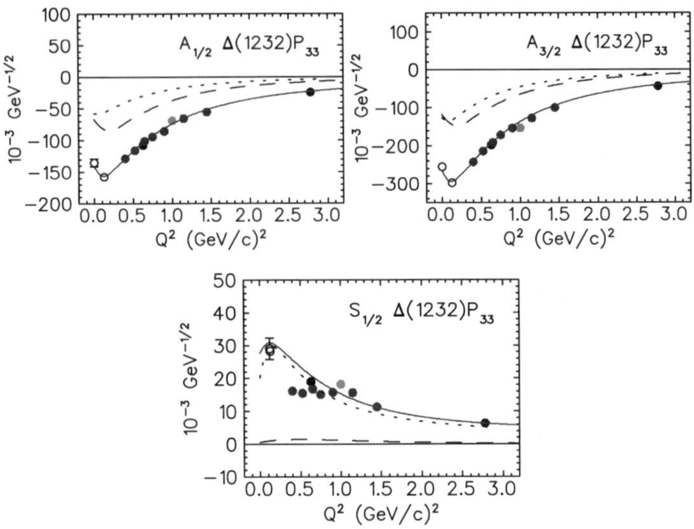

Fig. 4. The Q^2 dependence of the $N \to \Delta$ helicity amplitudes. The solid and dashed curves are the results of the superglobal fit with MAID and the predictions of the hCQM, respectively. The dotted lines show the pion cloud contributions calculated with DMT [42]. The data points at finite Q^2 are the results of the Mainz single-Q^2 fits [41]. At $Q^2 = 0$ for $A_{1/2}$ and $A_{3/2}$ the photon couplings from PDG are shown [11].

it is not easy to understand how dynamics is working. Finally, results are obtained for high quark masses ($> 100MeV$ for u, d quarks, hence for a pion mass $> 350MeV$). The goal of this work is to combine LQCD calculations with accurate phenomenological models in order to interpret and eventually guide LQCD results. This is the Trento-MIT programme and more details will be presented by Cristoforetti [44]. Knowing how LQCD observables depend on the quark mass, on can perform extrapolations. There are two regimes, the chiral one, valid for low pion mass, in which the dependence on quark mass is determined by the chiral Perturbation Theory (χPT) and the quark model one, characterized by pion masses larger than the ρ meson. The transition between the chiral and quark regime is studied with the Interacting Instanton Liquid Model (IILM) [9]. This choice is motivated by the fact that instantons appear to be the dynamical mechanism responsible for the chiral symmetry breaking, the masses and the electroweak structure of nucleon and pion are correctly reproduced and there is only one phenomenological parameter, the instanton size (already known). The transition scale is related to the eigenvalue spectrum of the Dirac operator in an Instanton background; the quasi-zero mode spectrum is peaked at $m^* \approx 80 MeV$ and for $m_q < m^*$, chiral effects dominate. Now, χPT predicts that the three point correlator is a constant as a function of the quark mass, but it can be calculated independently with IILM. With IILM one can calculate also the nucleon mass for different values of the pion mass. The results agree with the lattice calculations by CP-PACS [45] if the instanton size is $0.32 fm$, as it is shown in Fig. 5 [43]. IILM is then able to reproduce results in the chiral and quark regime.

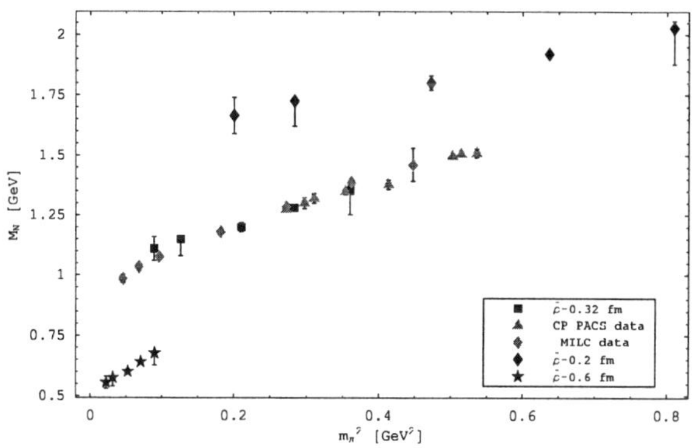

Fig. 5. Comparison [43] between nucleon masses obtained with the IILM for three different choices of the average instanton size $\bar{\rho}$ and the Lattice calculations by the CP-PACS collaboration [45].

3. Inclusive and semiinclusive reactions

3.1. *Nucleon structure functions*

From the recent Hall C high Q^2 experiments at Jlab, a systematic set of high quality data on the proton and the deuteron structure functions has been obtained [46]. From these data, one can extract the CN and Nachtmann moments of the nucleon structure function F_2 . The Nachtmann moments allow a cleaner extraction of dynamical higher twists related to multiparton correlations, since they are free from target mass corrections. One can write

$$M_n^{Nacht}(Q^2) = \mu_n(Q^2) + \frac{\alpha_n^{(4)}}{Q^2}[\alpha_s(Q^2)]^{\gamma_n^{(4)}} + \frac{\alpha_n^{(6)}}{Q^4}[\alpha_s(Q^2)]^{\gamma_n^{(6)}} + ... \quad (3)$$

where $\mu_n(Q^2)$ is the leading twist and the α and γ parameters are free and describe the effective strength and the anomalous dimensions of the

Higher Twist (HT). In this way one can separate the contributions of the HT up to the sixth order to the Nachtamann moments of the proton and the deuteron for $n = 2, 4, 6, 8$. In order to extract the neutron moments and perform a HT analysis, one has to take into acccount the nuclear effects in the deuteron at moderate and large $x (x > 0.1)$. The results for the neutron leading twists [47] are shown in Fig. 6.

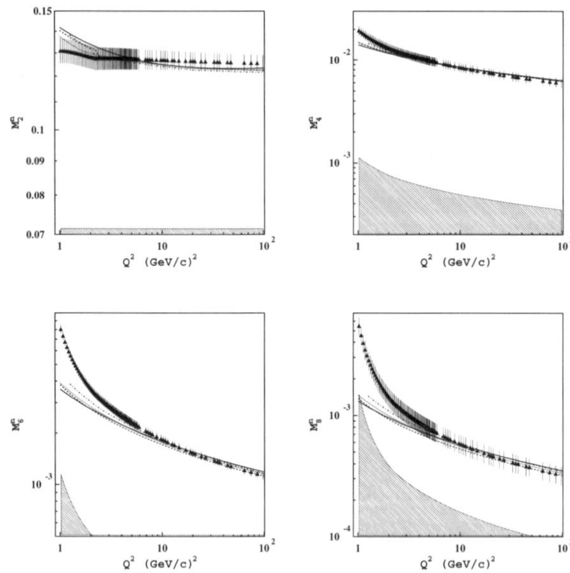

Fig. 6. The extracted moments of the neutron structure function for n = 2,4,6, and 8 as a function of Q^2 [47] in comparison to the moments of different parton distribution functions: GRV [48] (solid), Alekhin [49] (dashed), CTEQ [50] (dotted), MRST [51] (dot-dashed).

3.2. *Generalized Parton Distributions*

The GPDs can be used to describe various processes, such as the virtual photoproduction of mesons or the double virtual Compton scattering (DVCC). For high photon momentum Q^2 and moderate momentum transfer to the target, in the amplitudes of these reactions one can separate the hard process involving the virtual photon and the production mechanism from the soft scattering off the target. The latter can be described in terms of Generalized Parton Distributions (GPDs) [52], defined as follows:

$$GPDs \sim < P', S'| \int \frac{dz^-}{4\pi} \bar{q}(-\frac{z}{2}) \Gamma q(\frac{z}{2}) e^{ixP^+z^-} |P, S > \qquad (4)$$

in the above equation different types of GPDs (unpolarized, helicity and transverse) appear according to the polarizion conditions and the corresponding Dirac operator Γ. Of course, $P, S(P', S')$ are the initial (final) momentum and spin of the nucleon. If $P' \neq P$, the GPDs depend, besides the $t-$channel momentum transfer squared $t = (P - P')^2 = \Delta^2$, on two momentum fractions

$$x_{av} = \frac{(k + k')^+}{(P + P')^+}; \xi = \frac{(P - P')^+}{(P + P')^+} \qquad (5)$$

where x_{av} is the average fraction of the longitudinal momentum carried by the partons and ξ is the skewness parameter, describing the fraction of the longitudinal momentum transfer. In relation with the values of x_{av}, one can distinguish two separate regions, one for $|\xi| < |x_{av}| < 1$ (Dokschitzer-Gribov-Lipatov-Altarelli-Parisi, DGLAP [53]) and one for $|x_{av}| < |\xi|$ (Efremov-Radyushkin-Brodsky-Lepage, ERLB [54]). In the latter case, according to the parton interpretation of the process, a quark-antiquark pair is involved.

The Pavia-Trento collaboration has performed a systematic study of GPDs both in the unpolarized [55] and polarized case [56], using light cone wave functions and two different CQMs, the GBE [6] and the hypercentral one [5], the latter with relativistic kinetic energy but without hyperfine interaction [57]. Some results for the unpolarized case are reported in Fig. 7. Similar curves have been obtained for the longitudinally polarized GPDs. In the forward limit the GPDs of the H type converge to the usual unpolarized, helicity and transversity distributions (f_1, g_1, h_1, respectively), therefore the theoretical GPDs can be used also for the calculation of the measured asymmetries, after having performed a NLO evolution up to $Q^2 = 3GeV^2$ [56]; however the description is up to now only qualitative. The Pavia group has also performed a study of the chiral-odd GPDs [58], using an overlap representation, relativistic wave functions in the instant form and the hCQM in their relativized version without OGE. More details can be found in the talk by Pincetti [59].

One should also mention some recent work performed by the Perugia-Valencia collaboration. Starting from the assumption that quarks are complex systems containing partons of any type, a convolution of the quark GPD with the wave functions of the IK model is performed. This approach respects three important conditions, the forward, integral of $\tilde{H}$ and polynomial ones [60]. A further calculation of the H_T GPD has been done using simply the MIT bag model [61].

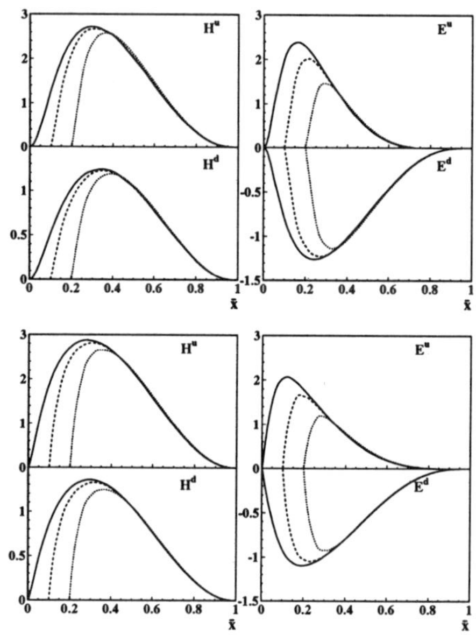

Fig. 7. The unpolarized GPDs H and E for u and d quarks at fixed $t = -0.5 GeV^2$ and various ξ values: 0 (solid), 0.1 (dashed) and 0.2 (dotted) [55], calculated with the GBE (left) and hCQM (right) models. In the latter case the relativistic kinetic energy is used but no hyperfine interaction is included.

A related work is given by the analysis performed recently by Bacchetta and Radici [62], aimed at the extraction of the transversity function h_1 in reactions with double production of hadrons starting from lepton scattering on polarized protons. From the observed off-plane asymmetry, one can obtain the so called dihadron fragmentation function which arises from the interference of the hadron pair produced either in $s-$ or $p-$ wave. The feasibility of this procedure has been shown by analyzing the double pair production in the Belle experiment [63] or in pp collisions at RICH-II [62]. the problem has been further analyzed with a direct calculation with a spectator model and a comparison with data from HERMES and COMPASS.

3.3. *Drell-Yan processes*

The lepton production by two polarized hadrons (polarized Drell-Yan) can also be used in order to investigate the nucleon transversity function. The spin asymmetries with transversely polarized hadrons, which have been first measured at BNL in 1976, are an interesting puzzle, since at high energies they reach values of about 40% and QCD is not able to explain them. The differential cross section for the polarized Drell-Yan production contains a series of terms linear in the structure functions of interest (transversity, Sivers function, Boer-Mulders function,...), weighted by some factors depending on the relative angle between the reaction and the production planes. Choosing accurately the kinemtical region of observation one can extract, at least in principle, the wanted structure function. In a series of papers [64], Bianconi and Radici have made a systematic study of various Drell-Yan processes performing a Monte-Carlo simulation and analyzing the measurability of the variouis quantities. The chosen reactions and kinematics are compatible with realistic experimental conditions. Their study regards in particular the muon pair production in different channels: polarized proton-antiproton (to be measured at PAX/ASSIA (GSI)), two protons (RICH-II) or pion-nucleon (COMPASS), leading to the extraction of the various strucure functions. Their results for the Sivers function are shown in Fig. 8, which illustrates the problem of its change of sign, predicted by general properties.

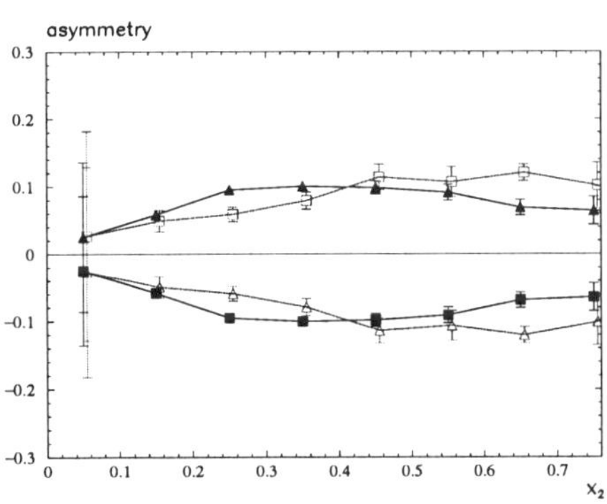

Fig. 8. Asymmetry for the Sivers effect in the Drell-Yan process [64].

The problem of extracting the Sivers function has also been addressed by studying the asymmetry in the proton-antiproton Drell-Yan process in the PAX kinematics; the dominant mechanism is $q\bar{q} \to c\bar{c}$ and Chiral models with vector mesons as gauge bosons can be used. [65]. The result is a direct access to the Sivers function, with the possibility of testing the relation $(Sivers)_u = -(Sivers)_d$, predicted by chiral models at leading order in $1/Nc$.

To conclude this section, it should be mentioned a simple result obtained recently [66]. Starting from a general parametrization of the correlator entering in the cross section (in particular the twist 2 T-even component) and making a comparison with the density matrix of a confined quark (which is interaction free but carries transverse momentum) one can get simple relations between structure functions and quark densities. In particular, the asymmetry for the pion-nucleon production of a muon pair turns out to be proportional to $1/Q^2$. This dependence can be directly tested and it is in agreement with data [66].

4. Quark-antiquark and/or meson cloud effects

The discussion on the missing strength in the electromagnetic excitation of the baryon resonances indicates that some fundamental mechanism is lacking in the dynamical description of the nucleon structure. This mechanism can be identified with the production of a quark-antiquark pair [37,67]. A strong interest on the possible influence of such quark-antiquark and/or meson cloud effects is emerging in the literature. The problem is how to include this feature into the description of hadrons. A possibility is to introduce higher Fock components in the hadron wave function. This means that, at the hadron scale, the baryon acquires a $4q - \bar{q}$ configuration

$$|\Psi> = \Psi_{qqq}|qqq> + \Psi_{3q\bar{q}q}|qqq\bar{q}q> \qquad (6)$$

This idea can be tested in the meson sector where states which are non $q\bar{q}$ candidates have been observed [11]. The spectroscopy of exotic $(q\bar{q}q\bar{q})$ meson states have been intensively studied by the Genoa group [68]. The states have been completely classified by means of the the group $O(3) \otimes SU_{sf}(6) \otimes SU_c(3)$ and explicitly constructed; moreover a mass formula have been obtained which gives predictions for the scalar nonets in agreement with the KLOE results. More details will be given in the talk by Galatà [69]. A similar test in the case of baryons is presently not feasible, since the observation of a pentaquark $(4q\bar{q})$ state has not been confirmed by high statistics remakes of previous experiments [11]. However, from the theoretical point of view, there is no reason against the existence of such exotic states.

The consistent inclusion of higher Fock components in the baryon wave function is a hard task, but, as a preliminary approach, one can introduce models in which the physical nucleon N is considered to be made of a bare nucleon dressed by a surrounding meson cloud

$$|N> \; = \Psi^N_{3q}|N(qqq)> + \sum_{B,M} \Psi^{BM}_{(3q)(q\bar{q})}|B(qqq)M(q\bar{q})> + ... \quad (7)$$

A model of this type have been used by the Pavia group for the calculation of GPDs [70]. The virtual photon can scatter either on a valence quark of the bare nucleon (N) (see top of Fig. 9) or on one of the constituents in the higher Fock state component (BM) (see bottom of Fig. 9). The soft scattering part is evaluated in the one meson approximation, using a light cone hamiltonian with hypercentral interaction [5] in the relativistic form [57] for the three quark state and a direct meson-baryon coupling. The GPDs are calculated by means of the convolution formalism; according to the interpretation quoted in Sec. 4.2, the process in the ERLB region (i.e. for $|\xi| \leq 1$), describes the emission of a quark-antiquark pair. The results for the isoscalar and isovector unpolarized and helicity GPDs are shown in Fig. 10.

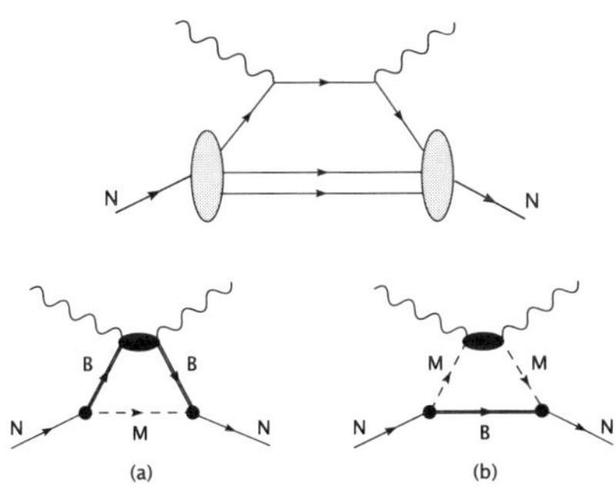

Fig. 9. Deeply virtual Compton scattering from the bare nucleon (up), from the virtual baryon (down part (a)) and meson (down part (b)) components of a dressed nucleon.

A similar approach has been followed using the hCQM [71]: to the model three quark state for the nucleon a pion cloud has been added by means of the meson-baryon coupling taken from [72]. The preliminary results show that the higher Fock components give relevant contributions both to the elastic from factors of the nucleon and to helicity amplitudes [71].

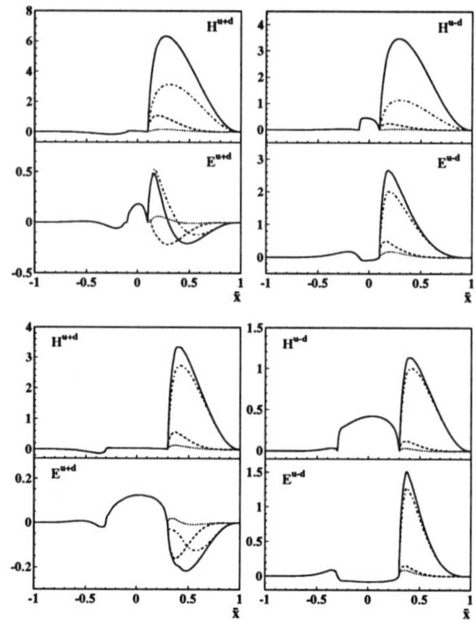

Fig. 10. Isoscalar $(u+d)$ and isovector $(u-d)$ of the averaged $H-$ and $E-$type GPDs calculated in the meson-cloud model [70] for $t = -0.5\ GeV^2$ and $\xi = 0.1$ (upper part) or $\xi = 0.3$ (lower part).

The contributions of the higher Fock components to the pion form factor has been recently calculated by the Rome group [73]. The photon vertex is dressed by means of a covariant vector meson dominance and a direct pion absorption by quarks is considered. Among the various terms contributing to the electromagnetic amplitude, the so called instantaneous terms seem to be of relevant importance. An important aspect of the approach is that a unified description of both time- and space-like form factors can be obtained. The theoretical results [73], in comparison to the experimental data, are shown on Fig. 11. The agreement is impressive. A similar calculation can be done also for the nucleon form factors [74]. This will be the topic of the talk by Pisano [75].

A consistent inclusion of quark-antiquark pair effects in CQMs is however not yet available. A promising approach is provided by a recent work aiming at unquenching the quark model [76]. The $q\bar{q}$ creation mechanism is introduced at the microscopic level within a stringlike approach. For the construction of the formalism, based on group theory, various problems have been solved. First of all the sum over all big towers of intermediate

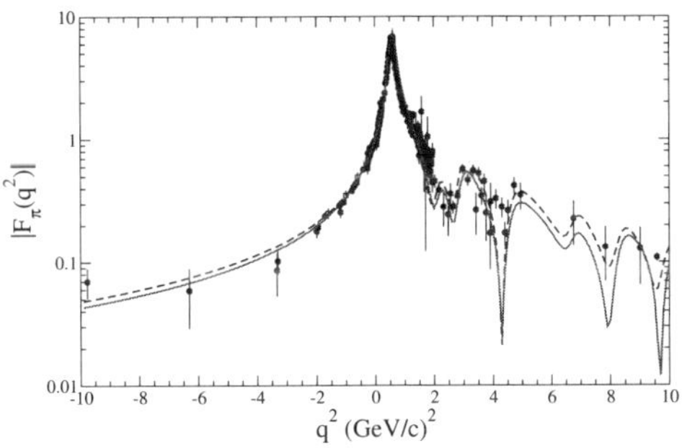

Fig. 11. The pion form factor in the time- and space-like regions (details and data in ref. [73].

states has been performed thanks to group theoretic methods, then the permutational symmetry for all identical quarks has been correctly taken into account and finally the determination of the pair creation vertex has been obtained. More details and some preliminary results will be given in the talk by Santopinto [77].

Acknowledgements

The author is indebted with the colleagues working in the Universities and/or INFN Sections of Ferrara, Genova, Roma1-2-3, Pavia, Perugia, Trento for supplying their results and sending materials useful for the preparation of this talk.

References

1. G. Morpurgo, 1965 Physics **2**, 95 (1965).
2. N. Isgur and G. Karl, Phys. Rev. D **18**, 4187 (1978); Phys. Rev. D **19**, 2653 (1979).
3. S. Capstick and N. Isgur, Phys. Rev. D **34**,2809 (1986).
4. R. Bijker, F. Iachello, and A. Leviatan, Ann. Phys. (N.Y.) **236**, 69 (1994).
5. M. Ferraris, M.M. Giannini, M. Pizzo, E. Santopinto, and L. Tiator, Phys. Lett. **B364**, 231 (1995).
6. L. Ya. Glozman and D.O. Riska, Phys. Rep. **268**, 263 (1996); L. Ya. Glozman, W. Plessas, K. Varga, and R.F. Wagenbrunn, Phys. Rev. D **58**, 094030 (1998).
7. U. Löring, K. Kretzschmar, B. Ch. Metsch, and H. R. Petry, Eur. Phys. J. **A10**, 309 (2001); U. Löring, B.Ch. Metsch, and H. R. Petry, Eur. Phys. J. **A10**, 395 (2001); *ibid.*, 447 (2001).
8. A. W. Thomas and W. Weise, The structure of the nucleon, Wiley-VCH (Berlin 2001).

9. T. Schaefer and E. Shuryak, Phys. Rev. D **53**, 6522 (1996); Rev. Mod. Phys. **70**, 323 (1998); D. Diakonov, Prog. Part. Nucl. Phys. **51**, 173 (2003).
10. A. De Rújula, H. Georgi, and S.L. Glashow, Phys. Rev. D **12**, 147 (1975).
11. W.-M. Yao *et al.* (Particle Data Group), J. Phys. **G33**, 1 (2006).
12. Gunnar S. Bali *et al.*, Phys. Rev. D **62**, 054503 (2000); Gunnar S. Bali, Phys. Rep. **343**, 1 (2001).
13. C. Alexandrou, P. de Forcrand, and O. Jahn, Nucl. Phys. Proc. Suppl. **119**, 667 (2003).
14. H. Suganuma, T.T. Takahashi, F. Okiharu, and H. Ichie, Nucl. Phys. Proc. Suppl. **141**, 92 (2000) (hep-lat/0407014).
15. M.M. Giannini, E. Santopinto, and A. Vassallo, Eur. Phys. J. **A12**, 447 (2001).
16. M.M. Giannini, E. Santopinto, and A. Vassallo, to be published.
17. E. Santopinto, Phys. Rev. C **72**, 022201 (2005).
18. M. K. Jones *et al.* [Jefferson Lab Hall A Collaboration], Phys. Rev. Lett. **84**, 1398 (2000); O. Gayou *et al.* [Jefferson Lab Hall A Collaboration], Phys. Rev. C **64**, 038202 (2001); Phys. Rev. Lett. **88**, 092301 (2002).
19. P.A.M. Dirac, Rev. Mod. Phys. **21**, 392 (1949).
20. G. Salmé, Invited Talk at this Conference.
21. E. Pace, G. Salmé, F. Cardarelli, and S. Simula, Nucl. Phys. **A666**, 33 (2000).
22. F. Cardarelli, E. Pace, G. Salmé, and S. Simula, Phys. Lett. **B357**, 267 (1995).
23. S. Boffi, L. Y. Glozman, W. Klink, W. Plessas, M. Radici, and R. F. Wagenbrunn, Eur. Phys. J. **A14**, 17 (2002).
24. G. Höhler, E. Pietarinen, I. Sabba Stefanescu, F. Borkowski, G. G. Simon, V. H. Walther and R. D. Wendling, Nucl. Phys. **B114**, 505 (1976); R. C. Walker *et al.*, Phys. Rev. D **49**, 5671 (1994); L. Andivahis *et al.*, Phys. Rev. D **50**, 5491 (1994); W. Bartel *et al.*, Nucl. Phys. **B58**, 429 (1973); A. F. Sill *et al.*, Phys. Rev. D **48**, 29 (1993).
25. M. De Sanctis, E. Santopinto, M. M. Giannini, and A. Vassallo, e-Print Archive:nucl-th/0506033.
26. B. D. Milbrath *et al.* [Bates FPP collaboration], Phys. Rev. Lett. **80**, 452 (1998) [*Erratum-ibid.* **82**, 2221 (1999)]; T. Pospischil *et al.* [A1 Collaboration], Eur. Phys. J. **A12**, 125 (2001); V. Punjabi *et al.*, Phys. Rev. C **71**, 055202 (2005); [*Erratum-ibid.* 069902 (2005)].
27. T. Melde, Contribution to this Confernce.
28. M. De Sanctis, E. Santopinto, and M. M. Giannini, Eur. Phys. J. **A1**, 187 (1998).
29. M. De Sanctis, M. M. Giannini, L. Repetto, and E. Santopinto, Phys. Rev. C **62**, 025208 (2000).
30. E. Tomasi-Gustafsson *et al.*, Eur. Phys. J. **A24**, 419 (2005).
31. A. Antonelli *et al.* [FENICE], Nucl. Phys. **B517**, 3 (1998).
32. B. Aubert *et al.* [BaBar], Phys. Rev. D **73**, 012005 (2006).
33. F. Ambrosino, M. Radici *et al.* , e-Print Archive:hep-ex/0603056.
34. A. Bianconi, B. Pasquini, and M. Radici, Phys. Rev. D **74**, 074012 (2006); hep-ph/0607277.

35. R. Koniuk and N. Isgur, Phys. Rev. D **21**, 1868 (1980).
36. L. A. Copley, G. Karl, and E. Obryk, Phys. Lett. **29**, 117 (1969).
37. M. Aiello, M. M. Giannini, and E. Santopinto, J. Phys. G: Nucl. Part. Phys. **24**, 753 (1998).
38. M. M. Giannini and E. Santopinto, to be published.
39. V. Burkert, in "Baryons 2002 (C.E. Carlson and B. A. Mecking eds.), World Scientific (Singapore), p. 29 (2003).
40. L. Tiator, D. Drechsel, S. Kamalov, M.M. Giannini, E. Sentopinto, and A. Vassallo, Eur. Phys. J. **A19** 55 (2004).
41. D. Drechsel, O. Hanstein, S.S. Kamalov, and L. Tiator, Nucl. Phys. **A645**, 145 (1999).
42. S. Kamalov, S.N. Yang, D. Drechsel, O. Hanstein, and L. Tiator, Phys. Rev. C **64**, 032201 (2001).
43. M. Cristoforetti, P. Faccioli, M. C. Traini, and J. W. Negele, e-Print arXiv:hep-ph/0605256.
44. M. Cristoforetti, Contribution to this Conference.
45. A. A. Khan *et al.*, Nucl. Phys. **B689**, 175 (2004).
46. M. Osipenko *et al.*, Phys. Rev. D **67**, 092001 (2003); Phys. Rev. C **73**, 045205 (2006).
47. M. Osipenko, W. Melnitchouk, S. Simula, S. Kulagin, and G. Ricco, Nucl. Phys. **A766**, 142 (2006).
48. M. Glück, E. Reya, and A. Vogt, Eur. Phys. J. **C5**, 461 (1998).
49. S. Alekhin, Phys. Rev. D **68**, 014002 (2003).
50. J. Pumplin *et al.*, JHEP **07**, 012 (2002).
51. A. D. Martin, R.G. Roberts, W. J. Stirling, and R. S. Thorne, Eur. Phys. J. **C39**, 155 (2005).
52. D. Müller *et al.*, Fortschr. Phys. **42**, 101, (1994); A. V. Radyushkin, Phys. Lett. **B380**, 417 (1996); *ibid.* **B385**, 333 (1996); Phys. Rev. D **56**, 5524 (1997); X. Ji, Phys. Rev. Lett. **78**, 610 (1997); M. Diehl, Eur. Phys. J. **C19**, 485 (2001).
53. V. N. Gribov and L. N. Lipatov, Sov. J. Nucl. Phys. **15**, 438 (1972); L. N. Lipatov, Sov. J. Nucl. Phys. **20**, 94 (1975); G. Altarelli and G. Parisi, Nucl. Phys. **B126**, 298 (1977); Y. L. Dokshitzer, Sov. Phys. JETP **46**, 641 (1977).
54. G. P. Lepage and S. J. Brodsky, Phys. Lett. **B87**, 359 (1979); A. V. Efremov and A. V. Radyushkin, Phys. Lett. **B94**, 245 (1980).
55. S. Boffi, B. Pasquini, and M. Traini, Nucl. Phys. **B649**, 243 (2003).
56. S. Boffi, B. Pasquini, and M. Traini, Nucl. Phys. **B680**, 147 (2004).
57. P. Faccioli, M. Traini, and V. Vento, Nucl. Phys. **A656**, 400 (1999).
58. B. Pasquini, M. Pincetti, and S. Boffi, Phys. Rev. D **72**, 094029 (2005).
59. M. Pincetti, Contribution to this Conference.
60. S. Scopetta and V. Vento, Phys. Rev. D **71**, 014014 (2005).
61. S. Scopetta, Phys. Rev. D **72**, 117502 (2005).
62. A. Bacchetta and M. Radici, Phys. Rev. D **67**, 094002 (2003); *ibid.* **69**, 074026 (2004); *ibid.* **70**, 094032 (2004); hep ph0608037.
63. D. Boer, R. Jakob, and M. Radici Phys. Rev. D **67**, 094003 (2003).
64. A. Bianconi and M. Radici, Phys. Rev. D **71**, 074014 (2005); J. Phys. **G31**, 645 (2005); Phys. Rev. D **72**, 074013 (2005); Phys. Rev. D **73**, 034018 (2006);

ibid., 114002 (2006).
65. A. Drago, Phys. Rev. D **71**, 0501282 (2005).
66. E. Di Salvo, hep-ph/0607184 .
67. M. Aiello, M. Ferraris, M.M. Giannini, M. Pizzo, and E. Santopinto, Phys. Lett. **B387**, 215 (1996).
68. E. Santopinto and G. Galatà. e-Print Archive: hep-ph/0612042 and hep-ph/0612044 .
69. G. Galatà, Contribution to this Conference.
70. B. Pasquini and S. Boffi, Phys. Rev. D **73**, 094029 (2006).
71. D. Y. Chen, Y. B. Dong, M. Giannini, and E. Santopinto, 5th International Conference on Perspectives in Hadronic Physics: Particle-Nucleus and Nucleus-Nucleus Scattering at Relativistic Energies, Trieste (Italy), May 22-26, 2006 (nucl-th/0611016).
72. S. Theberge and A. W. Thomas, Nucl. Phys. **A393**, 252 (1983).
73. J. P. B. C. De Melo, T. E. Frederico, E. Pace, and G. Salmé, Phys. Rev. D **73**, 074013 (2006).
74. J. P. B. C. De Melo, T. E. Frederico, E. Pace, S. Pisano, and G. Salmé, hep-ph/0607342.
75. S. Pisano, Contribution to this Conference.
76. E. Santopinto and R. Bijker, to be published.
77. E. Santopinto, Contribution to this Conference.

INSTANTONS AND THE NON-PERTURBATIVE QUARK DYNAMICS IN THE CHIRAL REGIME

M. CRISTOFORETTI, P. FACCIOLI and M.C. TRAINI

Dipartimento di Fisica and I.N.F.N., Università degli Studi di Trento, Via Sommarive 15, Povo (Trento) 38050 Italy.
European Centre for Theoretical Studies in Nuclear Physics and Related Areas, Strada delle Tabarelle 286, Villazzano (Trento), I-38050 Italy.

J.W. NEGELE

Center for Theoretical Physics Massachusetts Institute of Technology, NE25-4079, 77 Massachusetts Ave, Cambridge, MA 02139-4307, USA.

We use th Interacting Instanton Liquid Model to investigate the non-perturbative interaction associated to spontaneous chiral symmetry breaking. In particular, we study how hadrons' structure changes as a function of the quark mass. To support the validity of this model in the chiral regime, the quark mass dependencies of several properties are shown to agree with chiral perturbation theory, including the density of eigenmodes of the Dirac operator and the masses of the pion and nucleon. We identify a natural mass scale $m^* = 80$ MeV which sets the boundary of the mass regime where the non-perturbative QCD dynamics is determined by chiral effects.

Keywords: Instantons, Lattice QCD, Chiral extrapolation

1. Introduction

Although Lattice QCD provides the only available ab-initio tool for solving QCD non-perturbatively, such a framework does not allow a direct identification of the dynamical mechanism underlying hadron structure and of the relevant degrees of freedom at different scales. In this talk we present our recent attempt to extract complementary insight using the Interacting Instanton Liquid Model (IILM) [1].

In the IILM, the QCD path integral over all gluons is replaced by an effective theory in which instantons are the effective degrees of freedom, and the gauge fields of the theory are those generated by integrating over the positions, color orientations, and sizes of instantons. The QCD partition function reduced to:

$$\mathcal{Z}_{QCD} \simeq \mathcal{Z}_{ILM} = \sum_{N_+,N_-} \frac{1}{N_+!N_-!} \int \prod_i^{N_++N_-} \mathrm{d}\Omega_i d(\rho_i) e^{-S_{int}} \prod_i^{N_f} \det(i\not{D}+im_f).$$

Here, $\mathrm{d}\Omega_i = \mathrm{d}U_i \mathrm{d}^4 z_i \mathrm{d}\rho_i$ is the measure in the space of collective coordinates, color orientation, position and size, associated with the single instantons. Quantum fluctuations are included in Gaussian approximation, through the semi-classical instanton amplitude $d(\rho_i)$. In the formulation of the model we have considered, the only phenomenological parameter which has been introduced is the strenght of the instanton-instanton repulsion.

In the context of this model, we would like to understand the mass range in which instanton mediated chiral dynamics is manifested and the extent to which it is described by chiral perturbation theory. In particular, we would like to know the mass scale or scales at which the continuum fermion determinant is suppressed, at which zero modes become subdominant, and at which the density of quasi zero modes becomes independent of quark mass.

2. The chiral regime in the IILM

In order to investigate hadron structure in the chiral regime of QCD we check that IILM gives predictions in agreement with χpt. To test the IILM in the chiral regime we have analyzed the spectrum of the Dirac operator. In χpt there are clear statement on the distribution of eigenvalues of the Dirac operator $\rho(\lambda)$ depending on the mass of current quarks. When $m_q = 0$ from the Banks-Casher relation we have that $\rho(\lambda)$ have to be flat near the origin [2]

$$\lim_{\lambda \to 0} \lim_{m_q \to 0} \rho(\lambda) = \mathrm{Const} + O(\lambda^2) \qquad (N_f = 2), \tag{1}$$

In Fig. 1 we show the Dirac spectrum computed in the IILM for $m_q = 0$ where clearly appear that when λ goes to 0 the spectrum becomes flat.

When the quark mass is small but finite this flatness disappears and χpt predicts, for $N_f = 2$, a peak near the origin of the form $\rho(\lambda) - \rho(0) \propto m^\alpha \lambda^{1-\alpha}$. In particular the behavior of the Dirac spectrum in the small quark regime can be studied considering the 3-point scalar correlator [2]:

$$K^{abc} = \int \mathrm{d}^4x \mathrm{d}^4y \mathrm{d}^4z \langle 0|S^a(x) S^b(y) S^c(z)|0\rangle$$
$$S^a(x) = \bar{q}\tau^a q(x).$$

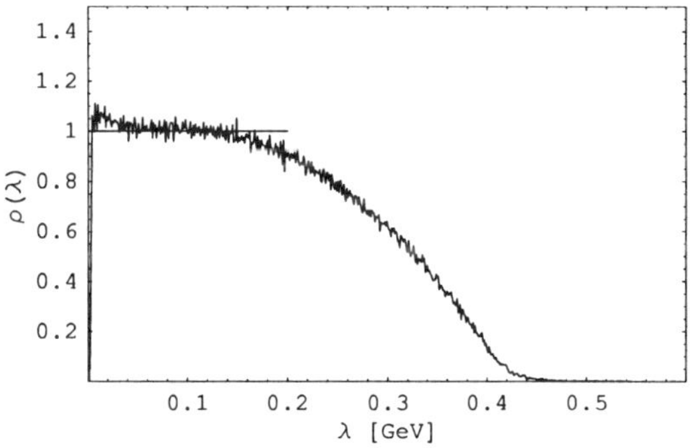

Fig. 1. Spectral density of the Dirac operator for $N_c = 3$, $N_f = 2$ and $m_q = 0$.

in particular χpt predicts that:

$$m_q \lim_{m_q \to 0} \frac{K^{abc}}{\rho_{m_q=0}(0)} = \frac{B(N_f^2 - 12)}{128\pi f_0^2 N_f} = \text{const.}$$

The same combination can be studied from the point of view of the spectral representation where we find

$$m_q \frac{K^{abc}}{\rho_{m_q=0}(0)} = -m_q^2 \int \mathrm{d}\lambda \frac{\rho_{m_q}(\lambda)}{\rho_{m_q=0}(0)} \frac{(m_q^2 - 3\lambda^2)}{(m_q^2 + \lambda^2)^3} d^{abc},$$

which can be computed directly in the IILM.

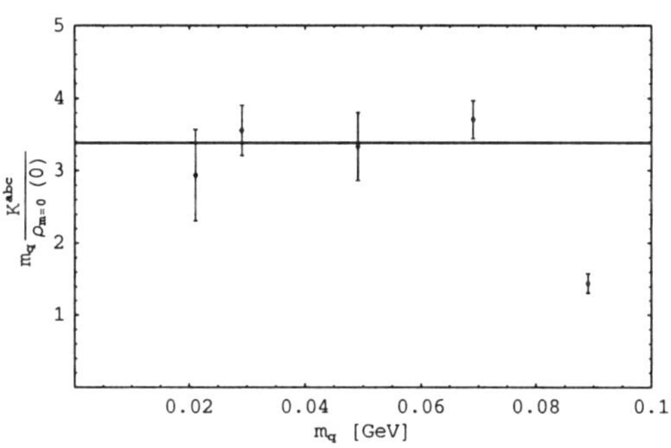

Fig. 2. $m_q \frac{K^{abc}}{\rho_{m=0}(0)}$ at different quark mass values in the IILM.

The results are plotted in Fig. 2, and we observe that the expected constant behavior with m is obtained for $m_q < 80$ MeV. Thus, in this case, the IILM reproduces the structure of the Dirac spectrum dictated by chiral symmetry.

3. Pion and nucleon mass in the IILM

The second step trying to understand the instantons contribution to the low-energy sector of QCD is verifying that the IILM effective theory can be used to calculate hadron masses and that these masses have the dependence on the quark mass expected from χpt. We have considered pion and nucleon because the mass of this two hadron depends to the first order from the instanton density. To calculate in the IILM the hadron masses we have performed the effective-mass plot analysis, which is routinely used to extract lowest-lying hadron masses from lattice simulations.

To compute the mass of pion and nucleon we have evaluated the correlation function:

$$G_\pi(\tau) = \int \mathrm{d}^3\underline{x}\langle 0|T[j_5(\underline{x},\tau)\bar{j}_5(\underline{0},0)|0\rangle,$$

$$G_N(\tau) = \int \mathrm{d}^3\underline{x}\langle 0|T[j_N(\underline{x},\tau)\bar{j}_N(\underline{0},0)P_+|0\rangle,$$

where the two currents are given by

$$j_N^a(x) = \varepsilon^{abc}\, u^a(x)C\gamma_5 d^b(x)u^c(x)$$
$$j_5^a(x) = \overline{q}(x)\gamma_5\tau^a q(x)$$

and $P_+ = \frac{1-\gamma_4}{2}$ is the positive-parity projector.

The mass of these hadrons can then be extracted from the plateau in the large Euclidean time limit of the effective mass, i.e. using

$$M_{\pi/N} = \lim_{\tau\to\infty} M^{eff}_{\pi/N}(\tau)$$
$$M^{eff}_{\pi/N}(\tau) = \frac{1}{\Delta}\ln\frac{G_{\pi/N}(\tau)}{G_{\pi/N}}(\tau+\Delta\tau).$$

We have computed the correlation function for five different values of the quark mass in the range $20 \lesssim m_q \lesssim 90$ MeV, finding the results reported in Table 1. We have found that all the pion effective mass plots display a very clean plateau, from which it is possible to unambiguously read-off the pion mass from a correlated chi-square fit. In the interpolation of the effective mass plots of the nucleon not only statistical but also systematic error has to be taken into account.

We have found that the pion data not only have a quark mass dependence consistent with the form expected from chiral perturbation theory, but we show that they yield prediction for the low energy constant f_0 and

Table 1. Pion and nucleon masses fitted by effective mass plot with $\chi^2/ndf \leq 1$. The quark masses are determined at a scale 2 GeV as in [3,4].

m_q [GeV]	Pion [GeV]	Nucleon [GeV]
0.021	0.300 ± 0.004	$1.11^{+0.05}_{-0.05}$
0.03	0.360 ± 0.004	$1.15^{+0.01}_{-0.07}$
0.05	0.460 ± 0.004	$1.20^{+0.02}_{-0.02}$
0.07	0.530 ± 0.004	$1.28^{+0.01}_{-0.01}$
0.09	0.600 ± 0.004	$1.35^{+0.02}_{-0.1}$

the chiral condensate consistent with those known from QCD:

$$f_0 = 0.085 \pm 0.003 \text{ GeV}$$
$$\langle \overline{q} q \rangle \approx -(0.259 \text{ GeV})^3 .$$

In order to assess the accuracy of the IILM prediction for the nucleon at different quark masses, we have verified that these are consistent with the corresponding lattice QCD calculation [5,6] (see Fig. 3). Note that the fit of the lattice data with extrapolation formulas based on χpt yields sensible low energy constants. Hence, since the IILM data are consistent with lattice data, we can conclude that the IILM data is also consistent with such low-energy constant. This implies that the Baryon chiral perturbation theory of the IILM is not very different from that of QCD.

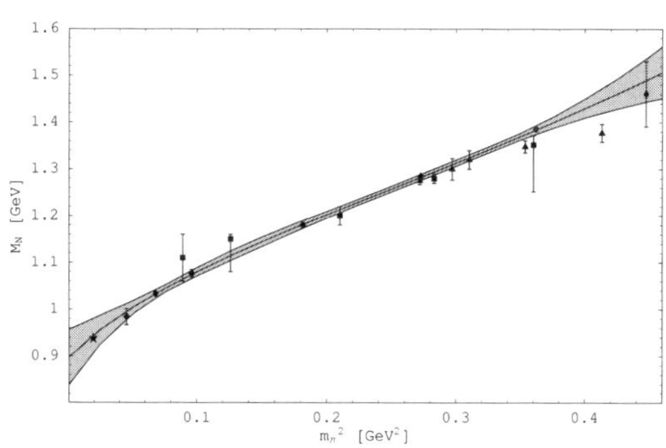

Fig. 3. Nucleon mass chiral extrapolation at $\mathcal{O}(p^4)$ in relativistic Bχpt. The error band are determined according to the method outlined in [7]. The squared points are the IILM data, the triangle are CP-PACS and the diamonds are MILC data. The star represents the physical point.

4. Instanton-induced dynamics in the chiral regime

Having carefully verified the consistency of the IILM results with lattice results and with the behavior expected from χpt, we conclude that the low energy effective theory indeed gives a useful approximation to the QCD path integral and hence use this model to explore the role of instanton induced dynamics. An essential quantity in the IILM is the overlap matrix $T_{ij} = \int d^4z \psi_I^{0\dagger}(z) i D_\mu \gamma_\mu \psi_j^0(z)$ specifing the probability that a quark hops from instanton i to instanton j, this matrix plays the main role both in the evaluation of the fermionic determinant and of the quark propagator. In the IILM these two quantities can be factorized in a zero mode part and a non-zero mode part

$$\mathrm{Det}_A(\not{D} + m_q) = \mathrm{Det}_{A\,zm} \times \mathrm{Det}'_{A\,nzm},$$
$$S(x,y)_A = S_{zm}(x,y)_A + S_{nzm}(x,y)_A \ ,$$

where

$$\mathrm{Det}_{A\,zm} = \mathrm{Det}(T + m_q),$$
$$S_{zm}(x,y)_A = \sum_{IJ} \psi_I^0(x) \psi_J^{0\dagger}(y) \left[\frac{1}{\hat{T} - i m_q}\right]_{IJ}.$$

The non-zero mode part of the determinant is approximated with the product of the non-zero mode contributions of each individual instanton:

$$\mathrm{Det}'_{A\,zm} = \prod_i^{N_+ + N_-} (1.34 m_q \rho_i) \tag{2}$$

while the non-zero part of the propagator results subleading in the low-energy regime. From these consideration it follows immediately that in order to understand the instanton-induced dynamics a carefully analysis of the dependence of the T_{ij} from the quark mass is needed. In Fig. 4 we have plotted the distributions of the *maximum* overlap matrix elements T_{IA} obtained at different values of the quark mass. We note that such distributions are peaked around the value $m^\star \simeq$ 80 MeV. The value $m^\star$ plays a central role in specifying the scale for instanton induced chiral symmetry breaking in the IILM. If we see the zero mode part of the propagator and the fermionic determinant, we understand that the scale $m^\star$ determines the regime where the quenched approximation is expected to become reliable. For quark masses $m_q \gg \max[T_{IJ}] \simeq m^\star$ the contribution of the overlap matrix elements to the propagator and the fermionic determinant becomes negligible. On the other hand, in this model, the mass contribution to the

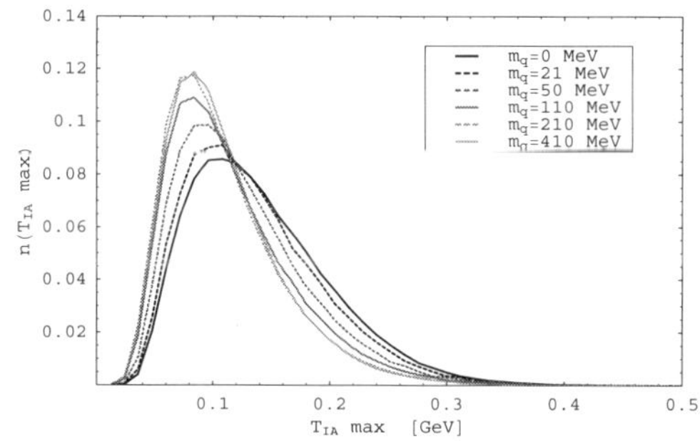

Fig. 4. Distribution of the maximum fermionic overlap matrix elements in the IILM for different quark masses.

ultraviolet sector of the spectrum factorizes and therefore cancels out in all ensemble averages.

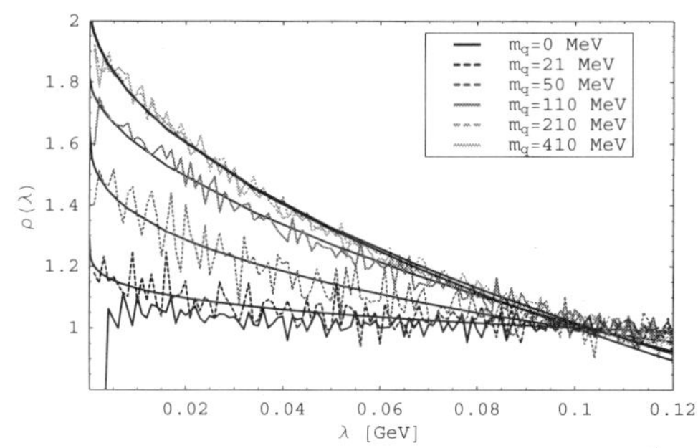

Fig. 5. Distribution of the maximum fermionic overlap matrix elements in the IILM for different quark masses.

Other two interesting features of the Instanton Model can be revealed from Fig 5: (i) for $m_q \lesssim m^\star$ the density of quasi zero-modes increases as the quark mass decreases and (ii) for $m_q \gtrsim m^\star$ the near zero-mode part of the spectrum becomes practically independent on the quark mass. These two non-trivial dynamical effects are in fact related and can be explained as follows. Quark loops are known to generate strong non-local correlations between pseudoparticles of opposite topological charge. Such correlations tend to suppress configurations in which one or more pseudoparticles are located far from all others. Hence, the contribution of the fermionic determinant leads to a reduction of the density of *nearly* zero-modes, as completely isolated instantons are known to have *exact* zero-modes. As the quark mass gets larger, such a topological screening becomes less and less effective and the population of near exact zero-modes increases, explaining the rise of the peak of eigenvalue density near the origin. On the other hand, we have seen that for $m_q \gtrsim m^\star$ the contribution of the fermionic determinant is suppressed and fermion-induced topological screening completely disappears. As a result, the low-virtuality sector of the Dirac spectrum stops depending on the quark mass. These are a interesting predictions of the instanton liquid model that can be checked with unquenched lattice calculations.

5. Conclusion

In this talk we havehave used the IILM as a tool to investigate the dynamical mechanisms which drive the dynamics and the structure of hadrons, in the light quark sector of QCD. We have shown that the model's predictions are consistent with QCD in the pion mass regime $\lesssim$ 500 MeV, where we expect chiral dynamics to play an important role. The model reproduces χpt prediction for the spectrum of the Dirac operator and is consistent with the lattice data for pion and nucleon masses in the range $m_\pi \approx 300 - 600$ MeV.

Having checked that chiral dynamics is correctly encoded in the instanton model, we have exploited our analytic handling on instanton-induced correlations to study the dynamical mechanisms involved in the transition into the chiral regime, at a microscopic level. We have identified a mass scale $m^\star = 80$ MeV above which we do not expect QCD correlators to be dominated by chiral dynamics and above which quenched calculations should begin to approximate full QCD.

References

1. M. Cristoforetti, P. Faccioli, J. Negele and M. Traini, hep-ph/0605256 (2006).
2. A. Smilga and J. Stern, Phys. Lett. **B318**, 531 (1993).
3. M. Musakhanov, *Seoul 2001, Hadrons and Nuclei* , 357 (2001).
4. P. Bowman, U. M. Heller, D. B. A. Leinweber, and G. Williams, Nucl. Phys. (Proc. Suppl.) **119**, 323 (2003).
5. A. A. Khan *et al.*, Nucl. Phys. **B689**, 175 (2004).
6. C. Aubin *et al.*, Phys. Rev. D **70**, 094505 (2004).
7. M. Procura, B. U. Munsch, T. Wollenweber, T. Hemmert, and W. Weise, Phys. Rev. D **73**, 114510 (2006).

ELECTROMAGNETIC FORM FACTORS OF THE NUCLEON IN SPACELIKE AND TIMELIKE REGIONS

J. P. B. C. DE MELO

Centro de Ciências Exatas e Tecnológicas, Universidade Cruzeiro do Sul, 08060-070, São Paulo, Brazil

T. FREDERICO

Dep. de Física, ITA, CTA, São José dos Campos, São Paulo, Brazil

E. PACE

Dipartimento di Fisica, Università di Roma "Tor Vergata" and Istituto Nazionale di Fisica Nucleare, Sezione Tor Vergata, Via della Ricerca Scientifica 1, I-00133 Roma, Italy

S. PISANO

Dipartimento di Fisica, Università di Roma "La Sapienza", P.le A. Moro 2, I-00185 Roma, Italy

G. SALMÈ

Istituto Nazionale di Fisica Nucleare, Sezione Roma I, P.le A. Moro 2, I-00185 Roma, Italy

An approach for a unified description of the nucleon electromagnetic form factors in spacelike and timelike regions is presented. The main ingredients of our model are: *i*) a Mandelstam formula for the matrix elements of the nucleon electromagnetic current; *ii*) a 3-dimensional reduction of the problem on the Light-Front performed within the so-called `Propagator Pole Approximation` (**PPA**), which consists in disregarding the analytical structure of the Bethe-Salpeter amplitudes and of the quark-photon vertex function in the integration over the minus components of the quark momenta; *iii*) a dressed photon vertex in the $q\bar{q}$ channel, where the photon is described by its spin-1, hadronic component.

1. Introduction

The electromagnetic form factors of the nucleon represent a powerful tool to investigate the nucleon structure. In particular, a detailed knowledge of the responses of the nucleon to an electromagnetic probe can give useful information on the quark distributions inside the hadron state, together with an insight on the role played by the non-valence Fock components. Presently, the experimental investigation of the nucleon electromagnetic form factors poses various problems in both the spacelike (SL) and timelike (TL) regions. In particular, in the SL region the most striking observation is the discrepancy in the results obtained for the proton ratio $G_E^p(Q^2)\mu_p/G_M^p(Q^2)$ between the data collected by the Rosenbluth separation method and by the polarization transfer one [1]. As to the TL region, a sizable difference occurs between the theoretical expectations from perturbative QCD [2] - that predicts, at the threshold, a ratio $\left(\frac{G_M^n}{G_M^p}\right)^2 \approx \left(\frac{q_d}{q_u}\right)^2 \approx 0.25$ - and the experimental data by Fenice [3], where $G_M^n \approx G_M^p$. From such a perspective, a unified investigation of both SL and TL regions is well motivated. In order to carry on the analysis for both these regions in the framework of the same Light-Front model, a reference frame with a non-vanishing plus component $q^+ \neq 0$ is needed, otherwise no pair-production mechanism - fundamental in the TL region - is allowed. In view of this, we will adopt a reference frame where $\mathbf{q}_\perp = 0$ and $q^+ = \sqrt{|q^2|}$.

2. The Model

The starting point of our model is the covariant formula *á la* Mandelstam [4] for the nucleon electromagnetic current. In the SL region, where the process under investigation is the scattering $e^- N \to e^- N$, we describe the matrix elements of the *macroscopic* current

$$\langle N;\sigma', P'_N|j^\mu\;|P_N,\sigma;N\rangle =$$
$$= \bar{U}_N(P'_N,\sigma')\left[-F_2(Q^2)\frac{P'_N{}^\mu + P_N{}^\mu}{2m_N} + \left(F_1(Q^2)+F_2(Q^2)\right)\gamma^\mu\right]U_N(P_N,\sigma) \tag{1}$$

where the Dirac (F_1) and Pauli (F_2) nucleon form factors are present, by the following *microscopical* approximation:

$$\langle\sigma', P'_N|j^\mu\;|P_N,\sigma\rangle = 3\;N_c\int\frac{d^4k_1}{(2\pi)^4}\int\frac{d^4k_2}{(2\pi)^4}\times$$
$$Tr_{\mathcal{T}(1,2)}\;Tr_{\mathcal{T}(3,N)}\;Tr_{\Gamma(1,2)}\left\{\;\bar{\Phi}_N^{\sigma'}(k_1,k_2,k'_3,P'_N)\;S^{-1}(k_1)\;S^{-1}(k_2)\;\mathcal{I}_3^\mu\times\right.$$
$$\left.\Phi_N^\sigma(k_1,k_2,k_3,P_N)\right\}\;. \tag{2}$$

Analogously, in the TL region we describe the annihilation matrix element $\langle N\ \bar{N}|j^\mu|0\rangle$:

$$\langle N;\sigma', P_N|j^\mu\ | - P_N, \sigma; N\rangle =$$
$$= \bar{U}_N(P_N,\sigma')\ \left[-F_2(q^2)\frac{P_N{}^\mu - P_{\bar{N}}{}^\mu}{2m_N} + \left(F_1(q^2)+F_2(q^2)\right)\ \gamma^\mu\right]\ V_N(P_{\bar{N}},\sigma) \tag{3}$$

by a *microscopical* approximation given by:

$$\langle N;\sigma', P_N|j^\mu\ | - P_{\bar{N}}, \sigma; N\rangle = 3\ N_c \int \frac{d^4k_1}{(2\pi)^4}\int \frac{d^4k_2}{(2\pi)^4}\ \times$$
$$Tr_{\tau_{(1,2)}}\ Tr_{\tau_{(3,N)}}\ Tr_{\Gamma_{(1,2)}}\ \left\{\bar{\Phi}_N^{\sigma'}(k_1,k_2,k_3',P_N)\ S^{-1}(k_1)\ S^{-1}(k_2)\ \mathcal{I}_3^\mu\ \times\right.$$
$$\left.\Phi_{\bar{N}}^\sigma(k_1,k_2,k_3,-P_{\bar{N}})\right\}\ . \tag{4}$$

In the previous expressions, $S(k) = (\not{k}+m)/(k^2-m^2+\imath\ \epsilon)$ is the quark propagator, the quantity $\Phi_N^\sigma(k_1,k_2,k_3,P_N)$, with its Dirac-conjugate $\bar{\Phi}_N^{\sigma'}(k_1,k_2,k_3',P_N')$, represents the nucleon Bethe-Salpeter amplitude (*BSA*), and $\mathcal{I}_3^\mu(k_3,q)$ indicates the q-γ vertex function, given by:

$$\mathcal{I}_3^\mu(k_3,q) = \left(\frac{1+\tau_z}{2}\right)\ \mathcal{I}_u^\mu(k_3,q)\ +\ \left(\frac{1-\tau_z}{2}\right)\ \mathcal{I}_d^\mu\ (k_3,q) =$$
$$= \frac{\mathcal{I}_u^\mu\ +\ \mathcal{I}_d^\mu}{2}\ +\ \tau_z\ \frac{\mathcal{I}_u^\mu\ -\ \mathcal{I}_d^\mu}{2} = \mathcal{I}_{IS}^\mu(k_3,q) + \tau_z\mathcal{I}_{IV}^\mu(k_3,q) \tag{5}$$

where $\mathcal{I}_u^\mu$ ($\mathcal{I}_d^\mu$) indicates the u (d) quark contribution, while $\mathcal{I}_{IS}^\mu(k_3,q)$ ($\mathcal{I}_{IV}^\mu(k_3,q)$) is the isoscalar (isovector) contribution.

To introduce a proper Dirac structure for the nucleon *BSA*, we describe the qqq-N interaction through an effective Lagrangian, which represents an isospin-zero and spin-zero coupling for the (1,2) quark pair, as in Ref. [5] but with $\alpha = 1$. Then, the nucleon *BSA* is approximated as follows

$$\Phi_N^\sigma(k_1,k_2,k_3,p) = \left[\ S(k_1)\ \imath\tau_y\ \gamma^5\ S_C(k_2)C\ S(k_3)\ +\ S(k_3)\ \imath\tau_y\ \gamma^5\ S_C(k_1)\right.$$
$$\left.\times C\ S(k_2)\ +\ S(k_3)\ \imath\tau_y\ \gamma^5\ S_C(k_2)C\ S(k_1)\ \right]\ \Lambda(k_1,k_2,k_3)\ \chi_{\tau_N}\ U_N(p,\sigma) \tag{6}$$

where: i) $\Lambda(k_1,k_2,k_3)$ describes the symmetric momentum dependence of the vertex function on the quark momentum variables, k_i, ii) $U_N(p,\sigma)$ is the nucleon Dirac spinor, iii) χ_{τ_N} the nucleon isospin state and iv) $S_C(k)$ is the charge conjugate quark propagator.

In order to apply phenomenological approximations inspired by the Hamiltonian language to our model, we need to project the matrix elements on the Light-Front. To this end, we integrate over the minus components of the

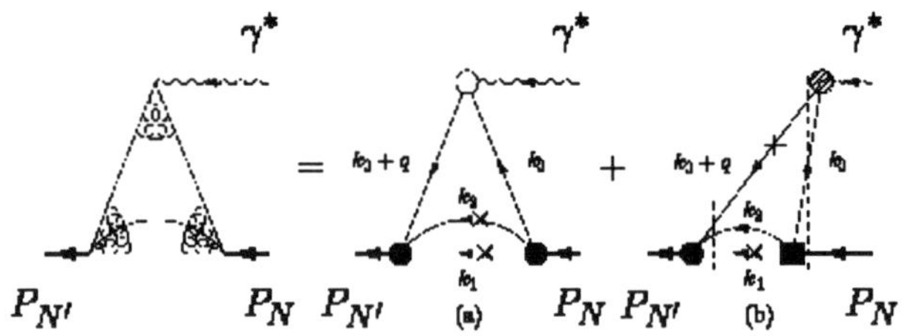

Fig. 1. Diagram (a) : valence, triangle contribution ($0 < k_i^+ < P_N^+, 0 < k_3^+ + q^+ < P_{N'}^+$). Diagram (b) : non-valence contribution ($0 > k_3^+ > -q^+$). The symbol $\times$ on a quark line indicates a quark on the mass shell, *i.e.* $k_{on}^- = (m^2 + k_\perp^2)/k^+$. (After Ref. [6]).

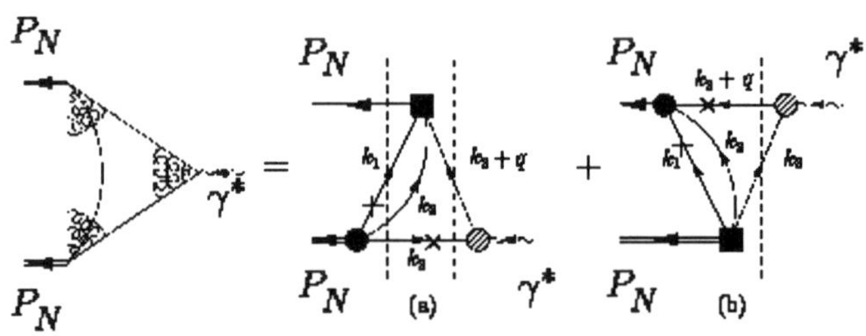

Fig. 2. Diagrams contributing to the nucleon EM form factors in the timelike region. The solid circles represent the on-shell amplitudes, the solid squares represent the non-valence ones and the shaded circles represent the dressed photon vertex.

quark momenta. We assume a suitable fall-off for the functions $\Lambda(k_1, k_2, k_3)$ and $\Lambda(k_1, k_2, k_3')$ appearing in the nucleon *BSAs*, to make finite the four dimensional integrations. Furthermore, we assume that the singularities of $\Lambda(k_1, k_2, k_3)$ and $\Lambda(k_1, k_2, k_3')$ give a negligible contribution to the integrations on k_1^- and on k_2^- and then these integrations are performed taking into account only the poles of the quark propagators.

The k_i^- integrations automatically single out two kinematical regions, namely a *valence* region, given by a triangle process (Fig. 1 (a)), with the spectator quarks on their mass shell and both the initial and the final nucleon vertexes in the valence sector, and the *non-valence* region (Fig. 1 (b)), where the $q\bar{q}$ pair production appears and only the final nucleon vertex is in the valence sector. This latter term can be seen as a higher Fock state contribution of the nucleon final state to the form factors. According to this kinematical separation, both the isoscalar and the isovector part of the quark-photon vertex $\mathcal{I}_3^\mu$ contain a purely bare valence contribution and a contribution corresponding to the pair production (Z-diagram), which can be decomposed in a bare, point-like term and a vector meson dominance (VMD) term (Fig. 3):

$$\mathcal{I}^\mu(k,q,i) = \mathcal{N}_i \; \theta(p^+ - k^+) \; \theta(k^+)\gamma^\mu + \\ +\theta(q^+ + k^+) \; \theta(-k^+) \, [Z_b \, \mathcal{N}_i \; \gamma^\mu + Z_V \Gamma^\mu(k,q,i)] \quad (7)$$

with $i = IS$, IV and $\mathcal{N}_{IS} = 1/6, \mathcal{N}_{IV} = 1/2$. The first term in Eq. (7) is the bare coupling of the triangle contribution, while Z_b, Z_V are renormalization constants to be determined from the phenomenological analysis of the data. As a consequence the SL form factors F^N are the sum of a valence term, F^N_Δ, plus a non-valence one, $F^N_\mathcal{Z}$.

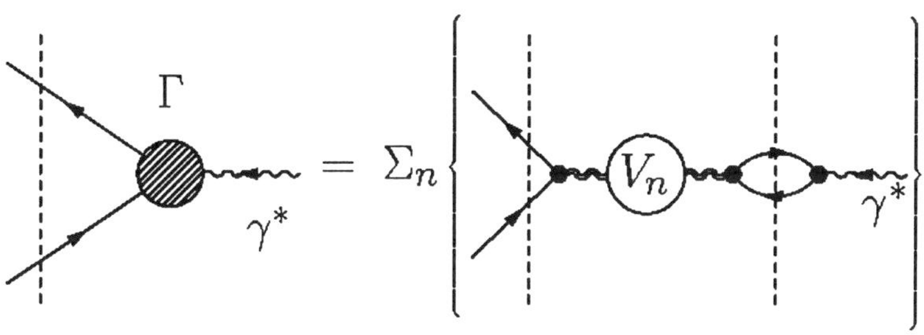

Fig. 3. Diagrammatic analysis of the VMD contribution to the quark-photon vertex, $\Gamma^\mu(k,q)$. (After Ref. [7]).

3. Phenomenological approximations

In order to overcome the lack of solutions for the 4-dimensional Bethe-Salpeter equation in the baryon case, we insert in our model some phenomenological approximations. In particular, we test an Ansatz for the momentum dependent part $\Lambda(k_1, k_2, k_3)$ of the nucleon BSA in Eq. (6), while the VMD description of the photon hadronic component, that appears in the quark-photon vertex, will be described by the eigenstates of a relativistic, squared mass operator introduced in Ref. [8].

In more details, the various amplitudes appearing in Figs. 1, 2 will be described as follows:

- for the solid circles, representing the valence, on-shell amplitudes, we adopt a power-law Ansatz *á la* Brodsky- Lepage [9];

- for the solid squares, representing the non-valence, off-shell amplitudes, we adopt a phenomenological form where the correlation between the spectator quarks is implemented;
- for the empty circle, representing a bare photon, a pointlike vertex, γ^μ, is used;
- for the shaded circles, representing a dressed photon, we use the sum of a bare term and a microscopical VMD model [7].

The actual forms are the following. For the on-shell nucleon amplitude, we have:

$$\Psi_N(k_1^+, \mathbf{k}_{1\perp}, k_2^+, \mathbf{k}_{2\perp}, P_N) = P_N^+ \frac{\Lambda(k_1, k_2, k_3)}{m_N^2 - M_0^2} \sim \frac{\mathcal{N}\; P_N^+\; \mathcal{F}(\xi_1, \xi_2, \xi_3)}{[\beta^2 + M_0^2(1,2,3)]^3} \tag{8}$$

where $\mathcal{F}(\xi_1, \xi_2, \xi_3)$ is a symmetric scalar function depending on the momentum fraction carried by each constituent and $M_0(1,2,3)$ is the light-front free mass for the three-quark system. At the present stage we take $\mathcal{F}(\xi_1, \xi_2, \xi_3) = 1$. As to the non-valence amplitude, it is described by:

$$\Lambda(k_1, k_2, k_3) \sim$$

$$\sim \frac{\mathcal{G}(\xi_1, \xi_2, \xi_3)}{\left[\beta_{off}^2 + M_0^2(1,2)\right]^p} \left\{ \frac{1}{\left[\beta_{off}^2 + M_0^2(3',2)\right]} + \frac{1}{\left[\beta_{off}^2 + M_0^2(3',1)\right]} \right\} \tag{9}$$

where $M_0(i,j)$ is the light-front free mass for a two-quark system and $\mathcal{G}(\xi_1, \xi_2, \xi_3)$ is a symmetric function of ξ_i. In this preliminary work $\mathcal{G}(\xi_1, \xi_2, \xi_3) = 1$ and $p = 2$ are chosen. As to the q-γ vertex, the term $\Gamma^\mu(k, q, i)$ in Eq. (7) is obtained through the same microscopical VMD model already used in the pion case with the same VM eigenstates [7]. Eventually, given the simplicity of the forms adopted for the nucleon amplitudes, a further dependence on the momentum transfer is introduced in the SL region through a modulation function $f(Q^2) = (1 + aQ^2)/(1 + bQ^2)$, that parameterizes all the effects not included in the naïve amplitudes we are using. This function acts in the following way on the total form factors F^N: $F^N(Q^2) = F^N_\Delta f^2(Q^2) + F^N_Z f(Q^2)$.

4. Results and perspectives

The preliminary results obtained with our model are reported in Figs. 4, 5 and 6. They show clearly the relevant role played by the pair production process - and then by the non-valence components - for the nucleon form factors. In particular, the interplay between the valence and non-valence

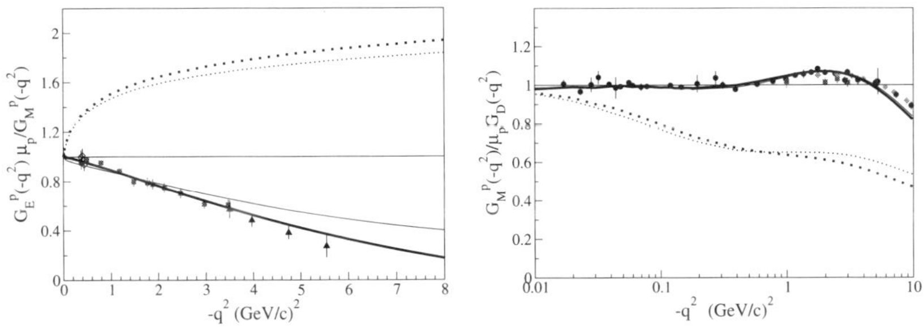

Fig. 4. **Left panel.** The ratio $G_E^p(Q^2)\mu_p/G_M^p(Q^2)$ vs Q^2. Thin (thick) solid line: full calculation for a quark mass $m_q = 190$ ($m_q = 200$) MeV, corresponding to the sum of all the contributions to $G_E^p(Q^2)$ and $G_M^p(Q^2)$, *i.e.* triangle plus pair production terms. Thin (thick) dotted line: triangle elastic contribution to $G_E^p(Q^2)$ and $G_M^p(Q^2)$ for a quark mass $m_q = 190$ ($m_q = 200$) MeV. **Right panel.** $G_M^p(Q^2)/\mu_p G_D(Q^2)$ vs Q^2, with the same notations as in the left panel, and $G_D(Q^2) = 1/(1+Q^2/0.71)^2$. For experimental data see Ref. [10].

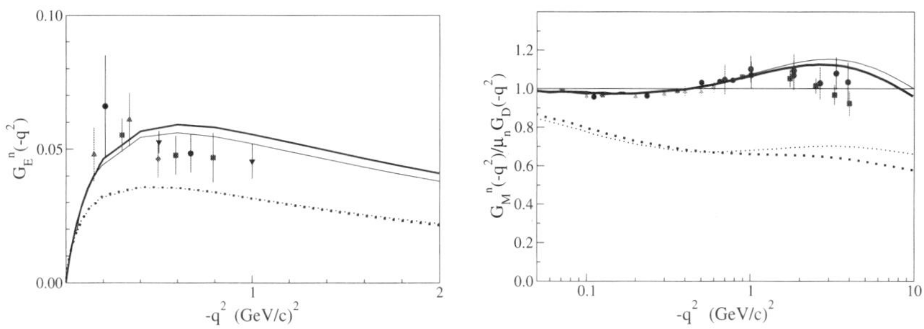

Fig. 5. **Left panel.** $G_E^n(Q^2)$ vs Q^2. Solid line: full calculation, corresponding to the sum of all the contributions to $G_E^n(Q^2)$, *i.e.* triangle plus pair production terms. Dotted line: triangle elastic contribution to $G_E^n(Q^2)$. **Right panel.** $G_M^n(Q^2)/\mu_n G_D(Q^2)$ vs Q^2 and $G_D(Q^2) = 1/(1+Q^2/0.71)^2$. The notation is the same as in Fig. 4. For experimental data see Ref. [10].

contributions generates the possibility to have a zero in the proton electric form factor. A reasonable description of the experimental data in the SL region is obtained, with a modulation factor $f(Q^2)$ which grows from 1 to 1.08, at most. As to the TL region, our preliminary results reproduce the behavior of the experimental data, but for a scale factor.

Various improvements can be applied to our model. First of all, a more elaborate form for the nucleon non-valence amplitude - that is tested for the first time in this model - could allow for a better description of the experimental data in the TL region. Another improvement can derive from

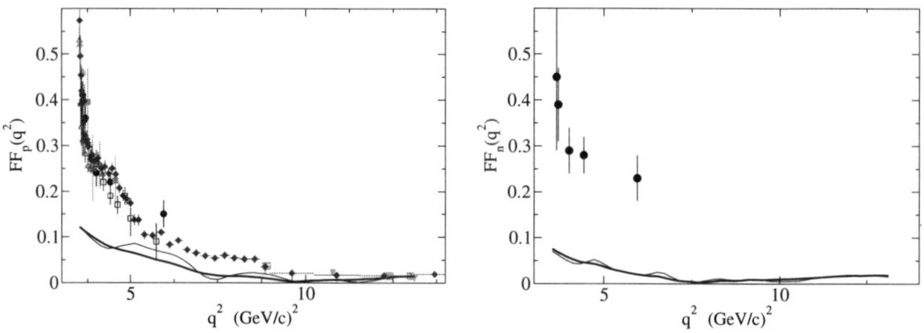

Fig. 6. **Left Panel**. Proton effective form factor in the timelike region. Thin (thick) solid line: calculation for a quark mass $m_q = 190$ ($m_q = 200$) MeV. For the experimental data see Ref. [11] and Refs. therein. **Right Panel**. Neutron effective form factor in the TL region, with the same notation of the Left Panel. Data by Fenice Collaboration [3].

a better description of the quark-photon vertex, and in particular for the VMD approximation. A new, fully covariant VMD model is presently under investigation.

References

1. J. Arrington, Phys. Rev. C **68**, 034325 (2003).
2. J. Ellis and M. Karliner, New J. Phys. **4**, 18 (2002).
3. Fenice Collaboration, Nucl. Phys. **B517**, 3 (1998).
4. S. Mandelstan, Proc. Royal Soc. (London) **A233**, 248 (1956).
5. W.R.B. de Araújo, E.F. Suisso, T. Frederico, M. Beyer, and H.J. Weber, Phys. Lett. **B478**, 86 (2000); Nucl. Phys. **A694**, 351 (2001).
6. E. Pace, G. Salmè, T. Frederico, S. Pisano, and J.P.B.C. de Melo, hep-ph/0607342, hep-ph/0611328.
7. J.P.B.C. de Melo, T. Frederico, E. Pace, and G. Salmè, Phys. Lett. **B581**, 75 (2004); Phys. Rev. D **73**, 074013 (2006).
8. T. Frederico, H.-C. Pauli, and S.-G. Zhou, Phys. Rev. D **66**, 054007 (2002); ibidem **66**, 116011 (2002).
9. G. P. Lepage and S. J. Brodsky, Phys. Rev. D **22**, 2157 (1980).
10. www.jlab.org/ cseely/nucleons.html and Refs. therein.
11. BaBar Collaboration, Phys. Rev. D **73**, 012005 (2006).

NUCLEON FORM FACTORS IN POINT-FORM SPECTATOR-MODEL CONSTRUCTIONS

T. MELDE

Theoretische Physik, Institut für Physik, Karl-Franzens-Universität, Universitätsplatz 5, A-8010 Graz, Austria

We discuss electromagnetic currents in the point-form formulation of relativistic quantum mechanics. The construction is along a spectator model and implies that only one quark is explicitly coupled to the photon, but nevertheless many-body contributions are present in the current operator. Such effects are unavoidable in relativistic constructions and resulting ambiguities are notably reduced by imposing charge normalization and time-reversal invariance. The residual theoretical indetermination introduces small but sizeable changes in the nucleon form-factors, particularly at higher Q^2 values, with the data generally centered in the middle of the theoretical band.

Keywords: Electromagnetic nucleon form-factors; Relativistic constituent quark model; Theoretical uncertainty estimation

1. Introduction

Nucleon form factors are of considerable importance, as they can give additional information on the internal structure of the nucleons. Renewed interest in the form factors is due to new polarization transfer measurements [1–4] for the proton, that produced data at variance with respect to earlier Rosenbluth data [5–7]. Accurate measurements [8,9] of new generation revealed the existence of a discrepancy that cannot be explained [10] with Coulomb distortion effects. It has been suggested that two-photon contributions could be the source of the disagreement [11,12], but to date the experimental situation is still under discussion (see also [13–15]). On the theoretical side, two-photon corrections [16–20] and additional Δ contributions [21] have been advocated to resolve this problem. A detailed discussion on the theoretical and experimental status of nucleon form factors can be found in two recent reviews [22,23].

Here, we reinvestigate the description of nucleon form factors within the Goldstone-boson exchange (GBE) constituent-quark model (CQM) [24].

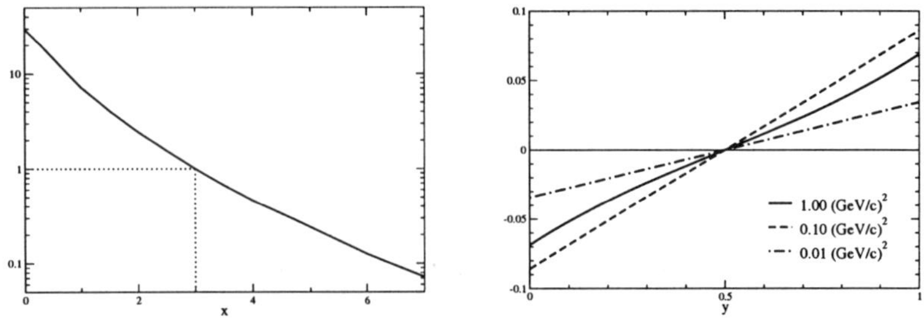

(a) (b)

Fig. 1. (a) Proton electric form factor G_E^p at momentum transfer $Q^2 = 0$ (i.e. proton charge) as a function of the exponent x in the normalization factor $\mathcal{N}$ of Eq. (5). (b) Expectation value of the electromagnetic current component $\hat{J}^{\mu=3}$ in the Breit frame as a function of the exponent y in the normalization factor $\mathcal{N}$ of (5) for three different values of the momentum transfer Q^2.

The approach is complementary to field-theoretic ones, like e.g. Refs. [25, 26], as it is valid for low and intermediate momentum transfers. The electromagnetic (EM) ground state properties of the nucleon already were reproduced quite well [27–30] with the point-form spectator model (PFSM) applied to the GBE CQM. With the advent of newer and more precise data for the nucleon form-factor, it becomes necessary to assess more carefully the predictive power of the PFSM construction and possible sources of indetermination [31]. In the last Cortona meeting [32], we presented arguments that translational invariance of the transition amplitude implies the presence of many-body contributions. In Ref. [33] it has been shown that in connection to these many-body effects certain ambiguities arise which cannot be fully constrained by Poincaré invariance alone. Here, we further constrain the PFSM construction by requiring charge normalization and time-reversal invariance of the elastic electromagnetic nucleon transition amplitude.

2. The electromagnetic current

The reduced matrix element of the PFSM EM current between three constituent quarks with individual momenta p_i and helicities σ_i reads [33]

$$\langle p_1', p_2', p_3'; \sigma_1', \sigma_2', \sigma_3' | \hat{J}^\mu_{\rm rd} | p_1, p_2, p_3; \sigma_1, \sigma_2, \sigma_3 \rangle =$$
$$3\mathcal{N} e_1 \bar{u}\left(p_1', \sigma_1'\right) \gamma^\mu u\left(p_1, \sigma_1\right) 2p_{20} \delta^3\left(\vec{p}_2 - \vec{p}_2'\right) 2p_{30} \delta^3\left(\vec{p}_3 - \vec{p}_3'\right) \delta_{\sigma_2 \sigma_2'} \delta_{\sigma_3 \sigma_3'} . \quad (1)$$

For the nucleon electromagnetic form-factors the transition amplitude between the incoming and outgoing baryon eigenstates is given by the follow-

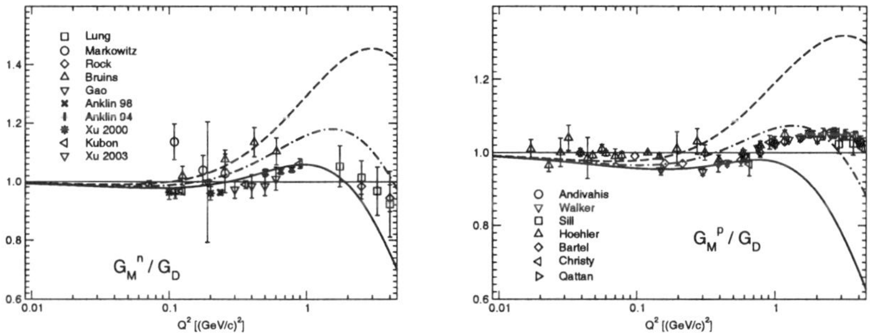

(a) (b)

Fig. 2. Comparison of magnetic form-factor to dipole ratios for the neutron (a) and proton (b) in the GBE CQM with PFSM currents. The full line denotes the theoretical results with $\mathcal{N}_{\text{geo}}$, the dashed line for $\mathcal{N}_{\text{ari}}$ and the dash-dotted line for $\mathcal{N}_{\text{fit}}$. All ratios are normalized to one at $Q^2 = 0$. Experiment given by Refs. [34–43] and Refs. [5–9,44,45]

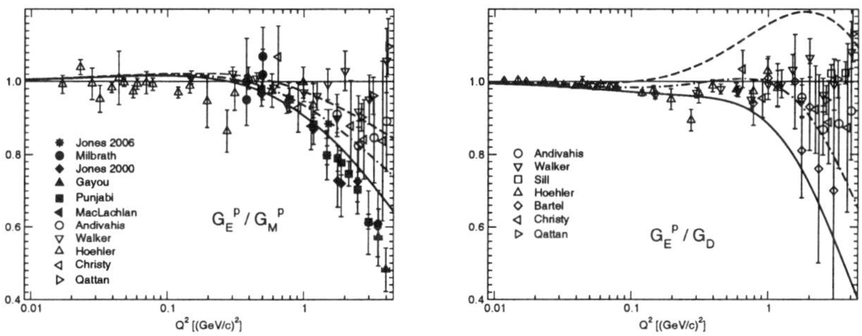

(a) (b)

Fig. 3. Notation same as Fig. 2 but for the proton form factor ratios; electric to magnetic (a) as well as electric to dipole (b). Experimental data from Refs. [1–9,14,44–46].

ing definition

$$F^{\mu}_{\Sigma',\Sigma}\left(Q^2\right) = \left\langle V', m_{\mathrm{N}}, \frac{1}{2}, \Sigma' \right| \hat{J}^{\mu}_{\mathrm{rd}} \left| V, m_{\mathrm{N}}, \frac{1}{2}, \Sigma \right\rangle . \tag{2}$$

In the Breit frame, the nucleon Sachs form-factors are related to the transition amplitude $F^{\mu}_{\Sigma',\Sigma}$ according to the expressions

$$F^{0}_{\Sigma',\Sigma}\left(Q^2\right) = 2m_{\mathrm{N}} G_{\mathrm{E}}\left(Q^2\right) \delta_{\Sigma'\Sigma} \tag{3}$$

$$\vec{F}_{\Sigma',\Sigma}\left(Q^2\right) = iQG_{\mathrm{M}}\left(Q^2\right) \chi^{\dagger}_{\Sigma'} \left(\vec{\sigma} \times \vec{e}_z\right) \chi_{\Sigma} , \tag{4}$$

where $\Sigma', \Sigma = \pm\frac{1}{2}$ are the spin projection along the z-axis of the nucleon and χ the corresponding Pauli spinors.

As has been observed in Ref. [33], a normalization factor $\mathcal{N}$ appears in the PFSM current of (1), which can be parametrized in the following way

$$\mathcal{N}(x,y) = \left(\frac{M}{\sum_i \omega_i}\right)^{xy} \left(\frac{M'}{\sum_i \omega'_i}\right)^{x(1-y)} . \tag{5}$$

In this form x and y are free parameters which we consider in the range $0, 0$, while M and M' are the mass eigenvalues of the baryon states (here, we restrict our investigation to the elastic case with $M = M'$). Conversely, $\sum_i \omega_i = \sum_i \sqrt{m_i^2 + \vec{k}_i^2}$ is the invariant mass of the free incoming three-quark system, where $\vec{k}_i$ are internal momenta, and similarly for $\sum_i \omega'_i$ for the outgoing. Obviously, the normalization factors so defined are all Lorentz invariant. Here, we further constrain the construction by adhering to additional global requirements for the EM amplitudes.

Charge normalization requires $x = 3$ (for any y values), because in Fig. 1(a) only for this value the proton charge is recovered at zero momentum transfer. This result substantiates the findings in Ref. [33], where one can see that only when the exponent takes the $x = 3$ value, one can recover a genuine one-body operator at zero momentum transfer. If we impose that the EM-current amplitude is time-reversal invariant, then it has been demonstrated (see Refs. [47,48]) that the third component of $\vec{F}_{\Sigma',\Sigma}$ has to vanish identically in the Breit frame. In Fig. 1(b) we present the corresponding transition amplitude exhibiting an antisymmetric structure around the value $y = \frac{1}{2}$. To obtain a vanishing amplitude the obvious choice clearly is the zero-crossing at $y = \frac{1}{2}$, leading to the symmetric (geometric) factor

$$\mathcal{N}_{\mathrm{S}} = \left(\frac{M}{\sum_i \omega_i}\right)^{\frac{3}{2}} \left(\frac{M'}{\sum_i \omega'_i}\right)^{\frac{3}{2}} . \tag{6}$$

However, the antisymmetric structure of the curves in Fig. 1(b) allows to consider also other, equally valid, PFSM normalizations. For example, physically valid constructions are

$$\mathcal{N}(z) = \frac{1}{2}\left[\left(\frac{M}{\sum_i \omega_i}\right)^{3z} \left(\frac{M'}{\sum_i \omega'_i}\right)^{3(1-z)} + \left(\frac{M'}{\sum_i \omega'_i}\right)^{3z} \left(\frac{M}{\sum_i \omega_i}\right)^{3(1-z)}\right] . \tag{7}$$

In Eq. (7), one can sort out two special cases, namely $\mathcal{N}_{\mathrm{S}}$ for $z = 0.5$ and $\mathcal{N}_{\mathrm{ari}}$ for $z = 0$ (or $z = 1$). Here, $\mathcal{N}_{\mathrm{S}}$ is given by Eq. (6), while $\mathcal{N}_{\mathrm{ari}}$ takes on the form of an arithmetic combination

$$\mathcal{N}_{\mathrm{ari}} = \frac{1}{2}\left[\left(\frac{M}{\sum_i \omega_i}\right)^{3} + \left(\frac{M'}{\sum_i \omega'_i}\right)^{3}\right] . \tag{8}$$

All forms implied by Eq. (7) lead to the correct charge normalization and fulfill time-reversal invariance[a]. In the $Q^2 \to 0$ limit all different expressions implied by Eq. (7) lead to the same transition amplitudes.

In Fig. 2 we show the ratio of magnetic nucleon to standard dipole form factors in comparison to experiment. As one moves to higher Q^2 momenta, one observes a broadening of the band between the curves $\mathcal{N}_{\rm S}$ and $\mathcal{N}_{\rm ari}$, indicating that the normalization factor effects mainly the higher momentum results. The predictions with $\mathcal{N}_{\rm S}$ (full line) exhibit a strong fall off above $Q^2 = 1{\rm GeV}^2$ following approximately the lower bounds of the experimental neutron data, but considerably underestimating the proton data. The dashed line, representing the results with $\mathcal{N}_{\rm ari}$, exhibit an overpredicton of the experimental data. We also have performed a one-parameter "best fit" (denoted by the dash-dotted line) to the experimental data of the dipole ratios G_M^n/G_D and G_M^p/G_D at $3{\rm GeV}^2$ momentum transfer. The value $z = 1/6$ in Eq. (7) produces a theoretical curve that lies still below the experimental data for the proton, while it is already above for the neutron. These reults are denoted as $\mathcal{N}_{\rm fit}$ and represented by the dash-dotted lines in Figs. 2,3. The two curves with $\mathcal{N}_{\rm geo}$ and $\mathcal{N}_{\rm ari}$ envelop the experimental data in most cases, while the curve with $\mathcal{N}_{\rm fit}$ lies in-between.

In Fig. 3 we show the ratio of the electric and magnetic proton form-factor (left) and the ratio of the electric proton form-factor to standard dipole parametrization (right). Again, the theoretical spread enlarges with increased momentum transfer. The dash-dotted line follows the experimental values and is congruent to latest data with the asymmetric beam-target experiment by Jones *et al.* [14]. Note that this line was obtained with the same $\mathcal{N}_{\rm fit}$, which was determined previously. To further decrease the theoretical spread, one needs to find additional conditions that further constrain the construction of PFSM operators. However, for the time being one could assume the band between the two curves as an estimate for the theoretical uncertainty of the PFSM construction of the nucleon EM form-factors.

3. Summary and conclusion

In this contribution we have reconsidered the construction of EM operators in the PFSM formalism. The spectator-like nature of the construction implies that only one quark is explicitly coupled to the current operator, but nevertheless the operator contains many-body effects. Such many-body effects appear in the kinematics and in the occurrence of a normalization

---

[a] a linear combination of the type $a\mathcal{N}_{\rm S} + (1-a)\,\mathcal{N}_{\rm ari}$ also is an allowed construction.

factor that is explicitly required by the theory. The specific expression for the factor, however, is not completely determined since Poincaré invariance and proton-charge normalization alone are not sufficient to uniquely determine the current-operator. Time-reversal invariance produces an important additional constraint that further restricts the possible choices, but still allows for a residual indetermination. This spread is zero at zero-momentum transfer and increases at higher Q^2-values. We have estimated the corresponding theoretical uncertainty and found that it remains reasonably small, at least for the EM form-factors of the nucleon. We have finally determined an optimal choice utilizing one single free parameter, obtaining results overall in good agreement with the experimental data.

Acknowledgements

The results reported in this contribution originated in a collaborative effort with L. Canton, W. Plessas and R. F. Wagenbrunn. The author is particularly thankful to L. Canton for his help in completing this manuscript and to R. F. Wagenbrunn for his expertise in the numerics. The author is grateful to the INFN Sezione di Padova and the University of Padova for their hospitality. This work was supported by the Austrian Science Fund (Projects P16945 and P19035). Finally, the author would like to thank the organizers for accepting this contribution to the Cortona meeting.

References

1. M. K. Jones *et al.*, Phys. Rev. Lett. **84**, 1398 (2000).
2. O. Gayou *et al.*, Phys. Rev. Lett. **88**, 092301 (2002).
3. V. Punjabi *et al.*, Phys. Rev. C **71**, 055202 (2005).
4. G. MacLachlan *et al.*, Nucl. Phys. **A764**, 261 (2006).
5. L. Andivahis *et al.*, Phys. Rev. D **50**, 5491 (1994).
6. R. C. Walker *et al.*, Phys. Lett. **B224**, 353 (1989).
7. G. Höhler, E. Pietarinen, I. S. Stefanescu, F. Borkowski, G. G. Simon, V. H. Walther, and R. D. Wendling, Nucl. Phys. **B114**, 505 (1976).
8. M. E. Christy *et al.*, Phys. Rev. C **70**, 015206 (2004).
9. I. A. Qattan *et al.*, Phys. Rev. Lett. **94**, 142301 (2005).
10. J. Arrington and I. Sick, Phys. Rev. C **70**, 028203 (2004).
11. J. Arrington, Phys. Rev. C **69**, 022201 (2004).
12. J. Arrington, Phys. Rev. C **71**, 015202 (2005).
13. V. Tvaskis *et al.*, Phys. Rev. C **73**, 025206 (2006).
14. M. K. Jones *et al.*, Phys. Rev. C **74**, 035201 (2006).
15. E. Tomasi-Gustafsson, hep-ph/0610108 (2006).
16. P. A. M. Guichon and M. Vanderhaeghen, Phys. Rev. Lett. **91**, 142303 (2003).

17. P. G. Blunden, W. Melnitchouk, and J. A. Tjon, Phys. Rev. Lett. **91**, 142304 (2003).
18. P. G. Blunden, W. Melnitchouk, and J. A. Tjon, Phys. Rev. C **72**, 034612 (2005).
19. Y. C. Chen, A. Afanasev, S. J. Brodsky, C. E. Carlson, and M. Vanderhaeghen, Phys. Rev. Lett. **93**, 122301 (2004).
20. A. V. Afanasev, S. J. Brodsky, C. E. Carlson, Y.-C. Chen, and M. Vanderhaeghen, Phys. Rev. D **72**, 013008 (2005).
21. S. Kondratyuk, P. G. Blunden, W. Melnitchouk, and J. A. Tjon, Phys. Rev. Lett. **95**, 172503 (2005).
22. J. Arrington, C. D. Roberts, and J. M. Zanotti, nucl-th/0611050 (2006).
23. C. F. Perdrisat, V. Punjabi, and M. Vanderhaeghen, hep-ph/0612014 (2006).
24. L. Y. Glozman, W. Plessas, K. Varga, and R. F. Wagenbrunn, Phys. Rev. D **58**, 094030 (1998).
25. M. Oettel, R. Alkofer, and L. von Smekal, Eur. Phys. J. A **8**, 553 (2000).
26. R. Alkofer, A. Holl, M. Kloker, A. Krassnigg, and C. D. Roberts, Few-Body Syst. **37**, 1 (2005).
27. R. F. Wagenbrunn, S. Boffi, W. Klink, W. Plessas, and M. Radici, Phys. Lett. **B511**, 33 (2001).
28. L. Y. Glozman, M. Radici, R. Wagenbrunn, S. Boffi, W. Klink, and W. Plessas, Phys. Lett. **B516**, 183 (2001).
29. S. Boffi, L. Glozman, W. Klink, W. Plessas, M. Radici, and R. Wagenbrunn, Eur. Phys. J. A **14**, 17 (2002).
30. K. Berger, R. F. Wagenbrunn, and W. Plessas, Phys. Rev. D **70**, 094027 (2004).
31. T. Melde, L. Canton, W. Plessas, and R. F. Wagenbrunn, nucl-th/0612013 (2006).
32. T. Melde, W. Plessas, R. F. Wagenbrunn, and L. Canton, Proceedings of the 10th Conference on Theoretical Nuclear Physics in Italy, Cortona, Italy, 6-9 Oct. 2004 (World Scientific, Singapore, 2004), p. 213.
33. T. Melde, L. Canton, W. Plessas, and R. F. Wagenbrunn, Eur. Phys. J. A **25**, 97 (2005).
34. A. Lung *et al.*, Phys. Rev. Lett. **70**, 718 (1993).
35. P. Markowitz *et al.*, Phys. Rev. C **48**, 5 (1993).
36. S. Rock *et al.*, Phys. Rev. Lett. **49**, 1139 (1982).
37. E. E. W. Bruins *et al.*, Phys. Rev. Lett. **75**, 21 (1995).
38. H. Gao *et al.*, Phys. Rev. C **50**, 546 (1994).
39. H. Anklin *et al.*, Phys. Lett. **B336**, 313 (1994).
40. H. Anklin *et al.*, Phys. Lett. **B428**, 248 (1998).
41. W. Xu *et al.*, Phys. Rev. Lett. **85**, 2900 (2000).
42. G. Kubon *et al.*, Phys. Lett. **B524**, 26 (2002).
43. W. Xu *et al.*, Phys. Rev. C **67**, 012201 (2003).
44. A. F. Sill *et al.*, Phys. Rev. D **48**, 29 (1993).
45. W. Bartel *et al.*, Nucl. Phys. **B58**, 429 (1973).
46. B. D. Milbrath *et al.*, Phys. Rev. Lett. **80**, 452 (1998).
47. L. I. Durand, P. C. DeCelles, and R. B. Marr, Phys. Rev **126**, 1882 (1962).

48. F. J. Ernst, R. G. Sachs, and K. C. Wali, Phys. Rev. **119**, 1105 (1960).

CHIRAL-ODD GENERALIZED PARTON DISTRIBUTIONS, TRANSVERSITY AND DOUBLE TRANSVERSE-SPIN ASYMMETRY IN DRELL–YAN DILEPTON PRODUCTION*

M. PINCETTI[‡], B. PASQUINI, and S. BOFFI

Dip. di Fisica Nucleare e Teorica, Università degli Studi di Pavia and INFN, Sezione di Pavia, Pavia, 27100, Italy

[‡] *E-mail: manuel.pincetti@pv.infn.it*

Within the framework of light-cone quantization we derive the overlap representation of generalized parton distributions for transversely polarized quarks using the Fock-state decomposition in the transverse-spin basis. We apply this formalism to the case of light-cone wave functions in a constituent quark model giving numerical results for the four chiral-odd generalized parton distributions in a region where they describe the emission and reabsorption of a quark by the nucleon. With the transversity distribution obtained in the forward limit of the generalized distribution, we provide some predictions for the double transverse-spin asymmetry in Drell-Yan dilepton production in the kinematics of the $\mathcal{PAX}$ experiment.

Keywords: Generalized Parton Distributions; Transversity; Drell–Yan; Double Transverse-Spin Asymmetry

1. Introduction

Quarks and gluons are the fundamental degrees of freedom in the study of strong interactions by means of QCD. Nevertheless, the objects experimentally observable are hadrons and, at present, we have not yet a complete description of how hadrons are built up in terms of quark and gluon fields. In hard scattering processes we parameterize this non perturbative information through a set of parton distribution (PDs), which are hadronic matrix elements of bilocal products of the light-front quark and gluon field operators. At leading twist a quark-parton model of the nucleon requires three PDs: the unpolarized distribution, $f_1(x)$, the helicity distribution, $g_1(x)$,

*This research is part of the EU Integrated Infrastructure Initiative Hadronphysics Project under contract number RII3-CT-2004-506078.

and the transversity distribution, $h_1(x)$. Whereas the first two distributions are well studied physical quantities, both from the experimental and theoretical point of view, so far, the last one is totally unknown. As a matter of fact, $h_1(x)$ does not contribute to inclusive deep-inelastic scattering because of its chiral-odd nature and is only experimentally accessible when coupled to another chiral-odd partner in the cross section. Notwithstanding, some theoretical activity has been developed to calculate it and to suggest some new experimental ways to extract $h_1(x)$ from data [1]. Among all the proposed experimental investigations, the favorite one is represented by the double transversely polarized Drell–Yan dilepton production [2]. In fact, this allows the study of the double transverse-spin asymmetry of lepton-pair production, giving direct access to the transversity distribution.

In recent years a novel class of parton distributions has been introduced, the generalized parton distributions (GPDs) [3]. The GPDs have attracted a considerable amount of interest, since it has been pointed out that they contain a wealth of information about the internal structure of hadrons, interpolating between the inclusive physics of parton distributions and the exclusive limit of form factors [4]. These new parton distributions functions are defined as non-diagonal hadronic matrix elements of bilocal products of the light-front quark and gluon field operators. A complete set of quark GPDs at leading twist include four helicity conserving, usually labeled H, E, $\tilde{H}$, $\tilde{E}$, and four helicity flipping chiral-odd GPDs, labeled H_T, E_T, $\tilde{H}_T$, $\tilde{E}_T$ [5,6]. In the forward limit, H, $\tilde{H}$ and H_T reduce to f_1, g_1 and h_1, respectively.

A variety of model calculations is available for helicity conserving GPDs. Less attention has been paid up to now to the chiral-odd sector. So far, there is only one proposal to access the chiral-odd GPDs in diffractive double meson production [7,8]. In this paper we study the chiral-odd GPDs by means of the overlap representation of light-cone wave functions (LCWFs) that was originally proposed by Diehl *et al.* [9]. In a fully covariant approach the connection between the overlap representation of GPDs and the non-diagonal one-body density matrix in momentum space has further been explored in Refs. [10] for the helicity conserving sector, making use of the correct transformation of the wave functions from the (canonical) instant-form to the (light-cone) front-form description. In this way the lowest-order Fock-space components of LCWFs with three valence quarks are directly linked to wave functions derived in constituent quark models (CQMs). Recently this approach has been extended to include the next-order Fock-state component, by developing a convolution formalism for the unpolarized GPDs which incorporate the sea quark distribution [11].

In this paper, following the approach discussed in Refs. [12,13], the four chiral-odd GPDs are derived in Sec. 2 by means of a Fock-state decomposition in the transverse spin basis and with LCWFs for three valence quarks. In Sec. 3 we present the numerical results obtained for the GPDs, while we show our estimations for the transversity distributions in Sec. 4 and some predictions for the double transverse-spin asymmetry in Drell–Yan dilepton production in Sec. 5.

2. The Overlap Representation

The chiral-odd GPDs are off-diagonal in the parton helicity basis. In the reference frame where the momenta $\vec{p}$ and $\vec{p}\,'$ of the initial and final nucleon lie in the $x-z$ plane, they become diagonal if one changes basis from eigenstates of helicity to eigenstates of transversity, i.e. states with spin projection $\uparrow$ ($\downarrow$) directed along (opposite to) the transverse direction $\hat{x}$. In such a basis they are obtained by means of the following relations

$$
\begin{aligned}
H^q_T &= \frac{1}{\sqrt{1-\xi^2}} T^q_{\uparrow\uparrow} - \frac{2M\xi}{\epsilon\sqrt{t_0-t}(1-\xi^2)} T^q_{\uparrow\downarrow}, & (1)\\
E^q_T &= \frac{2M\xi}{\epsilon\sqrt{t_0-t}} \frac{1}{1-\xi^2} T^q_{\uparrow\downarrow} + \frac{2M}{\epsilon\sqrt{t_0-t}(1-\xi^2)} \tilde{T}^q_{\uparrow\uparrow} \\
&\quad - \frac{4M^2}{(t_0-t)\sqrt{1-\xi^2}(1-\xi^2)} \left(\tilde{T}^q_{\downarrow\uparrow} - T^q_{\uparrow\uparrow} \right), & (2)\\
\tilde{H}^q_T &= \frac{2M^2}{(t_0-t)\sqrt{1-\xi^2}} (\tilde{T}^q_{\downarrow\uparrow} - T^q_{\uparrow\uparrow}), & (3)\\
\tilde{E}^q_T &= \frac{2M}{\epsilon\sqrt{t_0-t}(1-\xi^2)} \left(T^q_{\uparrow\downarrow} + \xi \tilde{T}^q_{\uparrow\uparrow} \right) \\
&\quad - \frac{4M^2\xi}{(t_0-t)\sqrt{1-\xi^2}(1-\xi^2)} \left(\tilde{T}^q_{\downarrow\uparrow} - T^q_{\uparrow\uparrow} \right), & (4)
\end{aligned}
$$

where $t = \Delta^2$ is the transferred momentum square, $-t_0 = 4m^2\xi^2/(1-\xi^2)$ is its minimum value for a given value of the skewness parameter $\xi = -\Delta^+/2P^+$ defined in terms of the momentum transfer $\Delta^\mu = p'^\mu - p^\mu$ and the average nucleon momentum $P^\mu = \frac{1}{2}(p+p')^\mu$, and $\epsilon = \mathrm{sign}(D^1)$, where D^1 is the x-component of $D^\alpha = P^+\Delta^\alpha - \Delta^+ P^\alpha$.

The transverse matrix elements are defined as follows

$$T^q_{\lambda'_t\lambda_t} = \langle p', \lambda'_t| \int \frac{dz^-}{2\pi} e^{i\bar{x}P^+z^-} \bar{\psi}(-z/2)\gamma^+\gamma^1\gamma_5\psi(z/2)|p, \lambda_t\rangle, \tag{5}$$

$$\tilde{T}^q_{\lambda'_t\lambda_t} = \langle p', \lambda'_t| \int \frac{dz^-}{2\pi} e^{i\bar{x}P^+z^-} \frac{i}{2}\bar{\psi}(-z/2)\sigma^{+1}\psi(z/2)|p, \lambda_t\rangle. \tag{6}$$

In the region $\xi \leq \bar{x} \leq 1$ of plus-momentum fractions, where the generalized quark distributions describe the emission of a quark with plus-momentum $(\bar{x}+\xi)P^+$ and its reabsorption with plus-momentum $(\bar{x}-\xi)P^+$, they can be written in the overlap representation of LCWFs as

$$T^q_{\lambda'_t\lambda_t} = \sum_{N,\beta=\beta'} \left(\sqrt{1-\xi}\right)^{2-N} \left(\sqrt{1+\xi}\right)^{2-N} \sum_{j=1}^{N} \mathrm{sign}(\mu^t_j)\delta_{s_j q}$$
$$\times \int [d\bar{x}]_N [d^2\vec{k}_\perp]_N \delta(\bar{x}-\bar{x}_j) \Psi^*_{\lambda'_t,N,\beta'}(\hat{r}')\Psi_{\lambda_t,N,\beta}(\tilde{r}), \tag{7}$$

$$\tilde{T}^q_{\lambda'_t\lambda_t} = \sum_{\beta,\beta',N} \left(\sqrt{1-\xi}\right)^{2-N} \left(\sqrt{1+\xi}\right)^{2-N} \sum_{j=1}^{N} \delta_{\mu^{t\prime}_j - \mu^t_j}\delta_{\mu^{t\prime}_i \mu^t_i} \mathrm{sign}(\mu^t_j)\delta_{s_j q}$$
$$\times \int [d\bar{x}]_N [d^2\vec{k}_\perp]_N \delta(\bar{x}-\bar{x}_j) \Psi^*_{\lambda'_t,N,\beta'}(\hat{r}')\Psi_{\lambda_t,N,\beta}(\tilde{r}), \tag{8}$$

where $\Psi_{\lambda_t,N,\beta}$ is the momentum LCWF of the N-parton Fock state, s_j labels the quantum numbers of the j-th active parton, with transverse initial (final) spin polarization μ^t_j ($\mu^{t\prime}_j$), and μ^t_i ($\mu^{t\prime}_i$) are the transverse spin of the spectator initial (final) quarks. The set of kinematical variables $\tilde{r}, \hat{r}'$ are defined according to Refs. [9,12].

3. Results

As an application of the general formalism we consider the valence-quark contribution ($N = 3$) to the chiral-odd GPDs with the LCWF derived starting from an instant-form of the proton obtained in the relativistic quark model of Ref. [14]. The structure of the nucleon wave function in this model is SU(6) symmetric for the spin-isospin components.

As an example we show in Fig. 1 the four calculated chiral-odd GPDs, H^u_T, E^u_T, $\tilde{H}^u_T$, $\tilde{E}^u_T$ at $t = -0.2$ (GeV)^2 and for different values of ξ. In all cases the GPDs vanish at $\bar{x} = \xi$ since in our approach they include the contribution of valence quarks only and we cannot populate the so-called ERBL region with $|\bar{x}| \leq \xi$ where quark-antiquark pairs and gluons are important. Therefore, at low $\bar{x}$ this gives a strong ξ dependence of the peak

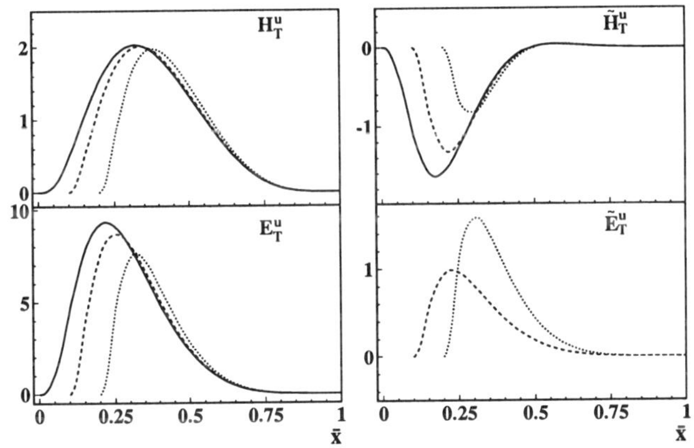

Fig. 1. The chiral odd generalized parton distributions calculated in the CQM of Ref. [14] for the flavour u at $t = -0.2$ (GeV)^2 and for different values of ξ: $\xi = 0$ (solid curves), $\xi = 0.1$ (dashed curves), $\xi = 0.2$ (dotted curves).

position of the distribution, but for large $\bar{x}$ the ξ dependence turns out to be rather weak.

Concerning the t dependence, it affects the low-$\bar{x}$ region and is more pronounced in the cases of E_T^q and $\tilde{H}_T^q$. For large $\bar{x}$ values the decay of all the distributions towards zero at the boundary $\bar{x} = 1$ is almost independent of t (see Ref. [12]).

4. The Forward Limit

In the forward limit $\Delta^\mu \to 0$ ($\bar{x} \to x$, with x being the usual Bjorken variable), only the quark GPD H_T^q can be measured and, in fact, becomes the quark transversity distribution $h_1^q(x)$. Although the quark GPDs E_T^q and $\tilde{H}_T^q$ do not contribute to the scattering amplitude, they remain finite in the forward limit, whereas $\tilde{E}_T^q$ vanishes identically being an odd function of ξ as already noticed in Ref. [6].

The transversity distribution h_1^q is the counterpart in the transverse-polarization space of the helicity parton distribution g_1^q which measures the helicity asymmetry. As it was stressed by Jaffe and Ji [2], in nonrelativistic situations where rotational and boost operations commute, one has $g_1^q = h_1^q$. Therefore the difference between h_1^q and g_1^q is a measure of the relativistic nature of the quarks inside the nucleon. In light-cone CQMs these relativistic effects are encoded in the Melosh rotations.

In Fig. 2 the helicity and transversity distributions, g_1 and h_1, obtained as a forward limit of the corresponding GPDs calculated with the CQM of

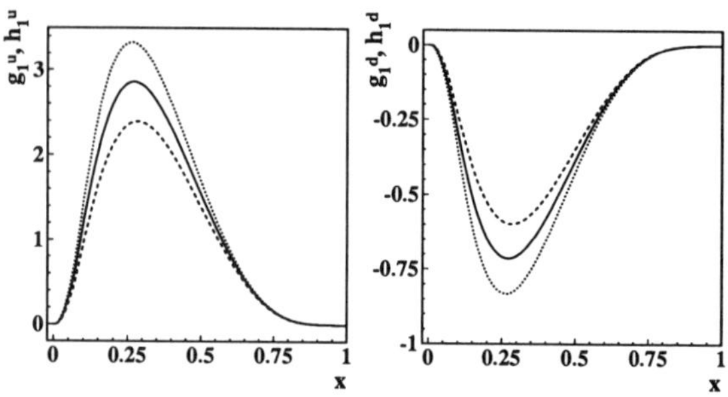

Fig. 2. Helicity and transversity distributions for the u (left panel) and d (right panel) quark. The solid lines correspond to h_1^q, the dashed lines show g_1^q, and the dotted lines are the nonrelativistic results when Melosh rotations reduce to the identity ($h_1^q = g_1^q$).

Ref. [14], are compared and plotted together with the nonrelativistic result when Melosh rotations reduce to identity. The large difference between g_1 and h_1 shows how big is the effect of relativity.

Recalling the expression for the unpolarized parton distribution f_1^q obtained in Ref. [10], it is easy to see that the following relations hold

$$2h_1^u(x) = g_1^u(x) + \frac{2}{3}f_1^u(x), \qquad 2h_1^d(x) = g_1^d(x) - \frac{1}{3}f_1^d(x), \tag{9}$$

which are compatible with the Soffer inequality [15]. In the nonrelativistic limit one obtains $h_1^u = g_1^u = 2/3 f_1^u$ and $h_1^d = g_1^d = -1/3 f_1^d$..

5. The Double Transverse-Spin Asymmetry

In order to directly access transversity via Drell-Yan lepton pair production one has to measure the double transverse-spin asymmetry A_{TT} in collisions between two transversely polarized hadrons:

$$A_{TT} = \frac{\mathbf{b}_c\sigma(\uparrow\uparrow) - \mathbf{b}_c\sigma(\uparrow\downarrow)}{\mathbf{b}_c\sigma(\uparrow\uparrow) + \mathbf{b}_c\sigma(\uparrow\downarrow)}, \tag{10}$$

with the arrows denoting the transverse directions along which the two colliding hadrons are polarized. The most favorable situation for a sizeable effect is the process $p^\uparrow\bar{p}^\uparrow \to l^+l^-X$ mediated by a virtual photon [16]. At LO, i.e. considering only the quark-antiquark annihilation graph, the double transverse-spin asymmetry for this process is given by

$$A_{TT}^{p\bar{p}} = a_{TT}\frac{\sum_q e_q^2[h_1^q(x_1,Q^2)h_1^q(x_2,Q^2) + h_1^{\bar{q}}(x_1,Q^2)h_1^{\bar{q}}(x_2,Q^2)]}{\sum_q e_q^2[f_1^q(x_1,Q^2)f_1^q(x_2,Q^2) + f_1^{\bar{q}}(x_1,Q^2)f_1^{\bar{q}}(x_2,Q^2)]}, \tag{11}$$

where e_q is the quark charge, Q^2 the invariant mass square of the lepton pair (dimuon), and a_{TT} the spin asymmetry of the QED elementary process $q\bar{q} \to l^+l^-$. Here, as in Ref. [13], we provide quantitative estimates of

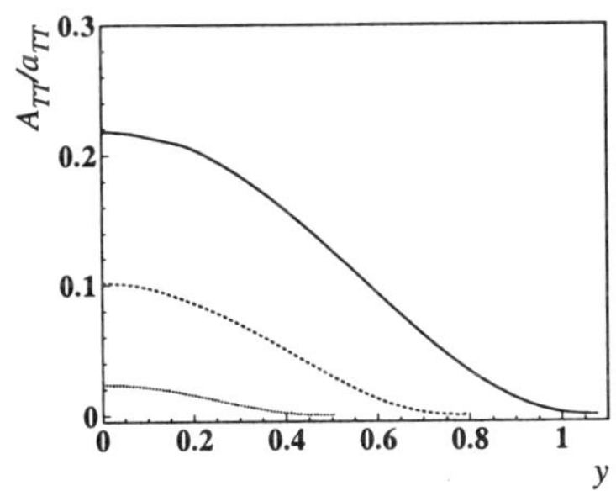

Fig. 3. The $A_{TT}^{p\bar{p}}/a_{TT}$ at a center of mass energy square of 45 GeV^2 and different values of Q^2: 5 GeV^2 (solid line), 9 GeV^2 (dashed line), 16 GeV^2 (dotted line).

$A_{TT}^{p\bar{p}}/a_{TT}$ for the kinematics of the proposed $\mathcal{PAX}$ experiment at GSI. In Fig. 3 our predictions for $A_{TT}^{p\bar{p}}/a_{TT}$ are plotted in terms of the rapidity $y = 1/2\ln(x_1/x_2)$, where the transversity distribution has been suitably evolved solving numerically the DGLAP equations [17] (see Fig. 4) and where the unpolarized distribution functions $f_1(x, Q^2)$ are taken from the GRV parameterization [18].

One may notice that our predictions for $A_{TT}^{p\bar{p}}/a_{TT}$ are much lower than the results obtained in others phenomenological analysis [16]. However, our study suggests the possibility of a measurable asymmetry, about 20%, at moderate Q^2 values, around 5÷10 GeV^2, giving direct access to the transversity distribution.

References

1. V. Barone, A. Drago, and P.G. Ratcliffe, Phys. Rep. **359**, 1 (2002); V. Barone and P.G. Ratcliffe, *Transverse spin physics* (World Scientific, Singapore, 2003).
2. J. Ralston and D. Soper, Nucl. Phys. **B152**, 109 (1979); R.L. Jaffe and X. Ji, Phys. Rev. Lett. **67**, 552 (1991); Nucl. Phys. **B375**, 527 (1992); J.L. Cortes, B. Pire, and J.P. Ralston, Z. Phys. C **55**, 409 (1992); X. Ji, Phys. Lett. **B284**, 137 (1992); P. Lenisa *et al.* ($\mathcal{PAX}$ Coll.), hep-ex/0505054; A. Bianconi and M. Radici, Phys. Rev. D **72**, 074013 (2005).
3. D. Müller *et al.*, Fortsch. Phys. **42**, 101 (1994); A.V. Radyushkin, Phys. Lett.

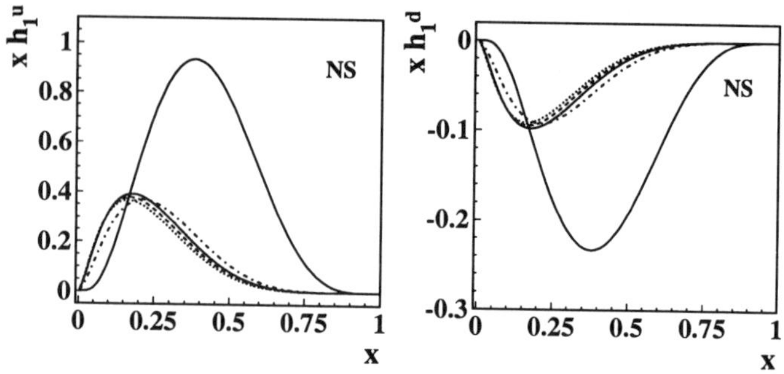

Fig. 4. Evolution of the transversity distribution for the u (left panel) and d (right panel) quark. Starting from the hadronic scale of the model $Q_0^2 = 0.079$ GeV^2 (upper curve), we plot the LO curves at different scales: $Q^2 = 5$ GeV^2, solid curves; $Q^2 = 9$ GeV^2, dashed curves; $Q^2 = 16$ GeV^2, dotted curves, and a NLO curve at $Q^2 = 5$ GeV^2 dashed-dotted curve.

B380 (1996) 417; Phys. Lett. **B385**, 333 (1996); X. Ji, Phys. Rev. Lett. **78**, 610 (1997); J. Phys. G **24**, 1181 (1998).

4. K. Goeke *et al.*, Prog. Part. Nucl. Phys. **47**, 401 (2001); M. Diehl, Phys. Rep. **388**, 41 (2003); X. Ji, Ann. Rev. Nucl. Part. Sci. **54**, 413 (2004); A.V. Belitsky and A.V. Radyushkin, Phys. Rep. **418**, 1 (2005).
5. P. Hoodbhoy and X. Ji, Phys. Rev. D **58**, 054006 (1998).
6. M. Diehl, Eur. Phys. J. C **19**, 485 (2001).
7. B. Pire *et al.*, Phys. Lett. **B550**, 65 (2002); Phys. Part. Nucl. **35**, 67 (2004).
8. R. Enberg, B. Pire, and L. Szymanowski, Eur. Phys. J. C **47**, 87 (2006).
9. M. Diehl *et al.*, Nucl. Phys. **B596**, 33 (2001).
10. S. Boffi, B. Pasquini, and M. Traini, Nucl. Phys. **B649**, 243 (2003); Nucl. Phys. **B680**, 147 (2004); Phys. Rev. D **71**, 034022 (2005).
11. B. Pasquini and S. Boffi, Phys. Rev. D **73**, 094001 (2006).
12. B. Pasquini, M. Pincetti, and S. Boffi, Phys. Rev. D **72**, 094029 (2006); M. Pincetti, B. Pasquini, and S. Boffi, hep-ph/0610051;
13. B. Pasquini, M. Pincetti, and S. Boffi, in preparation (2006).
14. F. Schlumpf, J. Phys. G **20** (1994) 237.
15. J. Soffer, Phys. Rev. Lett. **74**, 1292 (1995).
16. A. Efremov *et al.*, Eur. Phys. J. C **35**, 207 (2004); M. Anselmino *et al.*, Phys. Lett. **B594**, 97 (2004); V. Barone *et al.*, Phys. Lett. **B639**, 483 (2006).
17. Yu.L. Dokshitzer, JETP (Sov. Phys.) **46**, 641 (1977); V.N. Gribov and L.N. Lipatov, Sov. J. Nucl. Phys. **15**, 438 (1972); G. Altarelli and G. Parisi, Nucl. Phys. **B126**, 298 (1977).
18. M. Glück, E. Reya, and A. Vogt, Eur. Phys. J. C **5**, 461 (1998).

UNQUENCHING THE QUARK MODEL

E. SANTOPINTO
I.N.F.N. and Dipartimento di Fisica,
via Dodecaneso 33, Genova, I-16146, Italy
E-mail: santopinto@ge.infn.it

R. BIJKER
Departamento de Estructura de la Materia, Instituto de Ciencias Nucleares,
Universidad Nacional Autónoma de México, AP 70-543, 04510 México DF, Mexico
E-mail: bijker@nuclecu.unam.mx

We present the formalism for a new generation of unquenched quark models in which quark-antiquark pair effects are taken into account in an explicit form via a microscopic QCD-inspired quark-antiquark creation mechanism. No truncation in the sum over all the big tower of states is necessary since these states are automatically generated by means of powerful group-theoretical techniques. An important check on the formalism and the numerical results is provided by the closure limit. As an application, the effect of quark-antiquark pairs on the strange content of the proton spin, Δs, is discussed. The contributions of the up and down quarks, Δu and Δd, are also calculated. This has become possible after solving the difficult problem of permutational symmetry related to quark rearrangements. Finally, we present some preliminary results in the closure limit as well as an outlook for future applications.

Keywords: Constituent quark models, Meson and quark-antiquark pair effects, Group theory

1. Introduction

Since the QCD equations in the non-perturbative region are not solvable, many effective models of the hadrons have been developed, in particular many versions of the Constituent Quark Model (CQM). These are able to give good results for the static properties of the hadrons (like the spectrum and the magnetic moments), while they all fail to reproduce the dynamic ones, like the electromagnetic transition form factors at low Q^2 values. This seems to be a problem of degrees of freedom, since the region of low Q^2 means high distance, in which the creation of quark-antiquark-pair degrees has higher probability.

A priori, one would expect pair creation to be (after switching on a pair-creation mechanism) so enhanced in probability that a valence quark model would fail dramatically. On the other hand, it is known from the phenomenology that pair creation is suppressed, at least at the zeroth order in the static observables, since the observed spectrum is dominated by the valence quarks qqq for the baryons and by the valence $q\bar{q}$ for the mesons. Moreover, the resonances do not have decay widths of the order of the same resonance masses. Many years ago, Isgur tackled the intriguing and important question of why the CQMs work so well at an intermediate regime. The aim of this paper is to continue from where he left off and to develop the formalism for the light baryons with the linked problem of the permutation symmetry.

2. Degrees of freedom

Constituent Quark Models based on the effective degrees of freedom of three constituent quarks have been proposed in several versions: the Isgur-Karl model [1], the Capstick-Isgur model [2], the algebraic $U(7)$ model [3], the hypercentral model [4], the chiral boson exchange model [5,6] and the Bonn instanton model [7]. While these models display important and peculiar differences, they all have main features in common: they are all based on the effective degrees of freedom of three constituent quarks and on the $SU(6)$ spin-flavor symmetry; they also contain a long-range linear confining potential and an $SU(6)$-breaking term, even though the form and the advocated origin of this last term may be different, as in the one-gluon-exchange-inspired hyperfine interaction or the Goldston Boson Exchange $SU(6)$-breaking interaction or the instanton-induced-breaking term.

The photocouplings calculated by means of different constituent quark models (see for example Table 2 of Ref. [8], and references therein) have the same overall behaviour, having the same $SU(6)$ structure in common, but in many cases they show a lack of strength.

For each of those models for which calculations of the electromagnetic transition form factors are available, there is clearly the problem of missing strength at low Q^2, as is well illustrated in Fig. 1, which reports the transverse electromagnetic transition form factors for the $D_{13}(1520)$ resonance. The experimental data are compared with the predictions of different CQMs. The common problem of missing strength at low Q^2 can be ascribed to the lack of quark-antiquark effects, which are probably more important in the outer region of the nucleon. After understanding that the main prob-

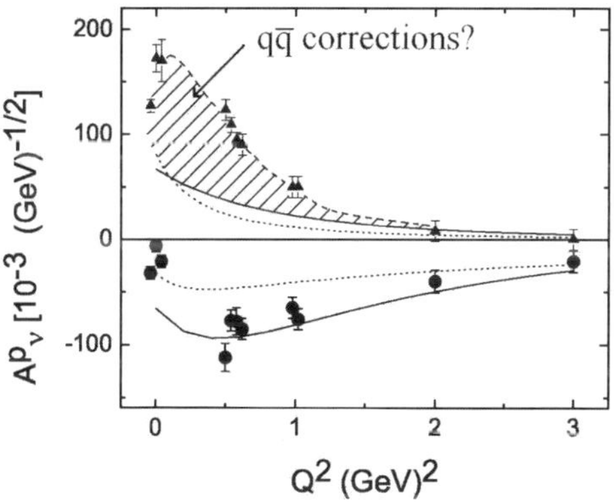

Fig. 1. Transition form factors for the $D_{13}(1520)$ resonance. Experimental data are compared with theoretical predictions from some CQMs, here the collective $U(7)$ [9], dotted line, and the hypercentral [10], unbroken line. The missing strength at low Q^2 is interpreted as due to $q\bar{q}$ pair effects.

lems of the CQMs are all linked to a problem of missing-degrees of freedom, and after identifing the key degrees of freedom as $q\bar{q}$ pairs, we are faced with two possibilities: the phenomenological parametrization or the microscopic explicit description. We have focused on the letter and tried to find a QCD-inspired pair-creation mechanism that at the same time should be encoded in such a way not as not to destroy the good CQM results. Moreover, it is important to find symmetry constraints in order to help to keep all the difficult calculations under control. In this respect, the closure limit, is very helpful, as will be explained in the following sections.

3. Flux-tube breaking model

In the flux-tube model for hadrons, the quark potential model arises from an adiabatic approximation to the gluonic degrees of freedom embodied in the flux tube [11]. The impact of $q\bar{q}$ pairs in meson spectroscopy has been studied in an elementary flux-tube breaking model [12] in which the quark pair creation occurs with $^3P_0$ quantum numbers. Subsequently, it has been shown by Geiger and Isgur [13] that a "miraculous" set of cancellations between apparently uncorrelated sets of mesons occurs in such a way to compensate each other and not to destroy the good CQM results for the mesons, and in particular in such a way that (i) the OZI hierarchy is preserved and (ii) there is a near immunity of the long-range confining potential. The change in the linear potential due to pairs bubbling in the string can be reabsorbed in a new strength of the linear potential, *i.e.* in a new string tension; thus the net effect of the mass shifts from pair creation is smaller than the naive expectation of the order of the strong decay widths. The important point is that it is necessary to sum over very large towers of intermediate states to see that the spectrum of the mesons is only weakly perturbed, after unquenching and renormalizing. No simple truncation of the set of meson loops can reproduce such results.

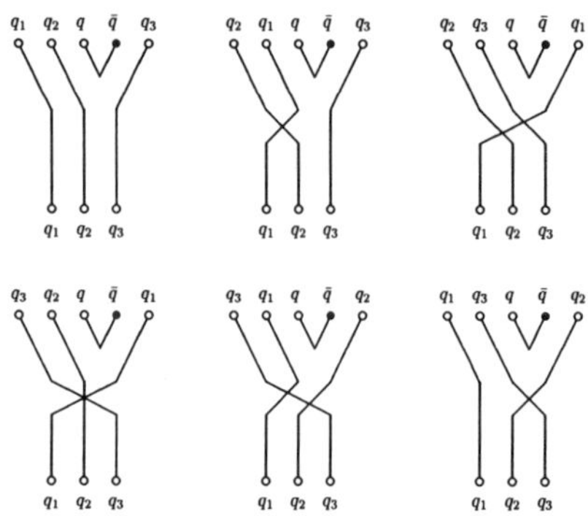

Fig. 2. Quark line diagrams for $A \to BC$.

This formalism set up for mesons could not be immediately extended to baryons, since there were extra problems, mainly linked with the permutation symmetry between identical quarks. In [14], Geiger and Isgur investigated the importance of $s\bar{s}$ loops in the proton by taking into account the contribution of the six different diagrams of Fig. 2 (in Fig. 3 of [14] only four diagrams are considered) and by using harmonic oscillator wave functions for the baryons and mesons.

In the present manuscript, we discuss some generalizations of this formalism for baryons. These new extensions make it possible to study the quark-antiquark contributions

- for any initial baryon resonance
- for any flavor of the quark-antiquark pair
- for any model of baryons and mesons, as long as their wave functions are expressed on the basis of the harmonic oscillator.

The problem of the permutation symmetry between identical quarks has been solved by means of group-theoretical techniques. In this way, we can evaluate the contribution of all the diagrams for any initial baryon $q_1q_2q_3$ (ground state or resonance) and for any flavor of the $q\bar{q}$ and not just for the $s\bar{s}$.

The pair-creation mechanism is inserted at the level of quarks and the one-loop diagram should be calculated, see Fig. 3, by summing over all the possible intermediate states. This is the crucial step, after solving all the problems linked with this infinite sum in the case of the baryons.

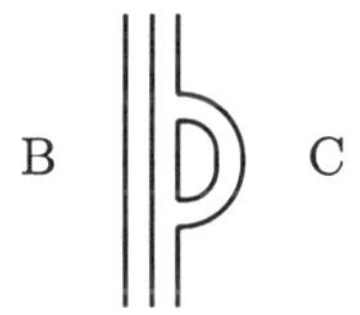

Fig. 3. One-loop diagram for the meson correction at the quark level.

Isgur has shown that as a first approximation we can think of the baryons as the sum of the zeroth order three-quark Fock component plus the sum over all the possibile intermediate states due to the creation of quark-antiquark pairs. Here we adopt a pair-creation model to incorporate the effects of $q\bar{q}$ pairs. Up to leading order in pair creation the baryon wave function is given by [14]

$$\mid \psi_A \rangle = \mathcal{N} \left[\mid A \rangle + \sum_{BClJ} \int d\vec{k} \mid BC\vec{k}\, lJ \rangle \frac{\langle BC\vec{k}\, lJ \mid h^\dagger_{q\bar{q}} \mid A \rangle}{M_A - E_B - E_C} \right] , \quad (1)$$

where $h^\dagger_{q\bar{q}}$ is the quark-antiquark pair-creation operator as in [14] , A is the initial baryon, B (C) denotes the intermediate baryon (meson), $\vec{k}$ and l the relative radial momentum and orbital angular momentum of B and C, and J is the total angular momentum $\vec{J} = \vec{J}_{BC} + \vec{l} = \vec{J}_B + \vec{J}_C + \vec{l}$. In general, matrix elements of observables $\hat{\mathcal{O}}$ are given by

$$\mathcal{O} = \langle \psi_A \mid \hat{\mathcal{O}} \mid \psi_A \rangle = \mathcal{O}_{\text{valence}} + \mathcal{O}_{\text{sea}} \quad (2)$$

where the first term denotes the contribution from the three valence quarks

$$\mathcal{O}_{\text{valence}} = \mathcal{N}^2 \langle A \mid \hat{\mathcal{O}} \mid A \rangle \quad (3)$$

and the second term the contribution from the quark-antiquark pairs

$$\mathcal{O}_{\text{sea}} = \mathcal{N}^2 \sum_{BClJ} \int d\vec{k} \sum_{B'C'l'J'} \int d\vec{k}' \frac{\langle A \mid h_{q\bar{q}} \mid B'C'\vec{k}'\, l'J' \rangle}{M_A - E_{B'} - E_{C'}} \langle B'C'\vec{k}'\, l'J' \mid \mathcal{O} \mid BC\vec{k}\, lJ \rangle \frac{\langle BC\vec{k}\, lJ \mid h^\dagger_{q\bar{q}} \mid A \rangle}{M_A - E_B - E_C} . \quad (4)$$

The sum is over a complete set of intermediate states, rather than just a few low-lying states. Not only does this have a significant impact on the numerical result, but it is necessary for consistency with the OZI-rule and the success of the CQMs in spectroscopy This is possible since the intermediate states are generated automatically by means of group-theory techniques. It has been implemented in a computer code which is able to construct states with the correct pemutational symmetry for any model and up to any shell. Therefore, the sum over intermediate states can be perfomed up to saturation and not just for the first few shells as in [14].

4. Closure limit

In this section, we discuss the closure limit which arises when the energy denominators do not depend strongly on the quantum numbers of the intermediate states in Eq. (4). In this case, the sum over the complete set of intermediate states can be solved by closure and the contribution of the quark-antiquark pairs to the matrix element reduces to

$$\mathcal{O}_{\text{sea}} \propto \langle A \mid h_{q\bar{q}} \, \mathcal{O} \, h^{\dagger}_{q\bar{q}} \mid A \rangle \ . \tag{5}$$

This means for example that, at least qualitatively, we can already say that the strange content of the proton should be very small, since within the closure limit it comes out to be exactly zero. The closure limit is very important for many reasons that are closely interconnected. It is a very good stringent test that a numerical calculation should fulfill. Moreover, and even more important, it also explains, when it comes out equal to zero, the success of the CQMs. The corrections due to the $q\bar{q}$ pairs are zero in the closure limit for many observables, while the sums over big tower of states are constrained by the closure limit in such a way that very differt meson and baryon states compensate one another.

5. Preliminary results

In this section, we discuss some preliminary results in the closure limit for the operator Δq to determine the fraction of the baryon's spin carried by each one of the flavors u, d and s

$$\Delta q \;=\; 2\langle S_z(q) + S_z(\bar{q})\rangle \ . \tag{6}$$

In the closure limit the relative contribution of the quark flavors from the quark-antiquark pairs to the baryon spin is the same as that from the valence quarks

$$\Delta u_{\text{sea}} : \Delta d_{\text{sea}} : \Delta s_{\text{sea}} \;=\; \Delta u_{\text{valence}} : \Delta d_{\text{valence}} : \Delta s_{\text{valence}} \ . \tag{7}$$

Table 1 shows the relative contributions of Δu, Δd and Δs to the spin of the ground state octet baryons in the closure limit. The closure limit, when

Table 1. Relative contributions of Δu, Δd and Δs to the spin of the ground state octet baryons with $^{2}8[56, 0^{+}]_{J}$ in the closure limit.

	Δu	:	Δd	:	Δs
p	4	:	-1	:	0
n	-1	:	4	:	0
Σ^{+}	4	:	0	:	-1
Σ^{0}	2	:	2	:	-1
Σ^{+}	0	:	4	:	-1
Λ	0	:	0	:	3
Ξ^{0}	-1	:	0	:	4
Ξ^{-}	0	:	-1	:	4

combined with symmetries, imposes other strong limits, as for example can be seen for the ground state decuplet baryons in Fig. 4. For the Δ resonances whose three-quark configuration does not contain strange quarks, the contribution of the $s\bar{s}$ pairs to the spin Δs and the magnetic moment μ_s vanishes in the closure limit. The same holds for the contribution of $d\bar{d}$ pairs to the Δ^{++}, Σ^{*+}, Ξ^{*0} and Ω^{-} resonances, and that of $u\bar{u}$ pairs to the Δ^{-}, Σ^{*-}, Ξ^{*-} and Ω^{-} resonances. These are very stringent require-

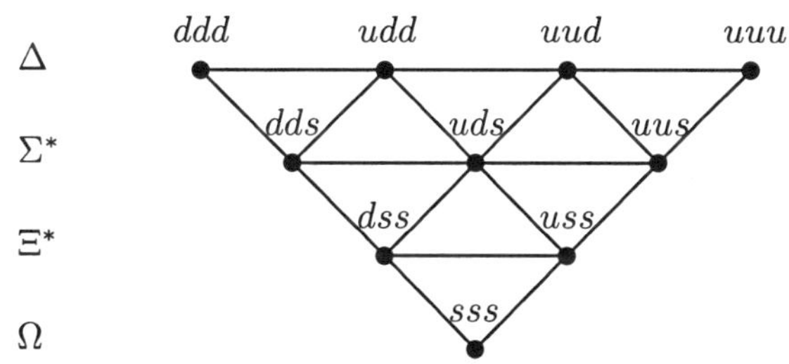

Fig. 4. Ground state decuplet baryons.

ments since each one involves the sum over a complete set of states of the intermediate baryon-meson system and provides a nontrivial test which involves the spin-flavor section, the permutation symmetry and the correct implementation of large towers of intermediate states.

6. Conclusions and outlook

The formalism described in this paper will be systematically used in order to study the still open problems regarding light baryons, such as the sea quark content and the spin problem of the proton [15] and the $q\bar{q}$ effects in the electromagnetic elastic and transition form factors [16].

Acknowledgments

This work was supported in part by the I.N.F.N. and in part by a grant from CONACyT, Mexico.

References

1. N. Isgur and G. Karl, Phys. Rev. D **20**, 1191 (1979).
2. S. Capstick and N. Isgur, Phys. Rev. D **34**, 2809 (1986).
3. R. Bijker, F. Iachello, and A. Leviatan, Ann. Phys. (N.Y.) **236**, 69 (1994); *ibid.* **284**, 89 (2000).
4. M. Ferraris, M. M. Giannini, M. Pizzo, E. Santopinto, and L. Tiator, Phys. Lett. **B364**, 231 (1995).
5. L. Ya. Glozman and D.O. Riska, Phys. Rep. C **268**, 263 (1996).
6. L. Ya. Glozman, Z. Papp, W. Plessas, K. Varga, and R. F. Wagenbrunn, Phys. Rev. C **57**, 3406 (1998); L. Ya. Glozman, W. Plessas, K. Varga, and R. F. Wagenbrunn, Phys. Rev. D **58**, 094030 (1998).
7. U. Löring, K. Kretzschmar, B. Ch. Metsch, and H. R. Petry, Eur. Phys. J. A **10**, 309 (2001); U. Löring, B.Ch. Metsch, and H. R. Petry, Eur. Phys. J. A **10**, 395 (2001); *ibid.* 447 (2001).
8. M. Aiello, M. Ferraris, M. M. Giannini, M. Pizzo, and E. Santopinto, Phys. Lett. **B387**, 215 (1996).
9. R. Bijker, F. Iachello, and A. Leviatan, Phys. Rev. C **54**, 1935 (1996).
10. M. Aiello, M. M. Giannini, and E. Santopinto, J. Phys. G **24**, 753 (1998).
11. N. Isgur and J. Paton, Phys. Rev. D **31**, 2910 (1985).
12. R. Kokoski and N. Isgur, Phys. Rev. D **35**, 907 (1987).
13. P. Geiger and N. Isgur, Phys. Rev. Lett. **67**, 1066 (1991); Phys. Rev. D **44**, 799 (1991); *ibid.* **47**, 5050 (1993).
14. P. Geiger and N. Isgur, Phys. Rev. D **55**, 299 (1997).
15. R. Bijker and E. Santopinto, to be published.
16. R. Bijker and E. Santopinto, work in progress.

TETRAQUARK STATES AND SPECTRUM

E. SANTOPINTO and G. GALATÀ

INFN and Università di Genova, Via Dodecaneso 33, 16146 Genova, Italy

A general classification of tetraquark states in terms of the spin-flavour, colour and spatial degrees of freedom has been constructed. The permutational symmetry properties of both the spin-flavour and orbital parts of the quark-quark and antiquark-antiquark subsystems are discussed in short. This classification is model indipendent and useful both for model-builders and experimentalists. An evaluation of the tetraquark spectrum is obtained from a generalization of a Iachello mass formula, originally developed for the $q\bar{q}$ mesons. The ground state tetraquark nonet is identified with $f_0(600)$, $\kappa(800)$, $f_0(980)$, $a_0(980)$.

1. Introduction

A new interest in light meson spectroscopy has been triggered recently by the KLOE, E791 and BES collaborations, which have provided evidence of the low mass resonances $f_0(600)$ [1–3] and $\kappa(800)$ [2,3]. Maiani *et al.* [4] have suggested that the lowest lying scalar mesons, $f_0(980)$, $a_0(980)$, $\kappa(800)$ and $f_0(600)$ could be described as a tetraquark nonet, in particular as a diquark and antidiquark system, since the quark-antiquark assignment to P-waves [5] has never worked in the scalar case [6]. In the traditional quark-antiquark scheme, the $f_0(980)$ is associated with non-strange quarks [5] and so it is difficult to explain both its higher mass respect to the other components of the nonet and its decay properties [4,6]. Already in the seventies Jaffe [6] suggested the tetraquark structure of the scalar nonet and proposed a four quark bag model. Other identifications have been proposed (for a complete review see [7–9] and references therein), in particular as quasimolecular-states [10] and as dynamically generated resonances [11].

We construct a complete classification scheme of the two quark-two antiquark states in terms of $SU(6)_{sf}$ [12]. An evaluation of the tetraquark spectrum for the lowest scalar meson nonet is obtained [12] from a generalization, to the tetraquark case, of the Iachello mass formula for normal mesons [13].

2. The classification of tetraquark states

The tetraquark wave function contains contributions connected to the spatial and the internal (colour, flavour and spin) degrees of freedom. We shall make use of symmetry principles without, for the moment, introducing any explicit dynamical model. In the construction of the classification scheme we are constrained by two conditions: the tetraquark wave functions should be a colour singlet, as all physical states, and the tetraquarks states must be antisymmetric for the exchange of the two quarks and the two antiquarks.

The allowed $SU(3)_f$ representations for the $qq\bar{q}\bar{q}$ system are obtained by means of the product $[3] \otimes [3] \otimes [\bar{3}] \otimes [\bar{3}] = [1] \oplus [8] \oplus [1] \oplus [8] \oplus [27] \oplus [8] \oplus [8] \oplus [10] \oplus [\overline{10}]$. The allowed isospin values are $I = 0, \frac{1}{2}, 1, \frac{3}{2}, 2$, while the hypercharge values are $Y = 0, \pm 1, \pm 2$. The values $I = \frac{3}{2}, 2$ and $Y = \pm 2$ are exotic, which means that they are forbidden for the $q\bar{q}$ mesons. The allowed $SU(2)_s$ representations are obtained by means of the product $[2] \otimes [2] \otimes [2] \otimes [2] = [1] \oplus [3] \oplus [1] \oplus [3] \oplus [3] \oplus [5]$. The tetraquarks can have an exotic spin $S = 2$, value forbidden for normal $q\bar{q}$ mesons. The $SU(6)_{sf}$-spin-flavour classification is obtained by $[6] \otimes [6] \otimes [\bar{6}] \otimes [\bar{6}] = [1] \oplus [35] \oplus [405] \oplus [1] \oplus [35] \oplus [189] \oplus [35] \oplus [280] \oplus [35] \oplus [\overline{280}]$. In Appendices A and B of Ref. [12] all the flavour and spin states in the $qq\bar{q}\bar{q}$ configuration are explicitly written. Moreover in the ideal mixing hypothesys, the flavour states of the tetraquarks are a superposition of the SU(3)-symmetrical states in such a way to have defined strange quark and antiquark numbers. Clearly the only states that can be mixed are those with the same good quantum numbers, i.e. same isospin and hypercharge, and they are explicitly constructed in Ref. [12]. The ideal mixing is essentially a consequence of the OZI rule and it is a hypothesys that has to be proved yet, but it is used by all the authors working on $q\bar{q}$ and tetraquarks. In a system made up o four objects, like the tetraquark, we have to define three relative coordinates that we choose as in Ref. [14]: two relative coordinates between two quarks and two antiquarks and a relative coordinate between their centers of mass to which are associated three orbital angular momenta, L_{13}, L_{24} and L_{12-34} respectively. We have four different spins and three orbital angular momenta and a total angular momentum J obtained by combining spins and orbital momenta. The parity for a tetraquark system is the product of the intrinsic parities of the quarks (+) and the antiquarks (-) times the factors coming from the spherical harmonics [14] as $P = P_q P_q P_{\bar{q}} P_{\bar{q}} (-1)^{L_{13}} (-1)^{L_{24}} (-1)^{L_{12-34}} = (-1)^{L_{13}+L_{24}+L_{12-34}}$. Only the states for which Q and $\bar{Q}$, where Q represents the couple of quarks and $\bar{Q}$ the couple of antiquarks, have opposite charges are C eigenvectors, with eigenvalues [14] $C = (-1)^{L_{12-34}+S}$. The eigenvalues of the G parity are in Ref. [12]. Tetraquark mesons do not have forbidden J^{PC} combinations. The tetraquark states must be antisymmetric for the exchange of the two quarks and the two antiquarks and so it is necessary to study the permutational symmetry (i.e. the irreducible representations of the group S_2) of the colour, flavour, spin and spatial parts of the wave functions of each subsystem. Moreover we have another constraint: only the singlet colour states are physical states. Regarding the two colour singlets allowed to the tetraquarks, it is better to write them by underlining their permutational S_2 symmetry, antisymmetric (A) or symmetric (S): (qq) in $[\bar{3}]_C$ (A) and $(\bar{q}\bar{q})$ in $[3]_C$ (A), or (qq) in $[6]_C(S)$ and $(\bar{q}\bar{q})$ in $[\bar{6}]_C$ (S). The following step is studying the permutational symmetry of the spatial part of the two quarks (two antiquarks) states and the permutational symmetry of the $\mathrm{SU}(6)_{sf}$ representations for a couple of quarks. The spatial, flavour, colour and spin parts with given permutational symmetry (S_2) must then be combined together to obtain completely antisymmetric states under the exchange of the two quarks and the two antiquarks. The resulting states are listed in Table III of Ref. [12]. In Table V, VI, VII and VIII of Ref. [12] the possible flavour, spin and J^{PC} values for different orbital angular momenta are studied.

In 1991 Iachello, Mukhopadhyay and Zhang developed a mass formula [13] for $q\bar{q}$ mesons,

$$M^2 = (N_n M_n + N_s M_s)^2 + a\,\nu + b\,L + c\,S + d\,J + e\,M'^2_{iji'j'} + f\,M''^2_{iji'j'}, \quad (1)$$

where N_n is the non-strange quark and antiquark number, $M_n \equiv M_u = M_d$ is the non-strange constituent quark mass, N_s is the strange quark and antiquark number, M_s is the strange constituent quark mass, ν is the vibrational quantum number, L, S and J are the total orbital angular momentum, the total spin and the total angular momentum respectively, $M'^2_{iji'j'}$ and $M''^2_{iji'j'}$ are two phenomenological terms which act only on the lowest pseudoscalar mesons. The first acts only on the octect and encodes the unusually low masses of the eight Goldstone bosons, while the second acts on the η and η' mesons and encodes the non-negligible $q\bar{q}$ annihilation effecs that arise when the lowest mesons are flavour diagonal. The flavour states are considered in the ideal mixing hypothesis, with the exception of the lowest pseudoscalar nonet. During the 15 years that have passed from the publication of Iachello's article the values reported by the PDG regarding the mesons are changed in a considerable way, so we have decided to updated the fit of the Iachello model using the latest values reported by the PDG [9] for the light $q\bar{q}$ mesons . The resulting parameters are reported in Ref. [12]. The Iachello mass formula was developed for $q\bar{q}$ mesons. In

Table 1. The candidate tetraquark nonet. Experimental data and quantum numbers.

Meson	$I^G(J^{PC})$	N_s	*Mass (GeV)*	Source
$a_0(980)$	$1^-(0^{++})$	2	0.9847 ± 0.0012	PDG [9]
$f_0(980)$	$0^+(0^{++})$	2	0.980 ± 0.010	PDG [9]
$f_0(600)$	$0^+(0^{++})$	0	0.478 ± 0.024	KLOE [1]
$k(800)$	$\frac{1}{2}(0^+)$	1	0.797 ± 0.019	E791 [15]

order to describe uncorrelated tetraquark systems by means of an algebraic model one should use a new spectrum generating algebra for the spatial part, i.e. U(10). We have not addressed this difficult problem, but we chose to write the internal degrees of freedom part of the mass formula in the same way as it was done for the $q\bar{q}$ mesons. The splitting inside a given flavour multiplet, to which is also associated a given spin, can be described by the part of the mass formula that depends on the numbers of strange and non-strange quarks and antiquarks. Thus we can use, with the only purpose of determining the mass splitting of the candidate tetraquark nonet, see Ref. [12],

$$M^2 = \alpha + (N_n M_n + N_s M_s)^2, \tag{2}$$

where α is a constant that includes all the spatial and spin dependence of the mass formula, and M_n and M_s are the masses of the constituent quarks as obtained from the previously discussed upgrade of the parameters of the Iachello mass formula. We set the energy scale, i. e. we determine the constant α, by applying Eq. (2) to the best-known candidate tetraquark, $a_0(980)$, see Ref. [12]. Thus, the masses of the other mesons belonging to the same tetraquark nonet, predicted with our simple formula, are $M(\kappa(800)) = 0.726\ GeV$, $M(f_0(600)) = 0.354\ GeV$ and $M(f_0(980)) = 0.984\ GeV$. These values seem in agreement with the experimental values, even if, before reaching any conclusion, new experiments are mandatory.

References

1. A. Aloisio *et al.*, Phys. Lett. **B537**, 21 (2002).
2. E. M. Aitala *et al.*, Phys. Rev. Lett. **86**, 770 (2001).
3. M. Ablikim *et al.* (2005), hep-ex/0506055.

4. L. Maiani, F. Piccinini, A. D. Polosa, and V. Riquer, Phys. Rev. Lett. **93**, 212002 (2004).
5. N. A. Tornqvist, Z. Phys. C **68**, 647 (1995).
6. R. L. Jaffe, Phys. Rev. D **15**, 267 (1977).
7. C. Amsler and N. A. Tornqvist, Phys. Rept. **389**, 61 (2004).
8. F. E. Close and N. A. Tornqvist, J. Phys. **G28**, R249 (2002).
9. S. Eidelman *et al.*, Phys. Lett. **B592**, 1 (2004), Particle Data Group collaboration.
10. J. D. Weinstein and N. Isgur, Phys. Rev. Lett. **48**, 659 (1982).
11. J. A. Oller and E. Oset, *Nucl. Phys.* **A620**, p. 438 (1997).
12. E. Santopinto and G. Galatà (2006), hep-ph/0605333.
13. F. Iachello, N. C. Mukhopadhyay, and L. Zhang, Phys. Lett. **B256**, 295 (1991).
14. F. J. Llanes-Estrada, ECONF **c0309101**, FRWP011 (2003).
15. E. M. Aitala *et al.*, Phys. Rev. Lett. **89**, 121801 (2002).

Quark-Gluon Plasma and Relativistic Heavy Ion Collisions

QUARK-GLUON PLASMA AND RELATIVISTIC HEAVY ION COLLISIONS

M. NARDI
INFN, Sezione di Torino
Via P.Giuria 1, 10125 Torino, Italy
E-mail: nardi@to.infn.it

1. Introduction

Lattice QCD calculations predict that, at high density and/or temperature, hadronic matter undergoes a transition towards a phase in which quarks and gluons are deconfined and weakly interacting. This new phase of matter is called "Quark-Gluon Plasma" (QGP). In the last twenty years, many efforts have been devoted to the search of the QGP in the laboratory, through ultra-relativistic heavy ion collisions. The first experiments of collisions with ions occurred at the AGS accelerator, in the "Brookhaven National Laboratory" (BNL), in 1986. In the '90s, a intense program research was performed at CERN-SPS, in fixed-target experiments with beam energies up to 200A GeV (A is the atomic number of the accelerated nucleus); the center of mass energy available in these experiments reaches about 20 GeV per nucleon pair. The SPS results are really remarkable and already suggest that in these collisions there is QGP formation. The year 2000, at BNL, saw the starting of the collider era: at RHIC, beams of Au nuclei were accelerated and collided at a maximum energy of $\sqrt{s_{NN}} = 200$ GeV. The RHIC machine is very versatile and is able to run at different incident energies: several intermediate energies, down to the top SPS, have been explored, and even experiments with asymmetric systems (deuteron-gold) were successfully performed. The ultimate machine in this field will be the Large Hadron Collider (LHC) which will start in 2007. The first heavy ion run (Pb-Pb) is expected to be in 2008 at $\sqrt{s_{NN}} = 5.5$ TeV.

Ultra-relativistic heavy ion collisions offer a unique possibility to investigate nuclear matter under extreme conditions of temperature and density in the laboratory. Such a collision is an extremely complicate event and its evolution goes through several different stages. The hope of the Scientific Community is that, during some intermediate stage, the QGP is formed. The challenge of both experimental and theoretical physicists is to identify some specific signatures of the new phase of matter and to extract them from the experimental data analysis. With increasing incident energy, new observables become available, other become of difficult interpretation because the background changes. Therefore, although at SPS energy we have already identified some concrete examples of deconfinement signatures, at RHIC and LHC energy it will be a different story, and new and more effective observables have to be studied.

Statistical QCD predicts color deconfinement for sufficiently hot, strongly interacting systems in full equilibrium. In the resulting QGP both the momenta and relative abundances of quarks (q), anti-quarks ($\bar{q}$) and gluons (g) are determined by the temperature of the medium. Whether and at what stage the matter produced in high energy nuclear collisions can reach such equilibrium is not evident from theoretical grounds. Experimental data seem to indicate that such an equilibrium is indeed realized, at least during the final stages of the reaction, and reveals itself in the distribution of particle species and momenta observed in the detector. Some recent developments in this field are discussed in Sec. 2.

For the initial stages of the collision a different description, not implying statistical equilibrium, is certainly more appropriate: the deconfined medium could be produced out-of-equilibrium and the thermalization could take place later (Sec. 3). On the other hand, some of the proposed QGP signatures do not require thermal equilibrium. A typical example is the J/ψ suppression, which depends on the presence of hard, free gluons, not on their statistical distribution (Sec. 4).

More recently, it has been suggested that some exotic particles, rarely produced in simple hadronic interactions, could be copiously produced in the decay of the deconfined phase and could therefore serve as signatures of the QGP production. An example is given in Sec. 5.

For an unambiguous interpretation of the experimental signals, it is of utmost importance the comparison of present and future data with the "standard" hadronic models which can be tested in other experiments, with small interacting systems (light nuclei, or proton-nucleus collisions) where the deconfined phase is not expected to occur. Very useful, in this respect, is also the comparison with collisions performed at much lower incident energy. These hadronic models give a firm "baseline" and any significant evidence of QGP formation should manifest itself as a departure from it.

Beside the phenomenological analysis of the experimental observables, many efforts have been made to produce a microscopic description of the deconfinement phase transition from s theoretical point of view. The fundamental QCD Lagrangian can not be used (apart from lattice calculations, yet with some limitations) due to the highly non-perturbative nature of the confined phase and of the transition regime, thus one relies on effective models, based on QCD inspired Lagrangians and mostly used in the mean field approximation. These approaches can help in understanding the properties of the deconfined versus confined phase, as a function of the temperature, energy density, baryonic density, isospin asymmetry, etc. This topic will not be touched here for lack of space; some aspects of it are covered in other contributions of these Proceedings [1].

Finally, one should note that the presence of a deconfined phase can affect the equation of state of neutron stars and, in general of astrophysical objects, with observable consequences on their properties (like mass and radius). Even the QGP-Hadron gas transition is believed to have occurred at the first stages of the Universe evolution, just after the Big Bang. Therefore, the study of the QGP is not only interesting by itself, but it is very important for its implications on other fields of Modern Physics.

2. Statistical hadron production

The statistical model successfully describes the multiplicities of different hadron species produced in high energy collisions of both elementary particles and heavy ions.

In the case of relativistic heavy ion collisions, with large multiplicities involved, the use of the grand canonical ensemble (GCE) is justified, and the model predicts that hadron yields are proportional to the volume of the system at hadronization. On the other hand, in high-energy elementary collisions the canonical ensemble (CE), enforcing the exact conservation of charges, must be used owing to the relatively low multiplicities. The exact charge conservation implies, for instance, that the mean multiplicity of heavy charged hadrons is smaller in the CE that in the GCE, this effect is known as "canonical suppression". At very low multiplicities, even the energy-momentum conservation should be explicitly imposed, and the micro-canonical ensemble description becomes necessary. Of course, the mean multiplicities calculated in the three approaches coincide in the thermodynamic limit ($V \to \infty$). The transition from MCE to CE and to GCE for small systems has been studied recently [2–4]. In particular, in Ref. [2] the comparison between the three descriptions is performed for a hadron gas at fixed energy density. As an example, Fig. 1 shows the ratios of the mean particle multiplicities in the MCE to those of the GCE as a function of the total energy of the system. The solid line shows the ratio for heavy, neutral particles of mass 3.1 GeV: the energy conservation forbids their production, in the MCE, if the energy is not high enough. The rise of the ratio of massless particles at low energies is due to the charge conservation: in the MCE charged particles are produced in pairs, while in the GCE there are no restrictions on their number which can become very small.

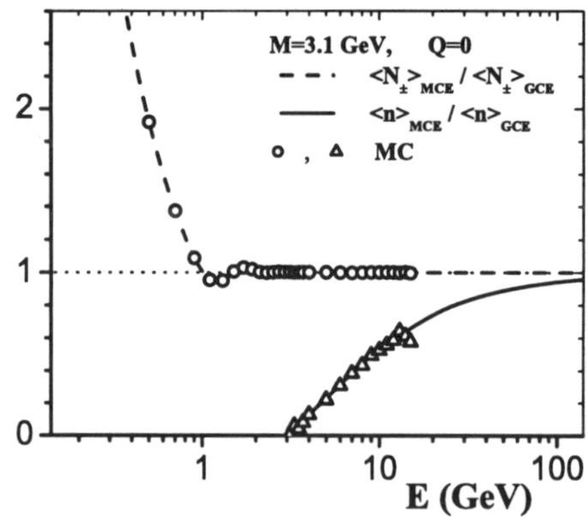

Fig. 1. The energy dependence of the ratios of the mean particle multiplicities in the MCE to those in the GCE. The lines show analytical results, the symbols are Monte Carlo results. Dashed lines (circles): ratio for massless particles. Solid lines (triangles): ratio for heavy neutral particles of mass 3.1 GeV. The total charge of the system is 0. Figure from Ref. [2].

In general, charge and momentum conservation constraints can cause non monotonic behaviors of ratios of particle multiplicities in the three statistical ensembles. As a more concrete example, let us consider the number of neutral pions and η mesons: their ratios as a function of the energy in the different statistical ensembles, are shown in Fig. 2. For both mesons, the mean multiplicities in the thermodynamic limit, here obtained for $E \to \infty$, coincide in the three statistical ensembles. For low energies (but still large compared to the pion mass) the total particle multiplicity in the MCE is 2 for pions, while in the GCE it is proportional to the energy and tends to 0 as E decreases: this explain why the ratio diverges in the low energy limit. The rise of the ratio CE/GCE for low energies is due to the higher temperature in the CE for small system at constant energy density. For η mesons, the peak in the MCE/GCE at $E \simeq 1$ GeV corresponds to the creation of a pair of mesons (the creation of a single meson is forbidden for momentum conservation). The following peaks correspond to the opening of new channels as the energy increases.

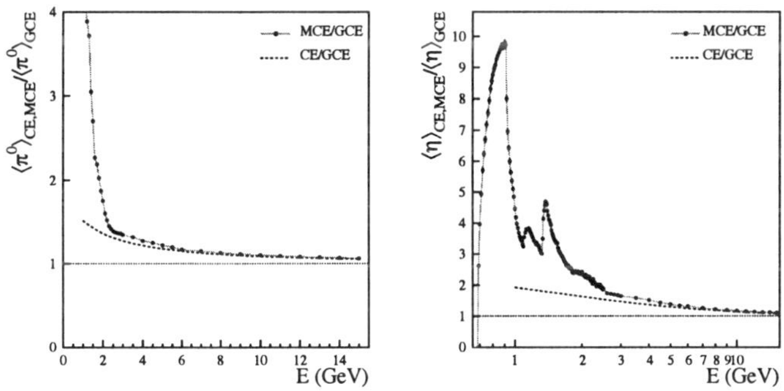

Fig. 2. The energy dependence of the ratio of mean π^0 (left) and η (right) mesons multiplicities in the MCE and CE to those in the GCE. Figures from Ref. [2].

It is evident from the previous discussion that for the statistical description of hadron production in a central collisions of heavy nuclei (Pb, Au) the GCE is adequate. In the GCE, the free parameters are : the temperature T, the volume V, the baryon chemical potential μ_B and the strangeness undersaturation parameter γ_s (or the Wroblewski factor λ_s, which measures the ratio of newly produced strange quarks to light quarks) of the system at the freeze-out. A study of the variation of these parameters with the energy or the system size is obviously interesting: a systematic analysis of all available data of particle multiplicity in proton-proton and central ion-ion collisions (from C-C to Pb-Pb) in a energy range from $\sqrt{s_{NN}} \gtrsim 4.5$ GeV (AGS) to $\sqrt{s_{NN}} \simeq 17.2$ GeV (SPS) is performed in Ref. [5]. The freeze-out temperature is not, by itself, a signature of phase transition: even in e^+e^- collisions, a distribution of secondary particles is found to be compatible with statistical production: an universal curve in the $T-\mu_B$ plane is found for both e^+e^- and $A-B$ collisions with different interacting nuclei. On the contrary, the change of γ_s from 0.4, at low energies and small systems, to 1 in central Pb-Pb at the highest SPS energy, signals a system approaching chemical equilibration (the quark s is suppressed with respect to u and d in normal hadronic matter because of its higher rest mass), a phenomenon related to chiral symmetry restoration, expected in a deconfined medium.

The energy dependence of the parameters is determined and extrapolated to RHIC energy ($\sqrt{s_{NN}} = 200$ GeV). Similarly the A-dependence (A is the atomic number of the interacting nuclei) of T and γ_s is determined. A better fit is obtained with a "two-component" model: in a central collision of two heavy nuclei (like Pb or Au), it is reasonable to assume that the deconfined phase is formed only in the central core of the produced system, where the density and the temperature are, presumably, higher. Near the surface, the conditions for deconfinement phase transitions are not reached and the system is in the hadronic phase, with γ_s comparable to the one found in light-ion collisions. Notice that the fit is performed only on data of central collisions.

Once the free parameters of the statistical hadronization model are known as a function of $\sqrt{s_{NN}}$ or A, it is possible to predict the average multiplicities and ratios of any particle species for any colliding system in central collision at any energy.

2.1. *Multiplicity fluctuations in the hadronic gas*

The study of fluctuations of particle multiplicities in heavy-ion reactions as drawn much attention in recent years because they can signal a phase transition. In Ref. [6] it is shown that enforcing the exact conservation of baryon number, strangeness, electric charge might cause large event-by-event fluctuations in the multiplicity distribution of produced particles; another source of fluctuations can be the decay of resonances, or some kinematical cuts in the detectors. It is extremely important to carefully study these "trivial" effects in order to isolate the contribution of critical fluctuations due to the phase transition.

2.2. *Production of of multiply heavy-flavored hadrons*

Charm and bottom quarks are expected to be abundantly produced in hadronic collisions at very high energies. In a central Pb-Pb collision at LHC energy, many pairs of $c\bar{c}$ and $b\bar{b}$ quarks are produced in the early stage of the collision. In the QGP phase they have the chance to reach thermal (not chemical, as their annihilation rate is very low, in comparison to the QGP lifetime) equilibrium [7] within the medium and, at the hadronization point, they will coalesce into hadrons. If the coalescence process occurs statistically, there is a finite chance that two or more uncorrelated (.i.e. not necessarily produced in the same hard interaction) heavy quarks coalesce in the same particle, thus forming a multiply heavy-flavored hadron. This effect is particularly interesting if compared to the production of such hadrons in more elementary interactions (proton-proton, e^+e^-, ...) : this analysis has been performed in Ref. [8], by calculating the mean multiplicity of heavy flavored particles in the framework of the Statistical Coalescence Model (SCM), with exact conservation of the number of c,$\bar{c}$,b and $\bar{b}$ quarks. The latter are produced in the hard collisions between primary nucleons (of the incident nuclei), therefore their multiplicity is expected to scale like the number of binary collisions N_{coll} (estimated with the Glauber model) : $\langle \nu_f \rangle = \langle \nu_{\bar{f}} \rangle = N_{\rm coll}\, \sigma_{f\bar{f}}/\sigma_{\rm inel}$ $(f = c, b)$. The ratio $\eta_f = \langle \nu_f \rangle / a_{f1}$ between the total number of quarks f and the number of hadrons containing one single quark of type f is shown to be proportional to the density of the heavy quarks f at the hadronization point, therefore it increases with the center-of-mass energy (since the cross section $\sigma_{f\bar{f}}$ grows much faster than the interaction volume): from RHIC to LHC energies it is expected to increase by one or two orders of magnitude. More quantitatively, the "primary" yields[a] expected for a few multi$-c$ and $-b$ hadrons for RHIC and LHC energies are given in Table 1. The feed-down from higher states increases these yields by a factor 4-5; an additional factor 2 comes from antiparticles yields.

Table 1. Primary yields at RHIC and LHC for central heavy-ion collisions given by the Statistical Coalescence Model.

	RHIC (Au-Au, $\sqrt{s_{NN}} = 200$ GeV)	LHC (Pb-Pb, $\sqrt{s_{NN}} = 5.5$ TeV)
Ω_{cc},Ξ_{cc}	$(0.07 - 7) \times 10^{-3}$	$0.019 - 0.38$
Ξ_{bc},Ω_{bc},B_c	$(0.04 - 6) \times 10^{-5}$	$(0.03 - 2.2) \times 10^{-2}$
Ξ_{bb},Ω_{bb}	$(0.2 - 3) \times 10^{-8}$	$(0.26 - 7) \times 10^{-5}$
Ω_{ccc}	$7 \times 10^{-7} - 10^{-4}$	$(0.1 - 3) \times 10^{-2}$

It is interesting to compare the ratio $\langle h_f \rangle / \langle \nu_f \rangle$ of the expected yields of hadrons with heavy flavor f to the total number of quarks f. For the case of multiply heavy-flavored baryons, this ratio in central Pb-Pb collisions at LHC according to the SCM will significantly exceed the ratio expected with QCD hard interactions: the most promising baryons are Ω_{ccc} and Ξ_{bc}, which are enhanced by two or even three order of magnitudes in the case of production by statistical coalescence. If such an enhancement is observed in heavy ion collisions, it would be another clear signature of QGP formation

[a] in this context "primary" denotes the production without including the feed-down from higher-state decays

It should be noted however, that the QCD predictions [9,10] are affected by large uncertainties, since these baryons have not been observed yet in hadronic scattering.

3. Kinetic description of the phase transition

Lattice QCD (LQCD) results are valid for a homogeneous, infinite system at thermal equilibrium. Obviously, a collision between heavy nuclei at very high energy is far from equilibrium and a realistic description can not be obtained with LQCD. Transport approaches [11,12] have been very useful in the recent past to describe many features of heavy ion collisions up to SPS energies. They follow the time evolution of each particle of the system; a collision occurs if two particles come to a distance less than or equal to $\sqrt{\sigma_{tot}/\pi}$ (σ_{tot} is the total cross section). It is therefore a microscopic description of the reaction. A recent work on this subject is presented in Ref. [13]. Here, the equilibrium feature of the system, starting from an out-of-equilibrium initial condition, are studied. Two clusters of pions are boosted at a given initial velocity and collide, elastically or inelastically according to the elementary cross-sections extracted from available experimental data. The evolution of the system is evaluated in two hypotheses: with a phase transition towards the QGP, implemented with the MIT Bag Model, or with a pure hadronic system, including all possible resonances formed in the collisions between the energetic pions. In the latter scenario, it is observed that an increase of the energy density of the system corresponds to a increase of the temperature only up to a limiting value T_H, the Hagedorn Temperature, because higher and higher mass resonances are excited, thus reducing the available kinetic energy. The initial state of pion gas can seem too simple, but it has the nice feature of generating a QGP with vanishing net baryonic density, very close to the conditions experimented at RHIC and, even more, at LHC energies. The equation of state obtained with this model is in qualitative agreement with LQCD results, as shown in Fig. 3. The model can be improved by adding a third quark flavor and some interaction among partons (neglected in the Bag Model). With this kind of model it is possible to look for new experimental observables that could signal the phase transition: the two working hypotheses (evolution with or without QGP) can be tested towards any possible experimental observable, if one observable is identified which is significantly different in the two scenarios, this observable is, by definition, a "QGP signature". An example is suggested in Ref. [14]: charmed mesons should show large momentum fluctuations in correspondence of the phase transition that should

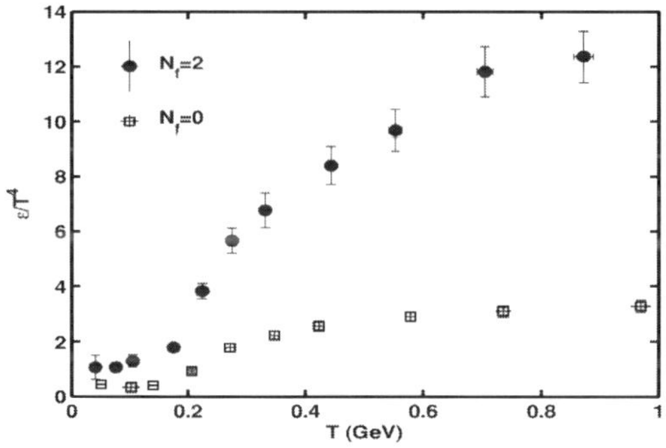

Fig. 3. Energy density divided by T^4 as function of the temperature, for 2 flavors ($N_f = 2$) or pure gluonic plasma ($N_f = 0$). From Ref. [13].

be experimentally visible as sudden increases of momentum and energy variances of D_c mesons. In the case of a normal hadronic gas, one should expect a regular, smooth behavior.

4. J/ψ suppression

The J/ψ suppression in heavy ion collisions has been proposed, twenty years ago, as a signature of QGP formation [15] : the J/ψ is strongly bound and its dissociation cross-section for inelastic interactions with hadrons in a gas at the temperature of about 200-300 MeV (the temperature expected for the dense, equilibrated medium in SPS experiments) is extremely small, therefore the hadronic gas is transparent for J/ψ's. On the other hand, in a deconfined medium the interaction between the c and $\bar{c}$ is screened and they can separate; when the QGP decays, they are far apart from each other and will hadronize by forming bound systems with other light quarks (the probability of meeting another c or $\bar{c}$ quark, independently produced, is negligible since their production is a very rare process at SPS). The situation is radically different at higher energies, like the ones available at RHIC and LHC, where the number of $c\bar{c}$ pairs initially produced is large enough that the probability of J/ψ formation by random recombination of two such quarks can, in principle, partially compensate the dissociation process in the deconfined phase.

In Refs. [16,17] the data on J/ψ suppression measured at SPS in Pb-Pb and S-U collisions are compared to the prediction of a theoretical model. The cross-section $J/\psi + h \to D^{(*)} + \bar{D}^{(*)}$, i.e. the dissociation of the J/ψ by interactions with hadrons, is assumed to be calculable within a "Constituent

Quark-Meson Model" [18–21]. The temperature of the hadronic fireball is determined from data at low centrality, and found to be $T \simeq 165-185$ Mev, close to the predicted temperature for the transition to QGP. Assuming that the energy density of the fireball is proportional to the average number of interacting nucleons per unit of overlapping area, the centrality dependence of the fireball temperature can be deduced. The centrality variable used in this work is the transverse size l of the fireball, related to the impact parameter b and the nuclear radius R by

$$l = 2R - b\ . \tag{1}$$

Figure 4 shows the geometry of the collision more clearly. The fireball is

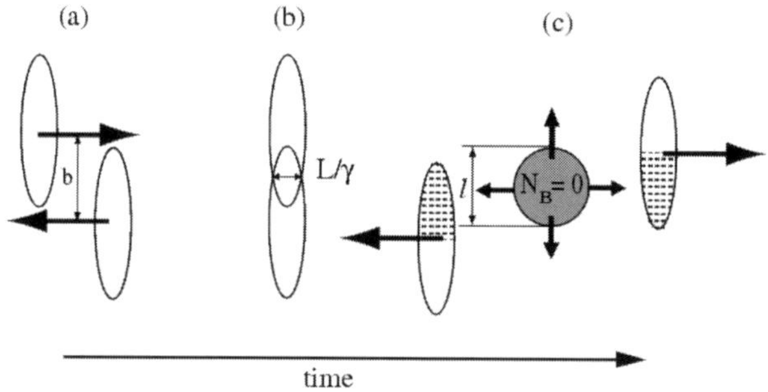

Fig. 4. Time evolution of the collision of two heavy nuclei at impact parameter b seen in the c.o.m. frame. From Ref. [17].

expected to quickly thermalize, giving rise to a hadron gas at temperature T. In this model, a boson gas made by the lowest-lying pseudoscalar and vector mesons is considered. The chemical potential is assumed to be 0 for simplicity. The particle density above the J/ψ-dissociation threshold (i.e. the density of particles with energy large enough to overcome the $2M_D$-$M_{J/\psi}$ gap) is calculated assuming a Bose-Einstein statistical distribution at temperature T. It can be seen that mesons like ρ and ω are much more effective in breaking the J/ψ than pions, which do not have, on average, enough energy.

The result for the J/ψ suppression is shown in Fig. 5. The dashed curve represents the normal nuclear absorption, observed in proton-nucleus collisions as well as in light nucleus-nucleus and peripheral Pb-Pb collisions and extrapolated to central Pb-Pb. The other three curves show the results for this model with different fireball temperatures. The curve correspond-

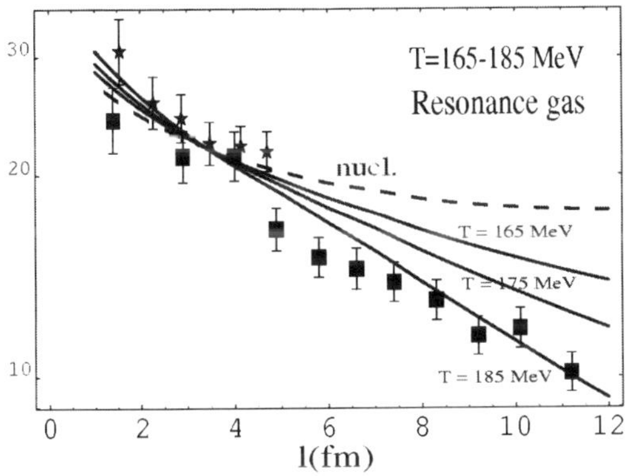

Fig. 5. J/ψ yields in S-U (stars) and Pb-Pb (boxes) collisions measured at SPS by NA50 Collab. [22,23]. See text for comments. Figure from Ref [17].

ing to $T = 185$ MeV seems to roughly reproduce the experimental data, however it should be noted that the quoted temperature refers to collisions at $l = 3.4$ fm, and it corresponds to a T well above 200 MeV for central collisions: this temperature is definitely not realistic for a hadron gas, since it is higher than the deconfining temperature. If one considers a Hagedorn gas (in which an increase of the energy density does not imply an increase of the temperature but rather favors the production of many more resonances resulting in a denser gas), one can not obtain a suppression for the J/ψ meson as strong as the observed one. The conclusion of the paper is therefore the following: a hadron gas in realistic conditions can not explain the J/ψ suppression measured at SPS and a new mechanism, like the formation the QGP, has to be invoked.

4.1. *J/ψ suppression and strangeness enhancement*

In Ref. [24] it has been observed that two well-known deconfinement signatures, namely the J/ψ suppression and strangeness enhancement, if studied as a function of the centrality variable l defined in Eq. (1) show a simultaneous departure from the normal nuclear baseline. The J/ψ suppression data are taken from Pb-Pb (NA50 [22,23]) and In-In (NA60 [25]) collisions at $\sqrt{s_{NN}} = 17.2$ GeV. The data for strangeness enhancement are taken by NA49 and NA57 in many interacting systems from p-p to Pb-Pb. The strangeness undersaturation factor γ_s is extracted by fitting the full-phase-space multiplicities of several strange particles with a simple form:

$$\rho_s(E,T) \equiv \frac{N_s}{V} = (2J+1)(\gamma_s)^{n_s} \exp\left(\frac{\vec{\mu}\cdot\vec{q}-E}{T}\right), \qquad (2)$$

where n_s is the number of valence strange quarks and/or antiquarks in the considered particle, $\vec{q}$ is a set of conserved charges and $\vec{\mu}$ the corresponding chemical potentials. The details of the fitting procedure are explained in Ref. [26]. The result is shown in Fig. 6. This is an even stronger evidence for the QGP formation in central heavy ion collisions at top SPS energy.

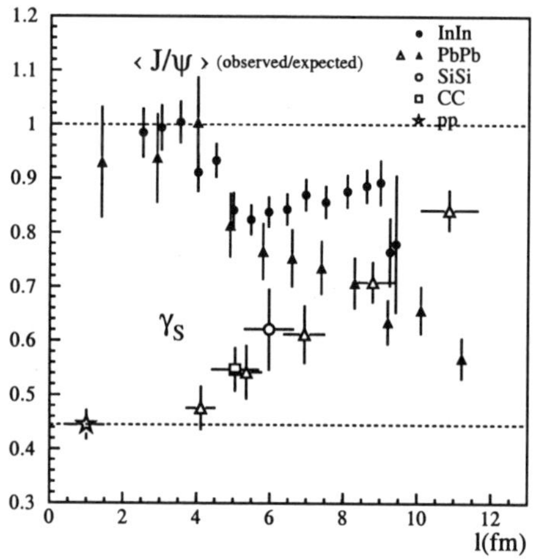

Fig. 6. Strangeness undersaturation factor γ_s (empty symbols) and J/ψ anomalous suppression ratio (full symbols) as a function of the transverse size of the interaction region. From Ref [24].

4.2. *New insights on J/ψ suppression*

In the last two years, the data on J/ψ suppression taken at RHIC, at energy much higher than SPS, became available. They caused a big surprise in the Community since the suppression observed in Au-Au collisions at $\sqrt{s_{NN}} = 200$ GeV is comparable with the one seen in Pb-Pb at $\sqrt{s_{NN}} = 17.2$ GeV : all the theoretical models, tuned to reproduce SPS data, predicted a much stronger suppression at RHIC. On one hand, hadronic models assume that the J/ψ breaks up in inelastic interactions with hadronic comovers; in these models, the density of comovers at RHIC is much larger than at LHC and, correspondly, the J/ψ suppression should be stronger. On the other hand, models which explain the SPS data with QGP formation, when extrapolated to RHIC energy simply shift the deconfinement onset to larger impact parameters (i.e. to more peripheral collisions), therefore they predict, in central collisions, a much stronger suppression that the one measured at LHC.

There are two possible explanations for having the same suppression at RHIC and SPS: the first possibility is that J/ψ's are indeed more suppressed in central Au-Au collisions at RHIC but they are also, partially, regenerated by statistical coalescence at the hadronization point, a process analogous to the one discussed in Sec. 2.2 for multi-heavy-flavored hadrons. The second explanation is inspired by recent Lattice QCD calculations on J/ψ's survival in a QGP: according to these results, at the deconfinement transition only the higher excited states of $c\bar{c}$ (namely χ_c and ψ') dissolve. The J/ψ itself survives up to a much higher temperature, estimated to be about 1.5-2 T_c (T_c is the critical temperature, at which deconfinement sets in). Such an higher temperature corresponds to a much higher energy density, certainly not attained at SPS and possibly not reachable even at RHIC (except, perhaps, in very central Au-Au collisions). As a consequence, the suppression observed so far at SPS and RHIC is only due to the dissociation of χ_c and ψ'. In normal conditions these mesons would undergo radiative decays, with their typical branching ratios, into J/ψ's: actually, about 40% of the observed J/ψ originate in this way (it is not possible, in the actual experimental setup to distinguish between "primary" and "secondary" J/ψ's). Therefore, a 40% suppression of J/ψ, the amount observed at SPS and RHIC, could be very well explained as the removal of the higher resonances contribution, the directly produced J/ψ's being unaffected. From the present status of the data it is not possible to choose between the two scenarios: it is necessary to have more precise experimental data at RHIC, possibly with different interacting nuclei.

The new LQCD results for the dissociation temperature of charmonium states aroused, at the beginning, some problem because they were in contrast with potential model calculations. The latter used the heavy quark free energy calculated on the lattice as a inter-quark potential: by solving the Schrödinger equation, the binding energy and radius of the charmonium states were deduced: in this way the dissociation temperature for J/ψ was found to be about T_c. The discrepancy has been solved in recent papers, like for instance Refs. [27,28]. Instead of using the free energy F as given by LQCD in the Schrödinger equation, the internal energy U is extracted by means of the thermodynamical equations:

$$F = U - TS \ , \qquad U = -T^2 \frac{\partial F/T}{\partial T} \ . \tag{3}$$

The internal energy U is a better candidate for the heavy quark potential to be used into the Schrödinger equation. In this way, the dissociation temperatures obtained for J/ψ and the other $c\bar{c}$ states are in good agreement with the LQCD results, as evident in Fig. 7, where it is shown that the binding energy for J/ψ vanishes between 1.7 and 2.2 T_c. The large uncertainty is due both to the uncertainty of the charm quark mass and to the different sets of LQCD results available, with 2 or 3 quark flavors or pure gauge.

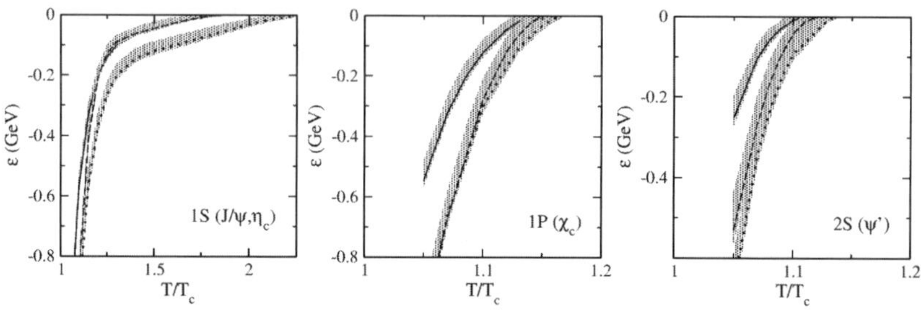

Fig. 7. Binding energies for charmonium states above T_c; the lines show the results for $m_c = 1.3$ GeV, the shadowed bands are the results for 1.15 GeV $< m_c <$ 1.35 GeV. The three different sets of results correspond to different LQCD data, with $N_f = 0, 2, 3$ respectively. From Ref. [27].

5. Exotic particles at RHIC and LHC

In addition to multi-heavy-flavored hadrons, discussed in Sec. 2.2, the high energies of RHIC and LHC should favor the production of exotic particles, like 4-quark mesons. An ideal candidate is the $f_0(980)$ meson, proposed to be a $qs\bar{q}\bar{s}$ bound state in Refs. [29–31] and already measured in heavy ion collisions [32]. The authors of Ref. [33] propose the measurement of p_T spectra to deduce the quark composition of $f_0(980)$, namely to decide whether $f_0(980)$ is formed by a $s\bar{s}$ pair or it is an exotic 4-quark state. The theoretical calculations rely upon a two component fragmentation/recombination model, validated by experimental information from STAR and PHENIX.

The soft part of the spectrum, at intermediate p_T, is assumed to be formed by the decay of a deconfined state of constituent quarks (gluon degrees of freedom are not included in the model) which recombine to form the observed hadrons; the consequence of this assumption is that the p_T distribution scales with the number of constituent quarks of a given hadron: this fact is well verified experimentally. The hard part is described by perturbative QCD with a simple implementation of the energy loss of the high energy partons traversing the medium. The p_T distribution of final particles is obtained by means of fragmentation functions which are tested by comparing the results with available experimental data.

The medium effects are studied through the ratios R_{CP} and R_{AA}, defined as follows:

$$R_{CP}(b) = \frac{N_{coll}(b)d^2N_{A-A}(b=0)/dP_T^2}{N_{coll}(b=0)d^2N_{A-A}(b=0)/dP_T^2} \tag{4}$$

$$R_{AA} = \frac{d^2N_{A-A}(b=0)/dP_T^2}{N_{coll}(b=0)d^2N_{p-p}/dP_T^2} \tag{5}$$

where $N_{coll}(b)$ is the mean number of nucleon-nucleon collisions at impact parameter b; the calculations are performed at $b = 12$ fm. More explicitly, R_{CP} is the ratio central/peripheral for a given observable in a nucleus-nucleus collision, whereas R_{AA} is the ratio for a given observable in central nucleus-nucleus to proton-proton collision at the same energy. The model is tested by calculating the ratios for many abundant particles: experimental spectra of pions, protons, kaons and lambdas are well reproduced. The spectrum for $f_0(980)$ is predicted with the two hypothesis about its composition and the results are shown in Fig. 8: $f_0 = s\bar{s}$ has a distribution very similar to the one of K_s (another quark-antiquark state) while the 4-quark state has a distribution more similar to $\Lambda + \bar{\Lambda}$. Similar properties are found by studying the R_{AA} ratios. The difference between the distributions in the two composition hypotheses for f_0 makes it possible, in principle, to discriminate between them by carefully analyzing the RHIC data. The practical difficulty is due to the limited statistics of these particles, which makes the p_T analysis very problematic. This problem could be overcome at LHC, where the larger energy available will, presumably, make the production of these particles more easy, and the analysis of their momentum distribution will be possible.

6. Particle correlations

Among the new experimental observable available at RHIC energies, the most interesting one is the so called "jet quenching", i.e. the strong modification of the spectrum of particles produced at high transverse momenta in ultrarelativistic heavy nucleus-nucleus collisions. At present its main ob-

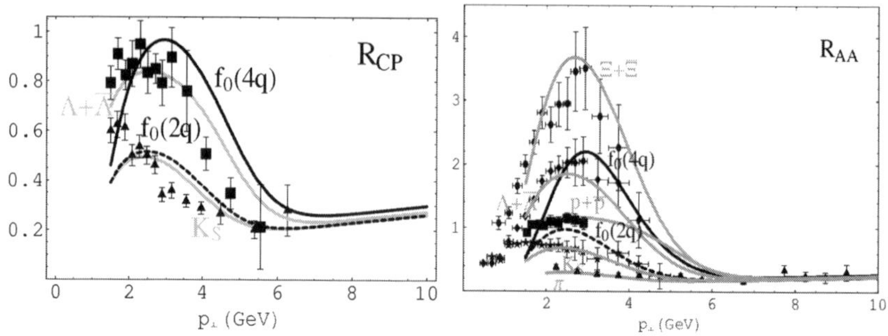

Fig. 8. Left: R_{CP} for K_s and $\Lambda + \bar{\Lambda}$ (gray lines), in Au-Au collisions at RHIC. Experimental data are from Ref. [34]. The peripheral collisions correspond to $b = 12$ fm. Black lines show the predictions for $f_0 = s\bar{s}$ (dashed) and $f_0 = qs\bar{q}\bar{s}$ (solid). Right: R_{AA} for different particles in Au-Au collisions at RHIC (gray lines), compared for the predictions for f_0 as a $s\bar{s}$ or as a 4-quark state, respectively dashed and solid black lines. Experimental data are from Ref. [35]. Figures from Ref. [33].

servable is the strong suppression of the inclusive particle yield in the whole available range of $p_T \lesssim 20$ GeV/c. This suppression can be understood as due to the energy loss of high-energy particles traversing the medium created in the collision. It is, of course, of the utmost importance to unravel where and how this energy was lost. The most promising experimental probes to investigate these issue are related to jet properties. Unfortunately, at RHIC energies, it is impossible to isolate a jet from the surrounding background: this will, hopefully, achievable only at LHC. Alternatively, with the RHIC data it is possible to study 2- or even 3-particle correlation, which should bring, at least partially, the desired information about parton-energy loss in the deconfined medium. The first RHIC data presented on this subject are already very intriguing. In particular, for some specific choice of the trigger particle (i.e. of the particle with maximum momentum in a given event) in central Au-Au collisions at the top RHIC energy, the distribution of accompanying particles is not peaked in the opposite direction (how it would be naively expected and how it occurs in proton-proton, deuteron-Au, peripheral Au-Au and even in central Au-Au with different trigger momenta) but is has a non-trivial structure with two maxima symmetrically displaced around 180°. A very fascinating explanation of this phenomenon has been given as a "sonic boom", the shock wave emitted by a particle traversing a medium with speed larger than the sound. However, before accepting this interpretation, one should explore all the alternative, and more traditional, explanations involving known properties of high density

QCD matter. There are two competing contributions that give the final p_T distribution of particles observed in the detector: initial state effects, which determine how many partons of a given p_T were produced, and final state interactions, with the medium produced during the reaction.

Initial state effects are dominated by parton saturation. At very high colliding energy the nucleus wave function is probed at very small Bjorken-x (fraction of longitudinal momentum carried by a parton), were the density of partons (mostly gluons) is so high that the normal evolution functions of perturbative QCD are not valid: the increase of the gluon distribution function for small x due to gluon splitting is compensated by the reverse process, gluon recombination; the gluon density, therefore, saturates. A simple model, based on this idea, is able to reproduce the global multiplicities and p_T distribution of hadrons produced at RHIC at different energies and for different interacting systems [36–39]. In Ref. [40] the same effect is found to influence also the angular correlation of produced particles. This work presents a quantitative study of angular correlation in the two particle spectrum produced by an energetic quark scattering off a dense hadronic target with sizable saturation momentum. The process studied in details is $qA \to q(\vec{k})g(\vec{p})X$: the vectors in parenthesis are the three-momenta of the quark and the gluon, X represents the unobserved fragments. The in-

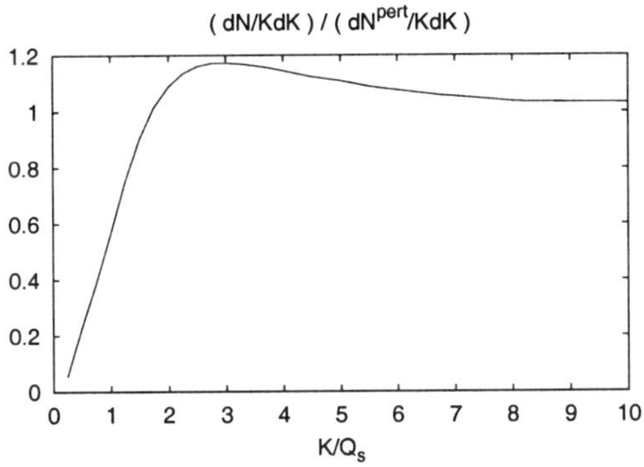

Fig. 9. The quark-gluon correlator normalized to its perturbative limit as a function of the total pair momentum K. From Ref. [40].

clusive process in which only the total transverse momentum $K = |\vec{k} + \vec{p}|$ of the qg pair in the final state is recorded serves as a check of the model: Fig. 9 shows the K distribution normalized to the perturbative (i.e. without saturation) results. It can be seen that the perturbative limit is recovered for momenta much larger than the saturation scale Q_s. Figure 10 shows the results for the two-particle distribution: for large momenta of the observed particles, the correlation is predominantly back-to-back, as it should be in the perturbative regime. As the trigger momentum gets closer to the saturation scale Q_s, the angular distribution of the accompanying particles broadens. The most interesting effect is observed when the momenta

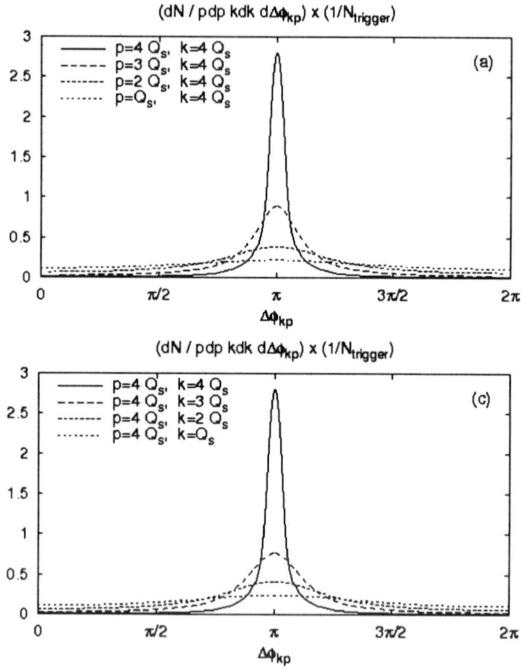

Fig. 10. The azimuthal dependence of the quark-gluon correlator for different values of quark (k) and gluon (p) transverse momentum. The trigger particle is either a quark (top panel) or a gluon (bottom panel) of momentum 4 Q_s. From Ref. [40].

are significantly smaller than Q_s: in general the azimuthal distribution is broad but still peaked back-to-back, however, in a narrow momentum range $0.5\ Q_s < p_{T trigger} < 1.5\ Q_s$, the azimuthal correlation splits into a double peak with maxima displaced away from 180°: an example is shown in Fig. 11. This behavior is in qualitative agreement with experimental data obtained at RHIC energies, but for a detailed and quantitative comparison more work is needed.

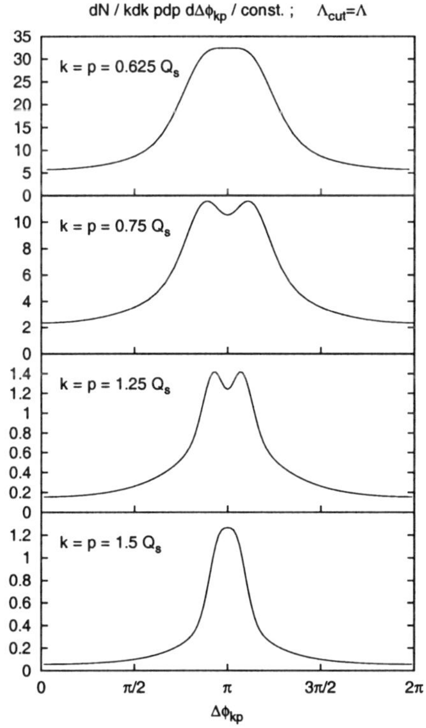

Fig. 11. The azimuthal dependence of the quark-gluon correlator for different values of quark (k) and gluon (p) transverse momentum close to the saturation scale. From Ref. [40].

References

1. See contributions by C. Ratti, M. Cristoforetti, M. Baldo, and G. Pagliara, in these Proceedings.
2. V. V. Begun, L. Ferroni, M. I. Gorenstein, M. Gazdzicki, and F. Becattini, J. Phys. **G32**, 1003 (2006).
3. F. Becattini and L. Ferroni, Eur. Phys. J. C **38**, 225 (2004).
4. F. Becattini and L. Ferroni, Eur. Phys. J. C **35**, 243 (2004).
5. F. Becattini, J. Manninen, and M. Gazdzicki, Phys. Rev. C **73**, 044905 (2006).
6. F. Becattini, A. Keranen, L. Ferroni, and T. Gabbriellini, Phys. Rev. C **72**, 064904 (2005).
7. H. van Hees and R. Rapp, Phys. Rev. C **71**, 034907 (2005).
8. F. Becattini, Phys. Rev. Lett. **95**, 022301 (2005).
9. A. V. Berezhnoi, V. V. Kiselev, A. K. Likhoded, and A. I. Onishchenko, Phys. Rev. D **57**, 4385 (1998).
10. M. A. Gomshi Nobary and R. Sepahvand, Nucl. Phys. **B741**, 34 (2006).

11. K. Geiger and B. Muller, Nucl. Phys. **B369**, 600 (1992).
12. K. Geiger, Phys. Rept. **258**, 237 (1995).
13. Z. G. Tan, D.-M. Zhou, S. Terranova, and A. Bonasera, Kinetic description of a finite temperature meson gas (2006), eprint: `nucl-th/0606055`.
14. S. Terranova, D. M. Zhou, and A. Bonasera, Eur. Phys. J. A **26**, 333 (2005).
15. T. Matsui and H. Satz, Phys. Lett. **B178**, 416 (1986).
16. L. Maiani, F. Piccinini, A. D. Polosa, and V. Riquer, Nucl. Phys. **A748**, 209 (2005).
17. L. Maiani, F. Piccinini, A. D. Polosa, and V. Riquer, Nucl. Phys. **A741**, 273 (2004).
18. V. Laporta, A. D. Polosa, F. Piccinini, and V. Riquer, J/ψ strong couplings to the vector mesons (2005), eprint: `nucl-th/0502009`.
19. A. Deandrea, G. Nardulli, and A. D. Polosa, Phys. Rev. D **68**, 034002 (2003).
20. A. Deandrea, N. Di Bartolomeo, R. Gatto, G. Nardulli, and A. D. Polosa, Phys. Rev. D **58**, 034004 (1998).
21. A. D. Polosa, Riv. Nuovo Cim. **23N11**, 1 (2000).
22. M. C. Abreu *et al.*, Phys. Lett. **B477**, 28 (2000).
23. M. C. Abreu *et al.*, Phys. Lett. **B521**, 195 (2001).
24. F. Becattini, L. Maiani, F. Piccinini, A. D. Polosa, and V. Riquer, Phys. Lett. **B632**, 233 (2006).
25. NA60 Collab, `http://na60.cern.ch/www`.
26. F. Becattini, M. Gazdzicki, A. Keranen, J. Manninen, and R. Stock, Phys. Rev. C **69**, 024905 (2004).
27. W. M. Alberico, A. Beraudo, A. De Pace, and A. Molinari, Phys. Rev. D **72**, 114011 (2005).
28. C.-Y. Wong, Phys. Rev. C **65**, 034902 (2002).
29. L. Maiani, F. Piccinini, A. D. Polosa, and V. Riquer, Phys. Rev. Lett. **93**, 212002 (2004).
30. L. Maiani, F. Piccinini, A. D. Polosa, and V. Riquer, Phys. Rev. D **71**, 014028 (2005).
31. R. L. Jaffe, Phys. Rept. **409**, 1 (2005).
32. J. Adams *et al.*, Phys. Rev. Lett. **92**, 092301 (2004).
33. L. Maiani, A. D. Polosa, V. Riquer, and C. A. Salgado, Counting valence quarks at rhic and lhc (2006), eprint: `hep-ph/0606217`.
34. H. Long, J. Phys. **G30**, S193 (2004).
35. S. Salur, System and energy dependence of strangeness production with star (2006).
36. D. Kharzeev and M. Nardi, Phys. Lett. **B507**, 121 (2001).
37. D. Kharzeev and E. Levin, Phys. Lett. **B523**, 79 (2001).
38. D. Kharzeev, E. Levin, and M. Nardi, Phys. Rev. C **71**, 054903 (2005).
39. D. Kharzeev, E. Levin, and M. Nardi, Nucl. Phys. **A730**, 448 (2004).
40. R. Baier, A. Kovner, M. Nardi, and U. A. Wiedemann, Phys. Rev. D **72**, 094013 (2005).

PHASES OF QCD: LATTICE THERMODYNAMICS, QUASIPARTICLES AND POLYAKOV LOOP

C. RATTI

ECT, I-38050 Villazzano (Trento) Italy and INFN, Gruppo Collegato di Trento, via Sommarive, I-38050 Povo (Trento) Italy*
E-mail: ratti@ect.it

S. RÖSSNER and W. WEISE

Physik-Department, Technische Universität München, D-85747 Garching, Germany
E-mail: sroessne@ph.tum.de

The thermodynamics of $N_f = 2$ quark flavors is investigated within an updated version of the PNJL model. Quarks interact through chiral four-point couplings and propagate in a homogeneous Polyakov loop background. The input is fixed exclusively by selected pure-gauge lattice QCD results and by pion properties in vacuum. Successful comparison with lattice QCD thermodynamics is obtained, both at zero and at finite quark chemical potential μ. Convergence properties of the Taylor series in μ/T are discussed within our model.

1. Introduction

In recent years, increasing activity has been devoted to the study of the QCD phase structure. Numerical simulations on the lattice are one of the major tool for this investigation. The equation of state of strongly interacting matter is now at hand as a function of temperature T and in a limited range of quark chemical potential μ. Improved multi-parameter reweighting techniques [1,2], Taylor series expansion methods [3–5] and analytic continuation from imaginary chemical potential [6–9] provide lattice data for the pressure, entropy density, quark density and selected susceptibilities.

The investigation of the phase diagram and thermodynamics of QCD within a field theoretical quasiparticle approach can be very useful in order to interpret lattice QCD results and extrapolate into regions not accessible by lattice computations. A promising ansatz of this sort is the PNJL model [10–14], a synthesis of Polyakov loop dynamics with the Nambu Jona-Lasinio model, combining the two principal non-perturbative features of low-energy QCD: confinement and spontaneous chiral symmetry breaking. This model has recently been used successfully in comparisons with a variety of lattice QCD data [14–22]. In this work, we summarize the most important results that we have obtained within the model, concerning $N_f = 2$ thermodynamics at zero and finite chemical potential*.

2. The model

The Euclidean action of the two-flavor PNJL model is

$$\mathcal{S}_E(\psi,\psi^\dagger,\phi) = \int_0^{\beta=1/T} d\tau \int d^3x \left[\psi^\dagger\,\partial_\tau\,\psi + \mathcal{H}(\psi,\psi^\dagger,\phi)\right] + \delta\mathcal{S}_E(\phi,T). \quad (1)$$

Here $\mathcal{H}$ is the fermionic Hamiltonian density given by:

$$\mathcal{H} = -i\psi^\dagger\left(\vec{\alpha}\cdot\vec{\nabla} + \gamma_4\,m_0 - \phi\right)\psi + \mathcal{V}(\psi,\psi^\dagger)\ , \quad (2)$$

where ψ is the $N_f = 2$ doublet quark field, $\vec{\alpha} = \gamma_0\,\vec{\gamma}$ and $\gamma_4 = i\gamma_0$ in terms of the standard Dirac γ matrices and $m_0 = diag(m_u, m_d)$ is the quark mass matrix. $\mathcal{V}(\psi,\psi^\dagger)$ is an $SU(2)\times SU(2)$ invariant four-fermion interaction acting in the pseudoscalar-isovector/scalar-isoscalar quark-antiquark channel†

$$\mathcal{V}\left(\psi,\psi^\dagger\right) = -\frac{G}{2}\left[\left(\bar{\psi}\psi\right)^2 + \left(\bar{\psi}\,i\gamma_5\vec{\tau}\,\psi\right)^2\right]. \quad (3)$$

Quarks move in a background color gauge field $\phi \equiv A_4 = iA_0$, where $A_0 = \delta_{\mu 0}\,g\mathcal{A}_a^\mu\,t^a$ with the $SU(3)_c$ gauge fields $\mathcal{A}_a^\mu$ and the generators $t^a = \lambda^a/2$. The matrix valued, constant field ϕ relates to the (traced) Polyakov loop as follows:

$$\Phi = \frac{1}{N_c}\mathrm{Tr}\left[\mathcal{P}\exp\left(i\int_0^\beta d\tau A_4\right)\right] = \frac{1}{3}\mathrm{Tr}\,e^{i\phi/T}\ , \quad (4)$$

In a convenient gauge (the so-called Polyakov gauge), one can choose a diagonal representation for the matrix ϕ,

$$\phi = \phi_3\,\lambda_3 + \phi_8\,\lambda_8\ , \quad (5)$$

*This presentation follows in large parts that of Ref. [14] where additional diquark degrees of freedom have also been treated explicitly.

†The scalar diquark interaction that has been considered in [14] is not relevant in the present discussion since the range of chemical potentials explored here is lower than the critical μ_q for diquark condensation.

which leaves only two independent variables, ϕ_3 and ϕ_8.

The piece $\delta\mathcal{S}_E = -\frac{V}{T}\mathcal{U}(\Phi,\Phi^*,T)$ of the action (1) controls the thermodynamics of the Polyakov loop. An improved form for the effective potential $\mathcal{U}(\Phi,\Phi^*,T)$ was introduced in our previous work [14]. It contains the logarithm of $J(\Phi)$, the Jacobi determinant which results from integrating out six non-diagonal group elements while keeping the two diagonal ones, $\phi_{3,8}$, to represent Φ:

$$\frac{\mathcal{U}(\Phi,\Phi^*,T)}{T^4} = -\frac{1}{2}a(T)\,\Phi^*\Phi + b(T)\,\ln\left[1 - 6\,\Phi^*\Phi + 4\left(\Phi^{*3} + \Phi^3\right) - 3\,(\Phi^*\Phi)^2\right] \tag{6}$$

with

$$a(T) = a_0 + a_1\left(\frac{T_0}{T}\right) + a_2\left(\frac{T_0}{T}\right)^2, \qquad b(T) = b_3\left(\frac{T_0}{T}\right)^3. \tag{7}$$

The logarithmic divergence of $\mathcal{U}(\Phi,\Phi^*,T)$ as $\Phi,\ \Phi^* \to 1$ automatically limits the Polyakov loop Φ to be always smaller than 1, reaching this value asymptotically only as $T \to \infty$. The parameters a_i and b_3 are taken from Ref. [14], where they were determined to reproduce lattice data for the thermodynamics of pure gauge lattice QCD up to about twice the critical temperature. At much higher temperatures, where transverse gluons begin to dominate, the PNJL model is not supposed to be applicable. The resulting parameters are

$$a_0 = 3.51\ ,\quad a_1 = -2.47\ ,\quad a_2 = 15.22\ ,\quad b_3 = -1.75\ .$$

The critical temperature T_0 for deconfinement in the pure gauge sector is fixed at 270 MeV in agreement with lattice results. The results of this combined fit are shown in Fig.1. The NJL part of the model involves three parameters: the bare quark mass which we take equal for u- and d-quarks, the coupling strength G and a three-momentum cutoff Λ. We take those from Ref. [13]:

$$m_{u,d} = 5.5\text{ MeV}\ ,\quad G = 10.1\text{ GeV}^{-2}\ ,\quad \Lambda = 0.65\text{ GeV}\ ,$$

which were fixed to reproduce the pion mass and decay constant in vacuum and the chiral condensate as $m_\pi = 139.3$ MeV, $f_\pi = 92.3$ MeV and $\langle\bar{\psi}_u\psi_u\rangle = -(251\text{ MeV})^3$.

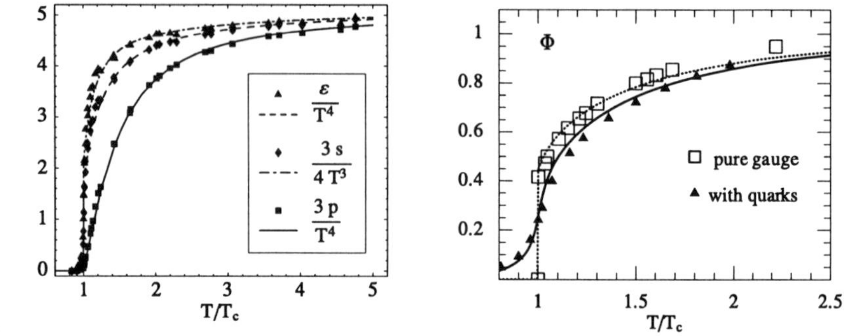

Fig. 1. Left: fit to scaled pressure, entropy density and energy density as functions of the temperature in the pure gauge sector, compared to the corresponding lattice data taken from Ref. [23]. Right: using the fit of the Polyakov loop (dotted line) to lattice results taken from Ref. [24] in the pure gauge sector (empty symbols), the PNJL model predicts the Polyakov loop behaviour as a function of temperature in the presence of dynamical quarks (solid line). This prediction is compared to lattice data in two flavours (full symbols) taken from Ref. [25].

3. Results

After performing a bosonization of the PNJL Lagrangian and introducing the auxiliary fields σ and $\vec{\pi}$ the thermodynamic potential results as follows:

$$\begin{aligned}\Omega\left(T,\mu,\sigma,\Phi,\Phi^*\right)=&\mathcal{U}\left(\Phi,\Phi^*,T\right)\\&+\frac{\sigma^2}{2G}-2N_f\int\frac{\mathrm{d}^3p}{(2\pi)^3}\left\{3E_p\theta\left(\Lambda^2-\vec{p}^{\,2}\right)\right.\\&+T\ln\left[1+3\Phi\mathrm{e}^{-(E_p-\mu)/T}+3\Phi^*\mathrm{e}^{-2(E_p-\mu)/T}+\mathrm{e}^{-3(E_p-\mu)/T}\right]\\&\left.+T\ln\left[1+3\Phi^*\mathrm{e}^{-(E_p+\mu)/T}+3\Phi\mathrm{e}^{-2(E_p+\mu)/T}+\mathrm{e}^{-3(E_p+\mu)/T}\right]\right\}\end{aligned} \tag{8}$$

where the quark quasiparticle energy is $E_p=\sqrt{\vec{p}^{\,2}+m^2}$ and the dynamical (constituent) quark mass is the same as in the standard NJL model: $m = m_0 - \langle\sigma\rangle = m_0 - G\langle\bar{\psi}\psi\rangle$.

In general, the fields Φ and Φ^* (and their thermal expectation values) are different at non-zero quark chemical potental [26]. We demonstrate in [14] that $\Phi \neq \Phi^*$ at $\mu \neq 0$ is primarily a consequence of quantum fluctuations of the fields around their mean field values. In the present work we stay at the mean-field limit which consistently implies $\Phi = \Phi^*$.

Minimizing $\Omega(T,\ \mu,\ \sigma,\ \Phi)$ determines the chiral condensate σ and the Polyakov loop Φ as functions of temperature and quark chemical potential. The right panel of Fig. 1 shows the temperature dependence of the

Polyakov loop for two situations: first for the pure-gauge case (dotted line), and secondly for the case including quarks at $\mu = 0$ (full line). While the pure-gauge lattice results have been used to fix the effective potential for the Polyakov loop as explained previously, the case with inclusion of quarks is a prediction of the model without any additional tuning of parameters. The resulting agreement with corresponding lattice data incorporating dynamical quarks [25] is striking.

Notice that, while the Polyakov loop shows a discontinuity at the critical temperature in the pure gauge system indicating a first order deconfinement phase transition, this turns into a smooth crossover when quarks are introduced. At the same time the critical temperature is reduced from $T_0 = 270$ MeV in the pure gauge case to around $T_c = 215$ MeV in the presence of quarks (not evident from Fig. 1 since the horizontal axis is scaled by T_c).

At non-zero chemical potential the lattice "data" are deduced from a Taylor expansion of the thermodynamic quantities in powers of μ/T around $\mu = 0$. For consistency reasons, we perform the same expansion in the PNJL model, and compare the lattice expansion coefficients to those calculated in our model. We have:

$$\frac{p\left(T,\mu\right)}{T^4} = \sum_{n=0}^{\infty} c_n\left(T\right)\left(\frac{\mu}{T}\right)^n ;$$
$$c_n\left(T\right) = \left.\frac{1}{n!}\frac{\partial^n\left(p\left(T,\mu\right)/T^4\right)}{\partial\left(\mu/T\right)^n}\right|_{\mu=0} \tag{9}$$

Our results for the coefficients c_2 and c_4 are shown in Fig. 2. Once a limited set of input parameters is fitted to Lattice QCD in the pure gauge sector and to pion properties in the hadronic sector, the PNJL model evidently provides a very good description of the lattice data. The results for c_6 (not shown here) are also in very good agreement with the lattice data and can be found in Ref. [14].

From these coefficients the Taylor-expanded pressure (9) is reconstructed and the quark number density follows as:

$$\frac{n_q}{T^3} = 2c_2\left(\frac{\mu}{T}\right) + 4c_4\left(\frac{\mu}{T}\right)^3 + 6c_6\left(\frac{\mu}{T}\right)^5 + ... \tag{10}$$

Fig. 3 displays results for the pressure difference, $\Delta p = p(T,\mu) - p(T,\mu = 0)$, and for the quark number density, both expanded around $\mu = 0$ up to fourth order in quark chemical potential, consistently with the corresponding lattice data. The full computation of the pressure and quark number density at finite chemical potential cannot be performed on the lattice,

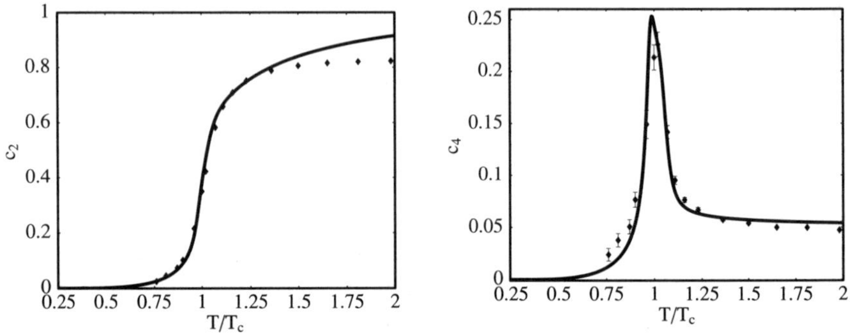

Fig. 2. Expansion coefficients c_2 and c_4: the PNJL model results are compared with the corresponding lattice results taken from Ref. [5].

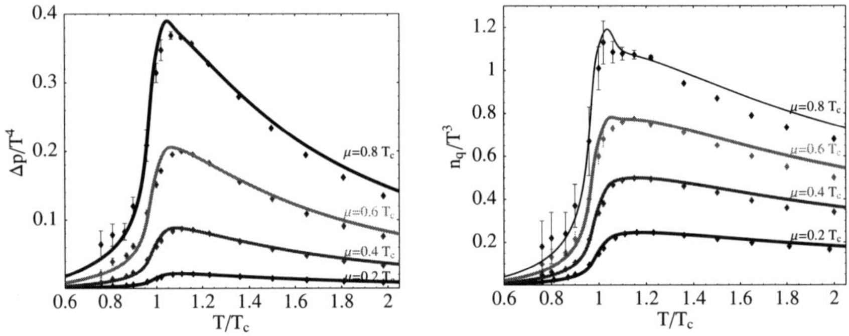

Fig. 3. Scaled pressure difference (left) and quark number density (right) as functions of T/T_c for different values of μ/T_c. In both cases the lattice data, obtained from a Taylor expansion to fourth order in quark chemical potential [4], are compared to the PNJL model results, also obtained from a Taylor expansion up to fourth order in quark chemical potential.

but it is instead possible in our model. We can thus investigate convergence properties of the Taylor series and study discrepancies between the full and truncated results. This study is shown in Fig. 4. The continuous lines (full results) are compared to second order (dashed) and fourth order (dotted) expansions in $\frac{\mu}{T}$. Evidently the series converges rapidly for small chemical potentials: the second order differs by less than 5% from the full result, and the fourth order basically coincides with the solid curve. It is nevertheless also evident that when μ/T_c becomes larger, of order 1, the Taylor expansion develops problems around T_c. In this region in fact, the expansion coefficient c_4 is rather large, and for this reason the fourth order contribution overshoots the full one around T_c. When T becomes larger, the coefficient c_4 drops and for this reason the agreement between the full result and the truncated one is again very good.

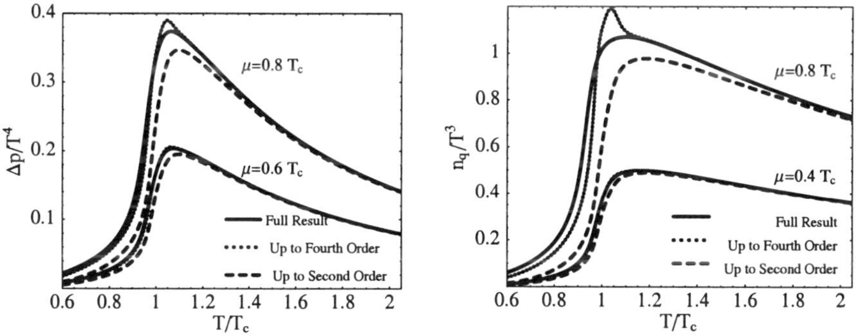

Fig. 4. Pressure difference (left) and quark number density (right) as functions of T/T_c for two different values of quark chemical potential. The full result (continuous line) is compared to the Taylor expansion up to second (dashed line) and fourth order (dotted line) in quark chemical potential.

4. Conclusions

The PNJL model represents a minimal synthesis of the two basic principles that govern QCD at low temperatures: spontaneous chiral symmetry breaking and confinement. The respective order parameters (the chiral quark condensate and the Polyakov loop) are given the meaning of collective degrees of freedom. Quarks couple to these collective fields according to the symmetry rules dictated by QCD itself. Once a limited set of input parameters is fitted to Lattice QCD in the pure gauge sector and to pion properties in the hadron sector, the quark-gluon thermodynamics above T_c up to about twice the critical temperature is well reproduced. In particular, the coefficients of the Taylor expansion in μ/T are in remarkably good agreement with the ones coming from lattice simulations. The PNJL model correctly describes the step from the first-order deconfinement transition observed in pure-gauge Lattice QCD to the crossover transition when $N_f = 2$ light quark flavours are added. The predicted behaviour of the Polyakov loop as a function of temperature in the presence of two quark flavours is in surprisingly good agreement with the corresponding lattice data: a highly non-trivial result.

The Taylor-expanded pressure difference and quark number density of the system have been calculated up to sixth order in quark chemical potential. The agreement with corresponding lattice data is again very good. Finally, a comparison between the full result for these two quantities and the truncated one has been performed. We observe fast convergence of the expansion in powers of μ/T at small chemical potential. At larger chemical potentials, comparable to T_c, limitations in the applicability of the Taylor expansion become apparent.

References

1. Z. Fodor, S. D. Katz, and K. K. Szabo, Phys. Lett. **B568**, 73 (2003).
2. Z. Fodor and S. D. Katz, JHEP **0203**, 014 (2002).
3. C. R. Allton *et al.*, Phys. Rev. D **66**, 074507 (2002).
4. C. R. Allton *et al.*, Phys. Rev. D **68**, 014507 (2003).
5. C. R. Allton *et al.*, Phys. Rev. D **71**, 054508 (2005).
6. E. Laermann and O. Philipsen, Ann. Rev. Nucl. Part. Sci. **53**, 163 (2003).
7. P. de Forcrand and O. Philipsen, Nucl. Phys. **B673**, 170 (2003)
8. M. D'Elia and M. P. Lombardo, Phys. Rev. D **67**, 014505 (2003).
9. M. D'Elia and M. P. Lombardo, Phys. Rev. D **70**, 074509 (2004).
10. K. Fukushima, Phys. Lett. **B591**, 277 (2004).
11. P. N. Meisinger and M. C. Ogilvie, Nucl. Phys. Proc. Suppl. **47**, 519 (1996), arXiv:hep-lat/9509050; Phys. Lett. **B379**, 163 (1996), arXiv:hep-lat/9512011.
12. E. Megias, E. Ruiz Arriola, and L. L. Salcedo, Phys. Rev. D **74**, 065005 (2006).
13. C. Ratti, M. A. Thaler, and W. Weise, Phys. Rev. D **73**, 014019 (2006); C. Ratti, M. A. Thaler, and W. Weise, nucl-th/0604025.
14. S. Roessner, C. Ratti, and W. Weise, arXiv:hep-ph/0609281.
15. S. K. Ghosh, T. K. Mukherjee, M. G. Mustafa, and R. Ray, Phys. Rev. D **73**, 114007 (2006).
16. C. Ratti, M. A. Thaler, and W. Weise, arXiv:nucl-th/0604025.
17. E. Megias, E. Ruiz Arriola, and L. L. Salcedo, arXiv:hep-ph/0607338.
18. H. Hansen *et al.*, arXiv:hep-ph/0609116.
19. C. Ratti, S. Roessner, M. A. Thaler, and W. Weise, Eur. Phys. J. C (2006) in press, arXiv:hep-ph/0609218.
20. S. Mukherjee, M. G. Mustafa, and R. Ray, arXiv:hep-ph/0609249.
21. Z. Zhang and Y. X. Liu, arXiv:hep-ph/0610221.
22. C. Sasaki, B. Friman, and K. Redlich, arXiv:hep-ph/0611147.
23. G. Boyd *et al.*, Nucl. Phys. **B469**, 419 (1996).
24. O. Kaczmarek, F. Karsch, P. Petreczky, and F. Zantow, Phys. Lett. **B543**, 41 (2002).
25. O. Kaczmarek and F. Zantow, Phys. Rev. D **71**, 114510 (2005).
26. A. Dumitru, R. D. Pisarski, and D. Zschiesche, Phys. Rev. D **72**, 065008 (2005).

Experimental Nuclear Physics in Italy

I LABORATORI NAZIONALI DEL SUD

R. Alba

Istituto Nazionale di Fisica Nucleare, Laboratori Nazionali del Sud, Via S. Sofia 62, 95123 Catania, Italy

A status report of the Laboratori Nazionali del Sud is provided. Instead of illustrating the details of the activities, focus is put on the more recent achievements mainly in the perspective of the future developments.

1. Introduction

The Laboratori Nazionali del Sud, located in the campus area of the Catania University, host two accelerators: a 15 MV terminal high voltage electrostatic Tandem and a K=800 Superconductive Cyclotron (CS) [1]. A large variety of ions is available, with energies from a few MeV to 80A MeV for the lightest ions accelerated by the CS.

Around the accelerators an intense experimental activity with heavy ion beams is performed, spanning from nuclear physics to applications and interdisciplinary research.

Fig.1 shows, as an example of the time distribution among the different fields, the situation of the accelerator operation in 2005.

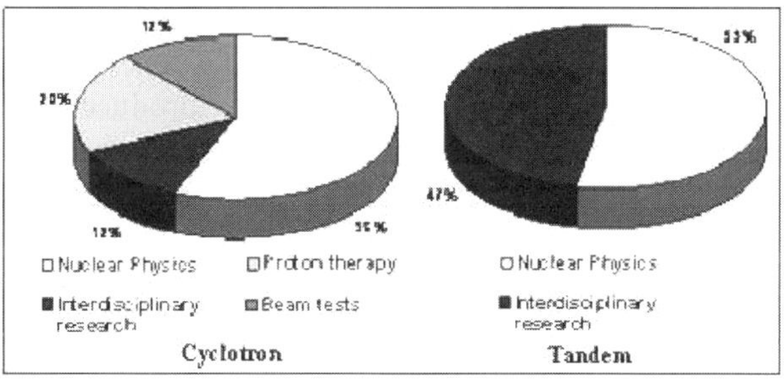

Fig. 1. Percentage of the total accelerator operation time devoted in 2005 to the different activities.

It can be seen that more than 50% of the total beam time is used for nuclear physics, but also interdisciplinary research, mainly aimed at the study of the interaction of ion beams with biological matter and of the ion induced material radiation damage, plays a relevant role. Moreover, a not negligible fraction of the CS operation time is devoted to the proton therapy (CATANA project).

In 2005 more than 10% of the CS beam has been used for tests. This originates in the attempt of increasing the intensity of the beam extracted, in order to reach the goals of the exotic ion beam facility EXCYT. Going through progressive improvements of materials and design of the electrostatic deflector, the extracted beam power has recently reached the value of about 200 Watt. To stress the excellent work done it is worth to note that, with the original deflector, the beam intensities of the Cyclotron were an order of magnitude lower when the EXCYT project planned to use 500 Watt of primary beam.

The activity related to the accelerators, including maintenance, new beam production and technological developments, represents a fraction of the LNS activities. With and without accelerators, the LANDIS laboratory, devoted to the study of cultural heritage, has reached excellent results and new perspectives are opened by the constant development of new techniques. Also, the researches in the field of high energy neutrino astrophysics have in the LNS a relevant reference point with the NEMO project, whose goal is the construction of a neutrino telescope in the Mediterranean Sea.

2. Exotic beams at LNS

2.1. *EXCYT*

EXCYT is the main facility for the production of exotic beams at LNS. It is based on the use of the two accelerators of LNS: the cyclotron as driver and the Tandem as final accelerator. The layout of the facility is shown in fig.2. The main feature of the beams produced are those of a Tandem accelerator, i.e. easily variable energy, good emittance and very low energy spread. Beams with energies from a few MeV till about 100 MeV can be produced. For special requirements beams of some hundreds of KeV, extracted before the Tandem, can also be used. The facility is now ready and fully tested. According with the requests of the researchers, the first beam chosen to start experiments is ^{8}Li, so it has also been used for the commissioning. First, with an intense stable ^{7}Li beam, produced in the target ion source complex, the efficiency of the charge exchange cell and the transmission through the different stages of the transport line to the Tandem have been measured and these values have been taken as a reference for the transport of the ^{8}Li. In July 2006 the ^{8}Li beam, produced via the reaction $^{1}3C+^{1}2C$ at 45A MeV using a 147 Watt primary beam, has been successfully extracted from the Tandem with an intensity of 2.1 10^{4} pps, intensity that, on the basis of the tests with the ^{7}Li, can still be improved of at least a factor 3 optimizing the transmission through the Tandem.

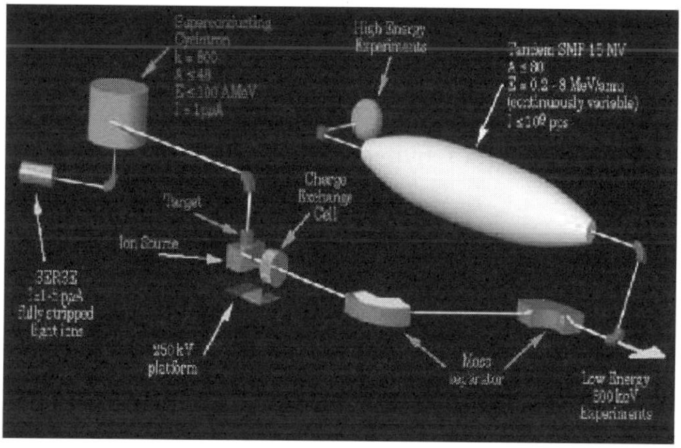

Fig. 2. Layout of the EXCYT facility.

An intense R& D activity is still in progress to improve the intensities. In particular, different projectile-target combinations, a new design and material for the target and strategies to increase the intensity of the primary beam are under study. However, the intensity already obtained is good enough to start the experimental program already approved by the LNS Scientific Committee:

- Measurement of the ^{8,9}Li + ^{28}Si reaction cross section at near barrier energies to determine the size of the unstable Li isotopes.
- Measurement of the α -^{8,9}Li elastic scattering excitation functions in reverse kinematics, aimed at studying backward angle resonances associated with cluster configurations of ^{12,13}B.
- Measurement of the ^{8}Li(αn)^{11}B cross section in the c.o.m. energy range from about 1.5 MeV down to the Gamow peak ($\simeq$0.5 MeV).

2.2. *FRIB's*

Born as an experiment on specific interests of a group of researchers, FRIBs is going to become a laboratory facility for in-flight production of exotic beams at intermediate energy. It is based on the fragmentation of projectiles accelerated by the CS. The layout is shown in fig.3. The production target is located close to the exit of the cyclotron and the exotic fragments produced are selected and transported by the same optical elements of the line normally used to transport the beams to the experimental areas, used, in this case, as a fragment separator. Since all the species produced with the same Bρ are transported along the beam line, it is necessary to identify event by event the ion impinging on the experimental target. This is done with a tagging detector located before the experimental point. In this way, it is also possible to perform different experiments at the same time, just selecting off line the beam required.

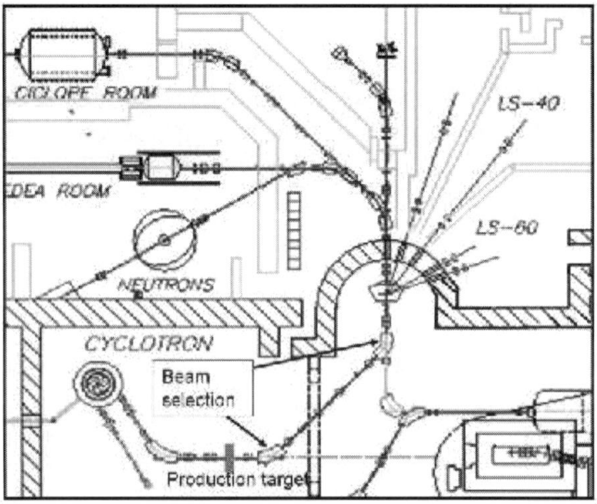

Fig. 3. Layout of the FRIBs facility.

Several tests have been made up to now to study not only the production of the exotic species for different projectile-target combination, but also to improve the transmission along the beam lines to the experimental points that host the large experimental devices of LNS. Rates of the order of some 10^4 have been obtained and already used in experiments aimed at the study of the fragmentation of exotic projectiles and of the diproton decay of ^{18}Ne.

3. Experimental nuclear physics at LNS

The experimental nuclear physics at LNS is mainly concentrated around the large devices already existing or under construction. In the following I will not present all the results obtained, that can be easily found in literature, but I will try to focus on the main research lines and perspectives, with only a few examples. It is to underline that, mainly in the difficult field of intermediate energies (20–100A MeV), where the comparison with theoretical models is crucial for the interpretation of the observed signals, the experimental activity takes great advantage of the collaboration of the LNS theoretical group.

The LNS are one the very few laboratories in the world where intermediate energies are available. The more exciting research line in this energy regime is the study of the nuclear Equation of State. In particular, in nuclear collisions induced by radioactive or stable neutron-rich nuclei, a transient state of nuclear matter with an appreciable isospin asymmetry as well as thermal and compressional excitation can be created. This offers the unique opportunity to study properties of highly isospin asymmetric nuclear matter. One of the most important goals of this study is to obtain information on the symmetry term of the nuclear equation of state (EOS). This is important, for instance, for understanding the explosion mechanism of supernova and the evolution of neutron stars. The study of nuclear multifragmentation is a very useful tool for this kind of investigations. The isotopic resolution of the products provides important information on the reaction dynamics and the observation of systematic trends, as, for instance, isoscaling, of isospin dependent observables can be related to the density dependence of the symmetry energy. The multidetector CHIMERA, fully operational at LNS since 2003, is mainly devoted to this kind of studies. With its about 1200 detectors, providing almost full solid angle coverage, and the capability of discriminating in charge and mass a large variety of the reaction products, Chimera represents at the moment the most powerful device in this field. An intense campaign of measurements has been done in the last years, calling many foreign users at LNS. Recently, the apparatus has also been upgraded to achieve very low identification thresholds and allow a complete r econstruction of the reaction events detecting also the emission from the target. For the future, some further improvements are foreseen, including the possibility of coupling part of the detector with other existing devices to explore particular physics topics. One possibility is to use the forward part of CHIMERA in connection with MEDEA+SOLE. MEDEA, made of 180 20 cm thick BaF_2 modules, allows a very efficient measurement of high energy photons and light charged particles while the superconducting solenoid SOLE collects the forward reaction products on a focal plane detector 15 meters from the target. The use of part of CHIMERA as focal plane detector opens, for instance, the possibility of a very good characterization of the evaporation residues in reverse kinematics (incomplete) fusion reactions. This could be very helpful in the studies on hot GDR suppression, that represent at the moment the main research line with MEDEA, mainly in the perspective of the determination of its isospin dependence. The main goal of this studies is the determination of the limiting excitation energy (or temperature) for the existence of the collective motion. The disappearance of this latter over a given temperature can be, in fact, interpreted as due to the transition of the nucleus from the liquid phase to the gaseous one. Indication for such behavior have already been found and further studies are in progress to map better the disappearance region for masses around 110–130.

Fig. 4. The multidetector CHIMERA.

An other well assessed research line at LNS is nuclear astrophysics, to which is devoted a large fraction of the Tandem beam time. Due to the very low energies involved, direct approaches to the study of nuclear reactions in astrophysics are very difficult because of the presence of Coulomb barrier. Usually, the experimental data at higher energies are extrapolated till the interesting ones. Also, the effect of the electron screening becomes important. Indirect methods provide a tool to overcome these problems. In the last years the Trojan Horse method has proven to be very successful. It consists in extracting a two body reaction cross section selecting the quasi-free contribution of an appropriate three body reaction. At LNS this method has been tested in several experimental condition, optimized and used for many measurements of bare nucleus astrophysical S (E) factor till very low energies. Important contributions on several astrophysics topics have been given, from the problem of the depletion of light elements in stellar environment to primordial nucleosynthesis, from electron screening to nucleosynthesis in AGB stars. Recently, the method has been tested also in cases in which the particle that participate in the reaction is a neutron. The results of the technique are very encouraging and this opens new perspectives since it would provide a virtual neutron source.

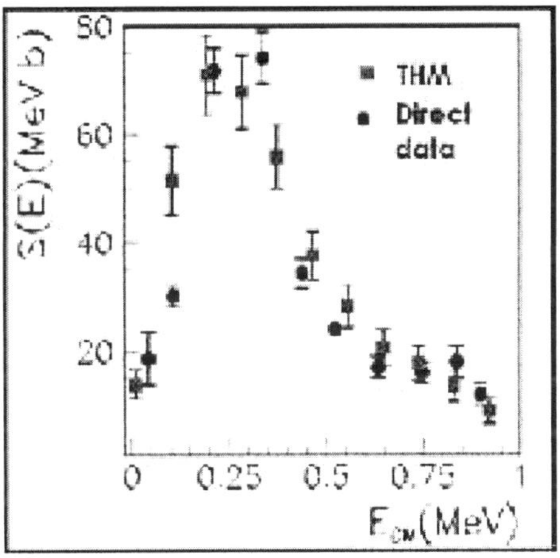

Fig. 5. The S(E) factor for the $^{9}Be(p,\alpha)^{6}Li$ reaction studied through the reaction $^{2}H(^{9}Be,\alpha\ ^{6}Li)n$.

The experimental programs for the next future will receive new hints from the availability of the EXCYT beams. The first experiment that will be performed is the measurement of the $^{8}Li(\alpha,n)^{11}B$ reaction cross section at Big Bang temperatures. The Standard Big Bang Model (SBBM) is able to predict, with very good agreement, the observed primordial abundances of the light elements ^{2}H, ^{3}He, ^{4}He and ^{7}Li. The basic assumption of the model is that during the phase of the primordial nucleosynthesis the universe had an homogeneous density distribution. However, theoretical studies have taken into account possible inhomogeneities in the baryon density arising from the quark-hadron phase transition, resulting in high-density proton rich and low-density neutron rich regions. In such an environment, heavy elements can be produced via the formation of unstable nuclei. One of the key unstable nuclei is ^{8}Li and the reaction ^{8}Li(α,n)^{11}B is a key one in the following reaction chain, very important for primordial nucleosynthesis leading to the formation of elements with A larger than 8, that represents the limit of the SBBM :

$$^{1}\mathrm{H}(\mathrm{n},\gamma)^{2}\mathrm{H}(\mathrm{n},\gamma)^{3}\mathrm{H}(\mathrm{d},\mathrm{n})^{4}\mathrm{He}(\mathrm{t},\gamma)^{7}\mathrm{Li}(\mathrm{n},\gamma)^{8}\mathrm{Li}(\alpha,\mathrm{n})^{11}\mathrm{B}(\mathrm{n},\gamma)^{12}\mathrm{B}(\beta,\nu)^{12}\mathrm{C}$$

At LNS the study of this reaction already started with a 8Li beam produced via direct reactions. With a ^{8}Li rate of about 150 pps a first measurement at 1.25 MeV c.o.m. energy has been made. But the relevant energy, corresponding to the Big Bang temperatures, ranges from 0.3 to 0.8 MeV. The aim of the experimental program with EXCYT is to explore this region and to study, eventually, the competing reactions ^{8}Li(d,t), ^{8}Li(d,p) and ^{8}Li(n,γ).

The advent of EXCYT at LNS will open new research fields involving the study of nuclear structure and reaction dynamics induced by nuclei with particular properties such as large values of isospin, haloes, large deformation. The MAGNEX spectrometer has been conceived in order to explore this physics, taking into account the RIB's relatively low intensities, wide energy (E=0.5 - 8 MeV/u) and mass ranges and the excellent optical properties of Tandem beams.

Of course, stable beams - including a limited range of those from CS - can also be used. MAGNEX is a large acceptance device (50 msr) based on a vertically focusing quadrupole and a bending dipole. A position sensitive start detector, located between the target and the quadrupole, gives both the angle of the particle and the start for the time of flight. The focal plane detector measures position and angles and provides the particle identification information. The spectrometer is now completely installed and commissioned. A program of ^{7}Li-induced charge exchange reactions is planned with MAGNEX at LNS. The (^{7}Li,^{7}Be) reaction at Tandem energies has been demonstrated to be a very useful tool to investigate spectroscopic properties of light nuclei, also far from the stability. Under these extreme conditions the standard representation of nuclei, based on the assumption of mean field dynamics, turns out to be often inadequate. In this context particular attention has been paid to the description of special correlations as the Dynamical Core Polarisation (DCP), i.e. the coupling of the loosely bound valence nucleon to the excited states of the core. Experimental signatures of the DCP correlations are found in transfer or charge exchange (CEX) reactions. Far from stability, in fact, also the core systems are more easily polarizable. Consequently the energy available in a binary reaction can be directly transferred to the core, being the valence neutron weakly influenced. One of the most striking observation is the appearance of very narrow resonances embedded in the continuum (called BSEC). Particularly interesting for this phenomenology is the exploration of the (Nα+3n) nuclei, for which an integer number of α particles is coupled to three extra neutrons. Examples are represented by ^{7}He, ^{11}Be, ^{15}C, 19O, ^{23}Ne, ^{27}Mg and heavier systems. For these nuclei one can assume that single particle excitations at low energy are mainly due to the t hree valence neutrons. Thus the three neutron dynamics in the nuclear medium can be studied from the measured energy spectra. The lighter system of this chain have already been studied in Orsay. The study of the heavier systems, even if difficult due to the strong decrease of cross sections, is very important in order to learn about the influence of mass and asymmetry parameter on these phenomena. The MAGNEX spectrometer at the LNS laboratory will allow to go on with the systematics by studying less favourable cases, fully exploiting its large solid angle and good energy resolution. Moreover, exploiting the EXCYT beam, spectroscopic studies of 8He are planned via the ^{7}Li(^{8}Li,^{7}Be)^{8}He reaction.

Fig. 6. The MAGNEX magnetic spectrometer.

4. The applications

4.1. *CATANA*

The facility for proton therapy CATANA, realized at LNS with the collaboration of the Catania University, started operation in 2002. It is based on the use of the 62 MeV proton beam of the CS and optimized for the treatment of the ocular diseases like uveal melanoma. The CS is the only machine in Italy and in the South of Europe used to this aim. Up to now 114 patients have been treated. Among them, 92 have been monitored after the treatment. Preliminary results of the follow-up for 56 patients at 6-8 months and 36 patients at 1 year indicate a size reduction, especially regarding the tumor thickness, in 75% of the patients, while in the remaining 25% a cessation of growth has been observed. The data evidence clearly a tumor control. For the next years, according to the recent agreement between INFN, Catania University, Centro Siciliano di Fisica Nucleare e Stuttura della Materia and Catania Azienda Policlinico, at least four irradiation sessions per year will be done.

Fig. 7. The CATANA treatment line.

Connected with this activity, a lot of experimental work, mainly on dosimetry and radiobiology, is done. Also, as a possible development of CATANA, the new Superconductive Cyclotron SCENT has been projected at LNS. This new machine is able to accelerate protons, lithium and carbon ions till a maximum energy of 300 MeV and is then particularly suitable to be used in a hadrotherapy center. Indeed, it has been proposed as the machine of the radiotherapy center of Cannizzaro hospital in Catania. Recently, INFN has approved a collaboration agreement with the IBA company to make this cyclotron available on the market. In case of sale from IBA to a final user, the first prototype will be realized under the responsibility of INFN.

4.2. *LANDIS*

The LANDIS (Laboratorio di Analisi Non Distruttiva In Situ), operational at LNS, is devoted to the applications of Nuclear Techniques to the Cultural Heritage field. The main scope is to develop new instruments, techniques and methods useful to solve problems posed by the art historians, archaeologists, museum conservators concerning the study, conservation and safeguard of Cultural Heritage masterpieces. The two main requirements are: the techniques must be non destructive; the instruments must be portable. To ensure the necessary link of LANDIS with the Cultural Heritage community, scientific collaborations with public institutions have been established; more specifically an agreement has been signed by INFN/LNS and the CNR/IBAM (Istituto per i Beni Archeologici e Monumentali).

Various new portable instruments have been realized at the LANDIS: the BSC-XRF spectrometer (a new system which enable to control the x-ray beam energy and intensity stability); an upgrading of a commercial XRD allowing to gain a factor 10 in statistics; the PIXE-alpha system, unique tool which allows to perform in situ PIXE analysis. The above systems have been widely used in several sites. Among the most relevant, Ravenna (Bibl. Classense), Rome (Istituto Centrale per la Patologia del Libro), Florence (Museo del Bargello), Heraklion (Archaeol. Museum), Atene (Benaki Museum), Creta (Festos and Prinias), Pylos (Nestor Palace) can be cited. Because of the specific applications and of the recent developments, some more details on the PIXE-alpha system are given. The system, realized in collaboration with the CEA/Saclay, is based on the use of a 210Po radioactive source, as charged particle emitter. With this system the light elements are well excited, but it has a low sensitivity for medium atomic number elements. The experience gained on the use of the PIXE-alpha system allowed to design and realize a new portable system XPIXE-alpha in which alpha particles and x-rays are used simultaneously as exciting beams. The radioactive source is ^{244}Cm. The new system permits to excite very well both light and medium atomic number elements. For its characteristics, this latter system can be also used for monitoring the atmospheric particulate matter. A new laboratory, the LAB-alfa, has been realised at the LNS to manipulate and realize the alpha emitter sources.

In parallel to the in situ activity, the deep proton activation analysis (DPAA) method for the determination of the chemical composition of roman coins has been developed. This non destructive method makes use of 22 MeV protons accelerated by the LNS Tandem to induce activation only inside the coins, minimizing the effects of the surface. Applied to ancient roman coins, DPPA has put in evidence, for instance, that in crisis periods, Romans used silver only for a thin layer at the surface of the coins while the interior was made with poorer materials.

5. NEMO

In the last years the scientific interests of LNS have been expanded in high energy neutrino astrophysics. High energy neutrinos may easily escape even from thick astrophysical sources, they are hardly absorbed during their propagation through the Universe and are not deflected by cosmic magnetic fields. They are, therefore, the ideal candidate particle to point back at their sources, but the quite low expected fluxes demand for the construction of km3-scale detectors. The only viable solution nowadays is to build a large array of light sensors in a natural transparent medium, such as water or ice.

Since 1998 the NEMO collaboration, with base at LNS, has undertaken a R& D program for the construction of an underwater km3 telescope for high energy neutrino astronomy. The NEMO collaboration research program focuses on the selection and characterization of a deep sea site in the Mediterranean, the accomplishment of a technological survey to investigate the project feasibility and the realization of prototypes to validate the technological solutions.

After many sea campaigns, aiming at measuring and monitoring for long time the water properties and current of many sites close to the Italian coast, a site at about 80 km off the SE coast of Sicily, in the Ionian Sea at a depth of 3500 m near the southernmost cape of Sicily (Capo Passero) has been selected as suitable to host a km3 telescope.

A general layout of the apparatus, which consists of a square array of structures called towers, connected each other by means of a network of submarine cables and junction boxes, has also been defined. The connection to the shore for power feeding and data transmission is assured by a single electro-optical cable. The tower consists of a flexible three-dimensional semi-rigid structure composed by a sequence of bars, which host the optical modules and the instrumentation, interlinked by a system of ropes and anchored on the seabed. The structure is kept vertical by an appropriate buoyancy on the top. In the layout proposed for the km3, each tower will consist of a sequence of 15 m long bars spaced vertically by 40 m, with the lowermost one placed at about 150 m above the sea bottom.

The validation of all the technical solutions proposed and a prototyping activity is the program of the NEMO Phase 1 and Phase 2 projects.

The NEMO Phase 1 is under way. It consists in the realization of an underwater test site located at 2000 m depth at a distance of about 25 km offshore Catania. When it will be completed, the site will host one instrumented tower connected by an underwater cable to a junction box, in order to reproduce the two typical main components foreseen for the full km3 apparatus. A sea campaign was conducted on January 2005 to install, on the deep sea ends of the 28 km electro-optical cable, two titanium termination frames with electro optical connectors. The installed connectors permit to plug and unplug underwater instrumentation by means of a Remotely Operated Vehicle (ROV). During the sea operation two experimental devices were installed. A prototype station equipped with four hydrophones for acoustic background measurements in deep underwater environment and the SN1 seismic monitoring station of the INGV (Istituto Nazionale di Geofisica e Vulcanologia). The two stations are fully operational and they are in data taking continuously since then. This operation has allowed to validate the technique of underwater connection by means of a ROV. This phase is expected to be concluded by the end of 2006 with the deployment and connection of the junction box and of a reduced dimension tower.

In order to validate the studied technologies at the depth of 3500 m and perform a long term site survey in Capo Passero, the NEMO Phase 2 has been already started. A 100 km electro-optical cable will be laid in 2007 to connect t a complete NEMO tower located at 3500 m depth to a shore station, whose building has already been bought.

The future developments are the design study (KM3NeT), supported by the European Community for 3 years (2006-2009) with the participation of 8 Countries, aimed to design a deep-sea km3-scale observatory for high energy neutrino astronomy and an associated platform for deep-sea science, including recommendation for the site choice. The next step should be the construction of the detector, if funded.

The realization of the KM3 in Capo Passero sea has been proposed by LNS as its future development in the INFN road map. This structure would be also an underwater multidisciplinary laboratory.

Acknowledgments

Thanks are due to all the LNS staff members and associated researchers that provided the material to make this presentation. A more complete and detailed review of the achievements, progresses and technological developments at LNS can be found in the activity reports [2] published each year, since 2000.

References

1. E. Migneco, Proceeding of the 9th Conference on Problems in Theoretical Nuclear Physics, World Scientific
2. INFN-LNS Activity Reports and references therein, www.lns.infn.it

NUCLEAR STRUCTURE AT THE LEGNARO NATIONAL LABORATORIES: FROM HIGH INTENSITY STABLE TO RADIOACTIVE NUCLEAR BEAMS

G. DE ANGELIS

INFN, Laboratori Nazionali di Legnaro, I-35020 Legnaro, Italy
email: giacomo.deangelis@lnl.infn.it

To understand the properties of a nucleus, apart from establishing the interaction between its components, it is necessary to determine the arrangement of the nucleons, i.e. the structure of a nucleus. So far our knowledge about the structure of nuclei is mostly limited to nuclei close to the valley of stability, or nuclei with a deficiency of neutrons, which can be produced in fusion-evaporation reactions with stable beams and stable targets. Future perspectives in nuclear structure rely on radioactive ion beams (RIB) as well as on high intensity beams of stable ions (HISB). A world wide effort is presently going on in order to built the next generation radioactive ion beam facilities like the FAIR and the EURISOL projects. The LNL are contributing to such development through the design study of the EURISOL project as well as through the design and construction of the intermediate facility SPES. Concerning the instrumentation, particularly powerful is the combination of large acceptance spectrometers with highly segmented γ-detector arrays. An example is the CLARA γ-ray detector array coupled with the PRISMA spectrometer at the Legnaro National Laboratories (LNL). The physics aims achievable with such device complement studies performed with current radioactive beam (RIB) facilities. With this set-up we have recently investigated the stability of the N=50 shell closure. Here the comparison of the experimental data with shell model calculations seems to indicate a persistence of the N=50 shell gap down to Z=31. Also the study of proton rich nuclei can strongly benefit from the use of high intensity stable beams using fusion evaporation reactions at energies close to the Coulomb barrier. Future perspectives at LNL are based on an increase in intensity as well as on the availability of heavy ion species. Moreover a new ISOL facility (SPES) dedicated to the production and acceleration of radioactive neutron rich species is now under development at LNL. Among the new instrumentation, the concept of γ-ray tracking has been recently introduced in nuclear spectroscopy. A new γ-ray detector array (AGATA) based on such technique is now under study in a wide european collaboration. The first sub-cluster of AGATA is foreseen to be installed at the PRISMA spectrometer.

1. Introduction

Despite the big progress made in the past, nuclear structure physics still holds many surprices. Basic questions remain to be answered as:
Where are the limits of nuclear existence ? What is the dependence of the nuclear force on the N/Z ratio? What is the relative importance of the motion of individual nucleons vs. nuclear collective motion? How are complex nuclei built from their constituents?

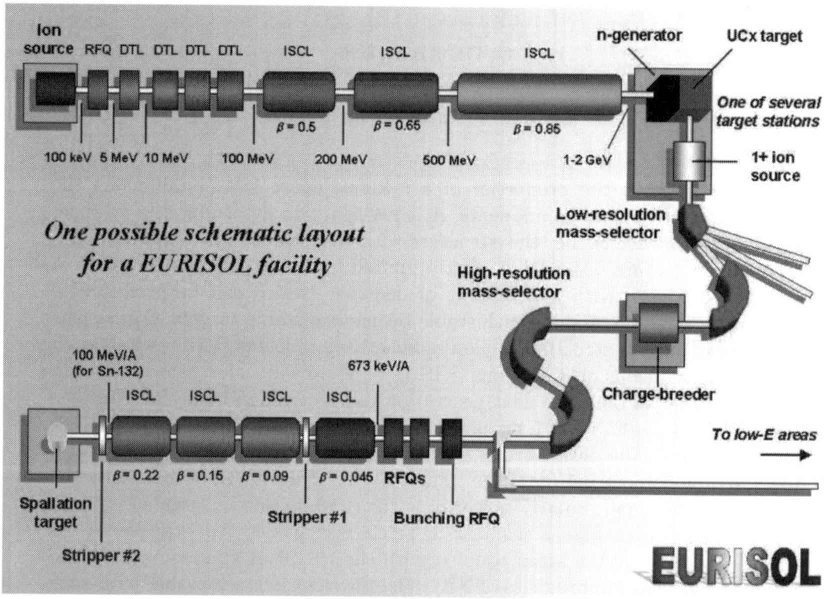

Fig. 1. Schematic lay-out of the EURISOL radioactive ion beam facility.

A tremendous effort is presently going on through a systematic exploration of new regions of the chart of nuclei as well as through the developments of high intensity stable and radioactive ion beam facilities like FAIR and EURISOL. The Legnaro National Laboratories are contributing to such process through the design study of the EURISOL project, the design and construction of an intermediate radioactive ion beam facility named SPES and the up-grading of the present accelerator complex to high intensity. In this contribution, after a short introduction on the status of the EURISOL and of the SPES projects, I will discuss some selected examples which show the big potential of stable beams using "unconventional" reactions for the study of the properties of the nuclear many body system at the limits of the isospin.

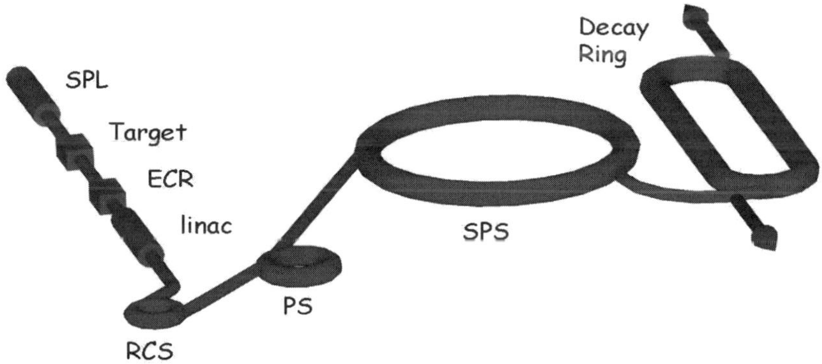

Fig. 2. Schematic lay-out of the EURISOL-Beta Beam conceptual design.

1.1. *The EURISOL Project*

The EURISOL project points to the design of the next generation European ISOL radioactive ion beam facility. The aim of such new facility is to provide, respect to the present ones, order of magnitudes more intense radioactive nuclear beams. The wide scientific programme related to the EURISOL project involves nuclear physics, nuclear astrophysics and fundamental interactions. The present design study includes a 5 MW proton linac accelerator, solid targets and proton-to-neutron converter targets (including a high power mercury jet target) a beam handling system (including charge breeding and cooling), a linac for post-acceleration and subsequent fragmentation [1]. The presently proposed driver accelerator is a 1 GeV, multi-MW, superconductive proton LINAC, figure 1. The ISOL facility would use both a 100-kW proton beam on solid target for direct RIBs production and a "converter" target to release high fluxes of spallation neutrons which would then produce RIBs by fission in a secondary target. For the high power "converter" target a windowless liquid mercury-jet target is being considered. For the EURISOL project synergies with the "Beta beam" neutrino project is now under study [2]. Such project aims to the production of a pure beam of electron neutrinos or their antiparticles through the decay of radioactive ions circulating in a storage ring.

This concept requires radioactive ions to be accelerated to as high Lorentz gamma as 100. The "beta beam" project proposes to use a thick ISOL target for the production of ^{6}He and ^{18}Ne. Acceleration could be performed first using a 100 MeV/u LINAC and later a new rapid cycling synchrotron (RCS), the PS and finally the SPS, figure 2. The isotopes would then be accumulated into a Decay Ring. The subsequend β^- and β^+ decaies would then produce beams of neutrinos and antineutrinos focused by the Lorentz boost. Research and development related to different tasks of the EURISOL

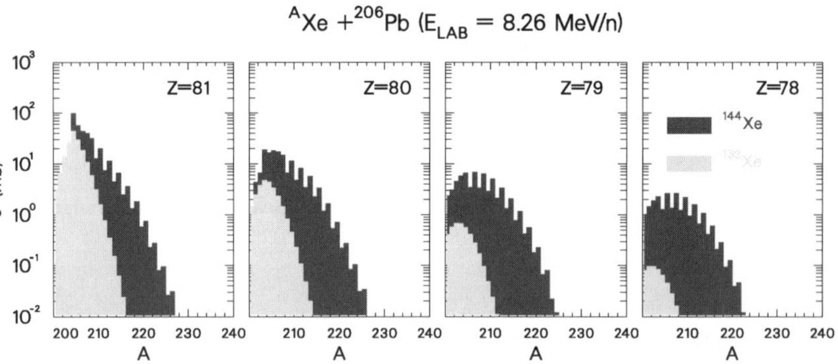

Fig. 3. Coupled channel calculations performed with the program GRAZING for the Xe + ^{206}Pb reaction. The production of the neutron rich reaction channels is strongly increased when going from ^{132}Xe to ^{144}Xe.

project are presently going on at LNL.

2. LNL future perspectives: from high intensity to radioactive nuclear beams

Nuclear structure perspectives mainly rely on the study of exotic nuclei. To accomplish such goal radioactive nuclear beams and high intensity beams of stable ions are complementary tools. Future perspectives of the Legnaro National Laboratories are therefore focused on the up-grading of the present accelerator complex to high intensity and on the construction of a new radioactive ion beam facility mainly devoted to beams of medium and heavy elements. In the following therefore I will shortly discuss:

a) the up-grading of the present superconductive LINAC to high intensity for medium and heavy ions.

b) the design and construction of the intermediate radioactive ion beam facility SPES.

2.1. *The up-grading of the superconductive LINAC ALPI*

The up-grading of the ALPI accelerator aims to the optimization of the present LINAC to intensities of the order of 100 pnA for medium and heavy elements. High intensity beams of stable heavy ions (like Xe or Pb) represent infact a very powerful tool, complementary to radioactive ion beams, for nuclear structure studies of neutron and proton rich nuclei. One of the best example of the potentialities of the high intensity stable beams in nuclear structure studies is related to the use of deep-inelastic and multi-nucleon transfer reactions to access nuclei with large neutron excess. Since the multinucleon flux moves from proton-stripping and neutron-pick-up to vice-versa when going from proton-rich to neutron-rich beams, heavy neutron-rich projectiles can be used to populate the most exotic final products. Figure 3 shows the production of neutron rich nuclei calculated using the program GRAZING [3]. One should notice that the relative intensity of the neutron rich target like channels, marked in black, strongly increases when using an heavy neutron-rich heavy beam or a radioactive ion beam.

The accelerator complex of LNL is presently based on a TANDEM XTU and on a new superconducting ion injector, both accelerating the ion beams into a superconductive LINAC. The new heavy ion injector, called PIAVE, consists of an ECR ion source coupled to two superconductive RFQs and to 8 low β resonators. It provides the optimum velocity matching to the ALPI superconductive LINAC.

The new accelerator complex, when completed, will allow to boost all stable ion beams, up to uranium, to energies above the Coulomb barrier. A recent up-grade of the superconductive resonators, using cavities based on spattered Nb, (see figure 4) has basically doubled all the accelerator fields, which now rich values of the order of 4.4 kV/m. Under such conditions higher energy as well as higher intensity beams can be achieved by the selection of more favorable charge states.

2.2. *The mid-term Isol facility SPES*

The use of beams of unstable ions will definitely allow to push the nuclear structure studies to the most exotic systems. A new mid-term radioactive nuclear beam facility dedicated to the production of neutron rich beams is now under study at LNL (see figure 5 and Ref. [4]). It is based on a

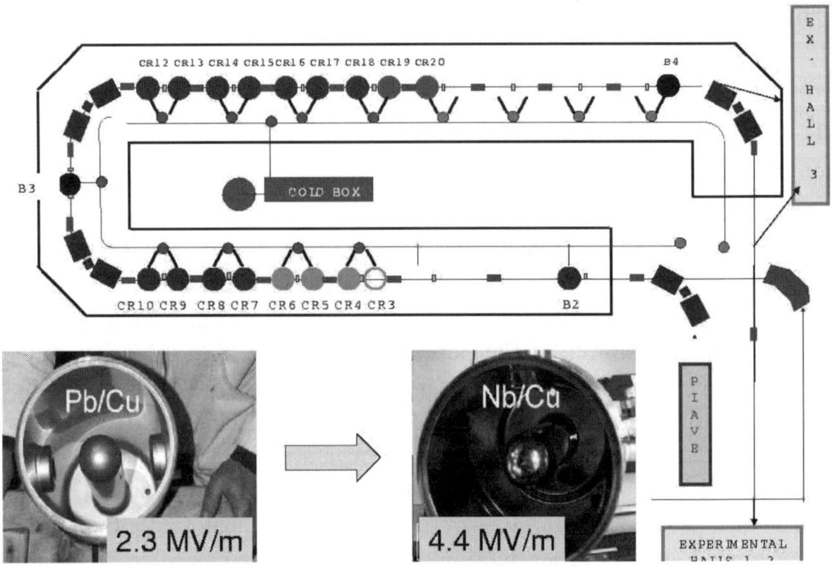

Fig. 4. Schematic lay-out of the LNL superconductive LINAC booster. The recent upgrading of the resonators, replacing the Pb cooting with a Nb cooting, has allowed to almost double the previous accelerating field, with a consisten increase both in energy and in intensity of the accelerated ion beams.

proton LINAC, the proton beam impinging directly on a U carbide target. Radioactive species are produced by proton induced fission. The fission products, after extraction from the Isol source and mass selection, will be post-accelerated by means of the present superconductive LINAC (PIAVE-ALPI). Prototypes of the RFQs and of the accelerating cavities are presently under development at LNL. A reference secondary beam intensity of 10^8 is foreseen for ^{132}Sn.

3. Nuclear Structure with (high intensity) stable ion beams

The most critical ingredient in determining the properties of a nucleus from a given effective interaction, is the overall number of nucleons and the ratio N/Z of neutrons to protons. It is the extremes in these quantities, which define the limits of existence for nuclear matter, that is going to be opened up for study with present and future radioactive ion beam accelerators. A presently available alternative approach, at least for systems not very far away from the valley of stability, is offered by high intensity beams of stable ions in combination with very efficient detectors. High intensity beams of

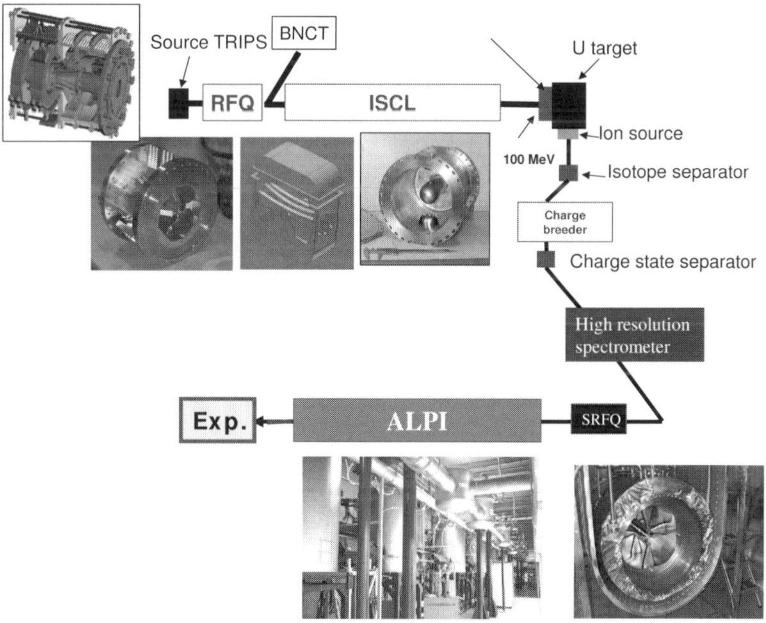

Fig. 5. Schematic lay-out of the SPES RIB-ISOL facility. It is based on an high intensity proton driver and uses an uranium carbide target for the production of the fission fragments. Post-acceleration is obtained using the present PIAVE-ALPI linac.

heavy ions can be used to reach proton and neutron rich systems where new exciting phenomena are expected to happen. In the following I will discuss few examples adressing the use, with unconventional reactions, of stable ion beams for nuclear structure studies of proton and neutron rich exotic nuclear systems.

3.1. *Exploring the changing of shell structure of neutron rich nuclei*

One aspect which is presently strongly discussed concerns the modification of the average field experienced by a single nucleon due to the changes in size and diffusivity for nuclei with strong neutron excess [5,6]. For large neutron excess the softening of the Woods-Saxon shape of the neutron potential is expected to cause a reduction of the spin-orbit interaction and therefore a migration of the high-l orbitals with a large impact on the shell structure

Fig. 6. View of the CLOVER array coupled to the PRISMA spectrometer.

of nuclei far from stability [7–10].

A different scenario has been recently suggested where the evolution of the shell structure, in going from stable to exotic nuclei, can be related to the effect of the tensor part of the nucleon-nucleon interaction [11]. The tensor-force, one of the most direct manifestation of the meson exchange origin of the nucleon-nucleon interaction, is responsible of the strong attraction between a proton and a neutron in the spin-flip partner orbits. A recent generalization of such mechanism foresees a similar behaviour also for orbitals with non identical orbital angular momenta. An attraction is expected for orbitals with antiparallel spin configuration whereas orbitals with parallel spin configuration should repell each other [8]. In most of the cases one is dealing with a combined effect of the attraction among orbitals with antiparallel spins and repulsion between orbitals with parallel spins. Those effects become particularly visible when moving away from the line of stability. In such cases the proton-neutron interaction is changed by empting of the partner orbit causing a modification into the effective single particle energies (evolution of the shell structure). The change of the shell structure based on such mechanism has been recently discussed in different mass regions of the nuclear chart [12,13] . In such contest neutron-rich nuclei close to shell gaps are particularly interesting since, when compared with the shell model prediction, they allow to search for anomalies into the shell structure [6,13]. It is predicted, for example, that the Z=28 gap for

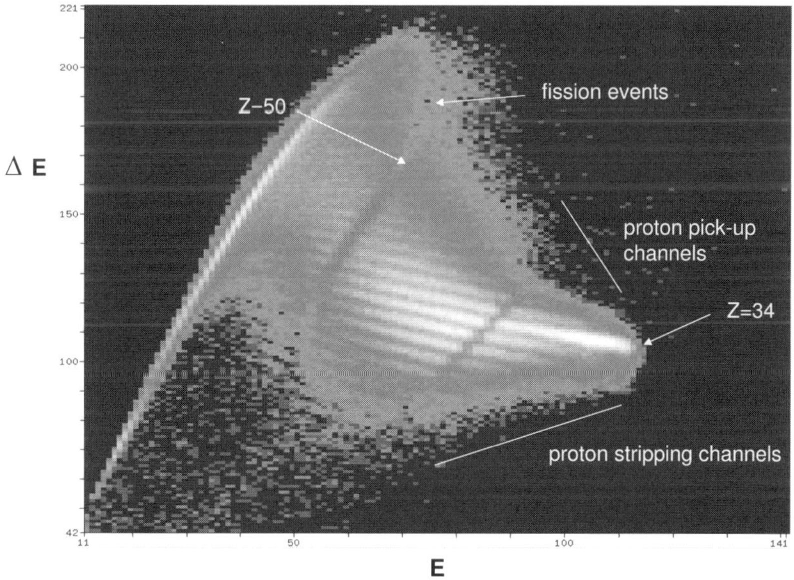

Fig. 7. Energy loss versus Total energy at the PRISMA focal plane for the ^{82}Se + ^{238}U reaction at 505 MeV of beam energy. Proton stripping reaction channels leading to more neutron rich nuclei are clearly visible.

protons in the pf-shell becomes smaller moving from ^{68}Ni to ^{78}Ni as consequence of the attraction between particles in the proton $f_{5/2}$ and neutron $g_{9/2}$ orbits and the repulsion between particles in the proton $f_{7/2}$ and the neutron $g_{9/2}$ configurations. The same argument also predicts a weakening of the N=50 shell gap when approaching the ^{78}Ni nucleus due to the attraction between particles in the neutron $g_{9/2}$ and $d_{5/2}$ configurations with the proton $f_{5/2}$ state and the repulsion between particles in the neutron $g_{7/2}$ with the proton $f_{5/2}$ states.

Experimental information has been obtained on a wide range of nuclei close to the N=50 shell closure by means of multi-nucleon transfer and deep-inelastic collisions. The reaction mechanism allows the population of medium and high spin yrast states. In most of the cases previous information on the excited states comes mainly from β-decay studies and direct reactions [14].

Excited states of previously unknown isotopes have been identified using the CLARA γ-detector array [15] in coincidence with the PRISMA spectrometer [16] - see figure 6. The N≈50 nuclei have been populated using the reaction ^{82}Se + ^{238}U. The combination of the Tandem-XTU and the

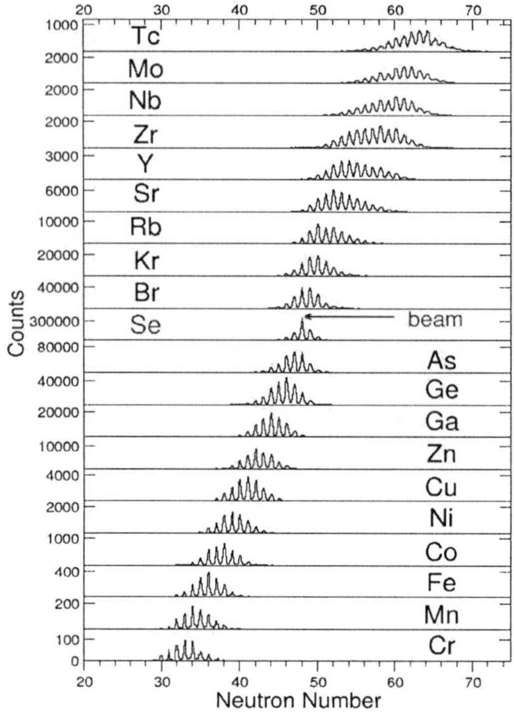

Fig. 8. Mass distributions for the different elements populated in the ^{82}Se + ^{238}U reaction at 505 MeV of beam energy.

superconductive LINAC ALPI accelerators at the Laboratori Nazionali di Legnaro, Italy, was used to accelerate a beam of ^{82}Se ions at an energy of 505 MeV. The target, isotopically enriched, was of a thickness of 400 μg/cm2. Projectile-like nuclei, produced following multinucleon transfer, were detected by the PRISMA spectrometer [13,14,17], placed at an angle of 64 degrees, covering an angular region around the grazing angle of the reaction. In figure 6 it is shown the PRISMA spectrometer with the CLOVER array CLARA positioned at the PRISMA target position. PRISMA is a magnetic spectrometer of large angular (≈80 msr) and energy (20%) acceptance, consisting of a quadrupole singlet followed by a dipole magnet. The position of the incoming ions entering the spectrometer were measured using a position sensitive MCP detector placed at 25 cm from the target. After the magnetic elements, the position of the ions were determined again at the focal plane of the spectrometer using a 10 elements, 100 cm long, multi-wire PPAC detector. Finally, the ions were stopped in a 10×4 elements ioniza-

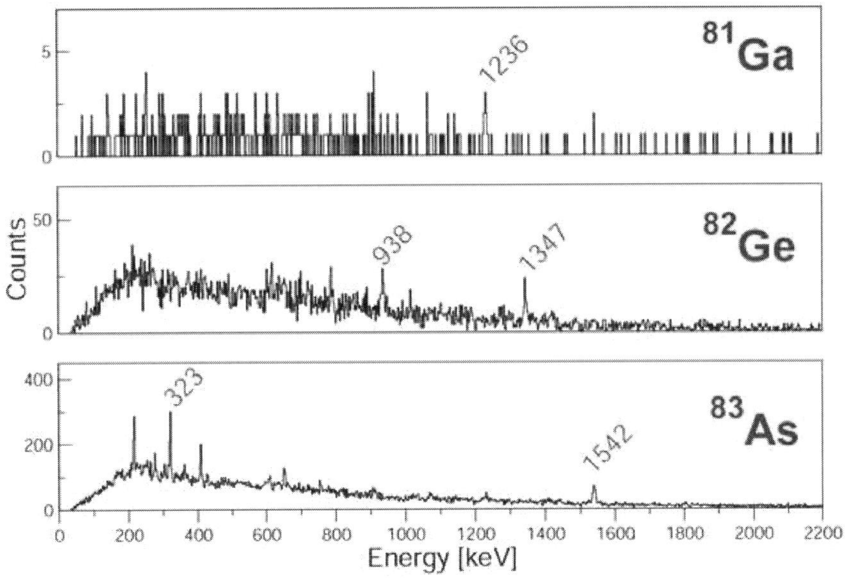

Fig. 9. Doppler corrected γ-ray spectra measured with the CLARA Ge-detector array for Z=33, 32, 31 and N=50 selected with the PRISMA Specrometer.

tion chamber measuring the time of flight (TOF). Such information allow a reconstruction of the trajectories, the identification of the atomic number Z, of the mass A and of the absolute value of the velocity - see figure 7. The mass resolution obtained in the present experiment, shown in figure 8, is of 1/180, allowing a good identification of all isotopes. Gamma ray spectra for selected N=50 isotones are shown in figure 9. For the construction of the level schemes as well as for the spin and parity assignment we have assumed that the population is mainly yrast and used, where possible, γ-coincidence data obtained in a second experiment at the GASP detector array [18]. The GASP apparatus consists of 40 Compton-suppressed large-volume germanium detectors and of an inner BGO ball acting as a total energy and multiplicity filter. Neutron-rich nuclei where populated using the reaction ^{82}Se + ^{192}Os at 460 MeV of beam energy. The Se beam was provided by the LNL accelerator complex. Triple and higher-fold γ-γ coincidences have been acquired in a "thick target" measurement. Here the target, isotopically enriched, was of a thickness of 60 mg/cm2, sufficient to stop all reaction fragments. Since all recoiling fragments were stopped in the target, Doppler broadening prevented the observation of transitions

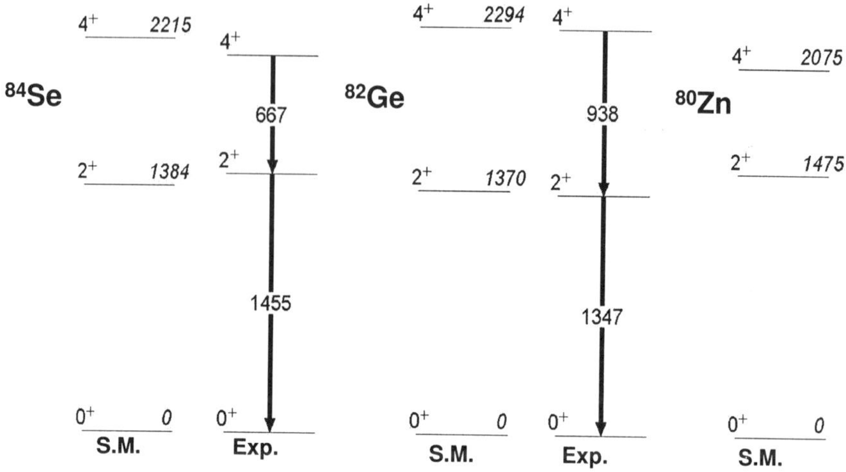

Fig. 10. Level scheme of the N=50 ^{84}Se, ^{82}Ge nuclei compared to Shell model calculations.

deexciting short lived states and only γ-decay with lifetime longer than the slowing-down time of the recoiling nuclei ($\approx$1 ps) could be resolved. The spin and parities of the levels were deduced, where possible, from angular distribution ratios from oriented states (ADO) as well as from the decay branchings. Details of the measurements as well as of the data reduction are reported in Ref. [17]. Results obtained for the N=50 isotones are shown in fugures 10 and 11 together with the predictions of shell model calculations.

Within the shell-model, the description of the nuclear mean field is conventionally obtained by considering the so-called monopole Hamiltonian constructed from the centroids of the two-body interaction. The eigenvalues of this Hamiltonian, usually referred as effective single particle energies, provide the average energies of the specific spherical configurations. They should reproduce the energies of single-proton or single neutron states in odd-A nuclei with Z or N equal to a magic number plus or minus one proton or one neutron. The nucleon-nucleon residual interaction, and in particular the tensor force part, changes the spherical single particle energies through the nuclear chart. As reported in Ref. [11,12,19] the monopole effect of the tensor force shifts systematically the single-particle levels as protons and neutrons fill certain orbits, the proton-neutron part being the dominant.

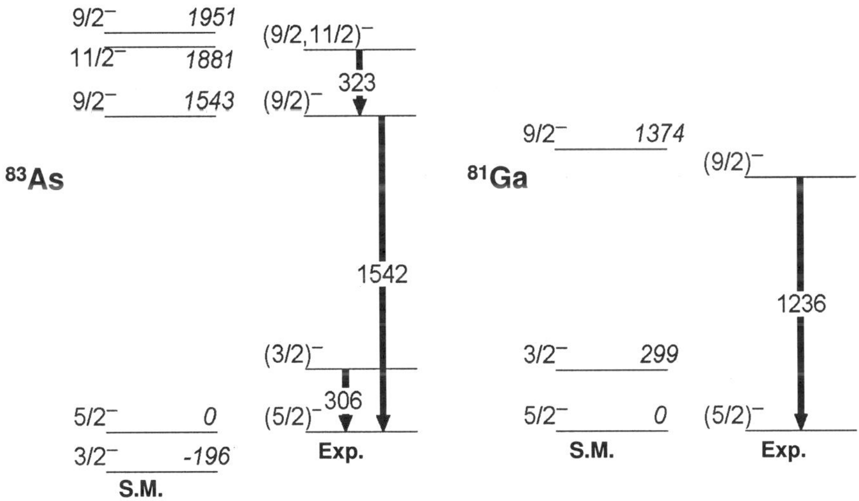

Fig. 11. Level scheme of the N=50 ^{83}As and ^{81}Ga compared to Shell model calculations.

This works attractively for anti-parallel spin configurations and repulsively for parallel spin configurations. The radial wave functions of the two orbits must also be similar in order to have a large overlap in the radial direction. For the same radial condition, larger orbital angular momenta enhance the tensor monopole effect [19]. For the N=50 isotones, due to the strong attraction between particles filling the $\pi f_{5/2}$ and the $\nu g_{9/2}$ orbits and the repulsion between the particles filling the $\pi d_{3/2}$ and the $\nu g_{9/2}$ orbits, one expects an inversion of the relative position of the $\pi f_{5/2}$ - $\pi d_{3/2}$ effective single particle states respect to the usual order where the $\pi d_{3/2}$ state is the lowest in excitation enery. Such inversion is confirmed by the experimental data (figure 11). It is interesting to mention here the good agreement obtained between the excitation energies of the newly identified states and the predicted values of the shell model calculation using the effective interaction reported in Ref. [20].

3.2. *Exploring Coulomb effects in N=Z nuclei*

Nuclei along the N=Z line or close to it display unique characteristics. The neutrons and protons occupy the same orbitals and have maximum spatial overlap of their wave functions, which leads to the development of strong

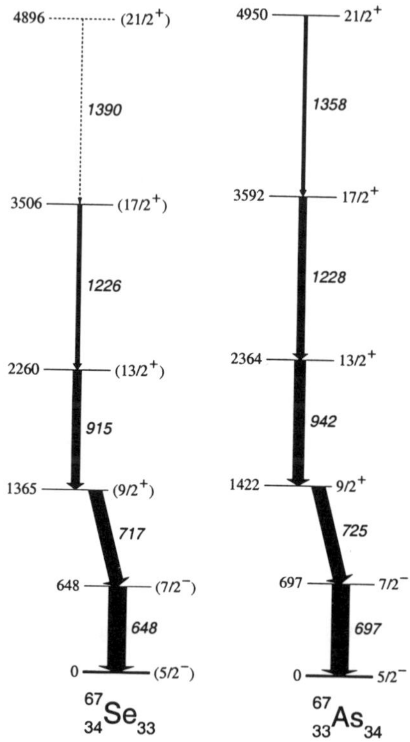

Fig. 12. Medium mass N≈Z nuclei can be populated using high intensity beams of stable ions at energies close to the Coulomb barrier. As an example I report here the A=67 mirror pair ^{67}Se-^{67}As identified using the EUROBALL [24] γ-ray spectrometer with the EUCLIDES charged particle detector array and the Neutron-wall. The mirror symmetry can be used to effectively cancel the effects of the strong force, making visible effects normally hidden as for example the electromagnetic spin orbit term.

collective effects when the number of valence nucleons increases [21]. Moreover the charge-independence of the nuclear force gives rise to a neutron-proton symmetry (represented by the isospin quantum number), which manifests itself in some structure features observable only on, or very near, the N=Z line. An example is the neutron-proton pairing, a new kind of superfluidity of nuclear matter. In nuclei far away from N=Z, the proton-

proton (pp) and neutron-neutron (nn) pairing dominate, whereas in nuclei with N=Z a sizable contribution of neutron-proton (np) correlations is expected due to the strong overlap of the wave functions. These correlations can be isovector (T=1,S=0) or isoscalar (T=0,S=1) [22]. Figuring out their character and whether they form a static pair condensate (an average field, made of np Cooper pairs) has been a continuous challenge since medium mass N=Z nuclei came into the reach of experiments [23]. The isospin quantum number leads also to the concepts of isobaric multiplets and mirror nuclei, whose structure differs only because of the Coulomb force. The increase in sensitivity and resolving power of the gamma arrays have allowed, during the last decade, more and more detailed studies of mirror energy differences, that is, differences in the energies at each yrast state of mirror nuclei (and in isobaric multiplets). Such differences give information on the nucleon alignment at the backbending, the evolution of the nuclear radii along the band, as well as on the isospin non-conserving terms in the nuclear interactions [25–28]. Large values of the mirror energy differences have for example revealed the contribution of the electromagnetic spin-orbit interaction [29]. Such force is well known from atomic physics but for nuclei it is normally hidden by the nuclear spin-orbit interaction. Only the cancellation of the effects of the strong force obtained comparing mirror nuclei allows its manifestation. The electromagnetic spin-orbit contribution to the single-particle energy for a specific orbital is given by [30]

$$E_{so} = (g_s - g_l)(e/2m_n^2c^2) < (1/r)d/drV_C(r) >< l.s >$$

Here g_s and g_l are the free-nucleon g factors. The spin-orbit energy has opposite sign for protons or neutrons and therefore the spin-orbit contribution to the single particle energy is sensitive to configuration changes. Moreover, it is large for configuration changes involving a spin flip. Using the uniformly charged sphere approximation for the radial term, the spin-orbit energy contributions for spin aligned or anti-aligned orbitals can be estimated as

$$\begin{array}{ll} E_{so} = (g_s - g_l)(e^2\hbar^2)/(16\pi\epsilon_0 r^3 m_n^2 c^2)(Z/A)(-l) & (\uparrow\uparrow) \\ E_{so} = (g_s - g_l)(e^2\hbar^2)/(16\pi\epsilon_0 r^3 m_n^2 c^2)(Z/A)(l+1) & (\uparrow\downarrow) \end{array}$$

Estimates of the spin-orbit contribution to the mirror energy differences for specific states in the A=31, 35, 39, 47, 61 mirror pair compare quite well with the experimental values, the maximum difference being less then 100

keV [31]. Interestingly for the heaviest mirror pair presently known, shown in figure 12, such estimate fails both in absolute value then in sign. The calculated ΔE_{so} value (^{67}Se-^{67}As) for the $9/2^+ - 5/2^-$ states is about 300 keV whereas experimentally one finds -57 keV. Such low value of the spin-orbit energy difference suggests a dominant configuration for the $9/2^+$ state based on simultaneous occupation (and with similar strength) of the same orbital by protons and neutrons. The main component can be symbolically expressed as an admixture of $^{66}Ge \times g_{9/2}(\pi)$ and $^{66}As \times g_{9/2}(\nu)$. For such configuration one expects a cancellation of the electromagnetic spin-orbit shift due to the opposite signs of the g factors for protons and neutrons. Such simultaneous occupation by protons and neutrons is a unique property for N=Z nuclei and it has been recently predicted by shell model calculations using an extended P+QQ Hamiltonian including T=0 monopole field, quadrupole-quadrupole and octupole-octupole force [32].
The use of mirror nuclei allows figuring out the interplay between the Coulomb potential and the isospin-breaking nuclear interaction. Electromagnetic transition probabilities are also crucial probes of the isospin symmetry, and its validity with increasing values of A and Z. If the isospin symmetry is valid, there are two important consequences: E1 (ΔT=0) transitions in N=Z nuclei are strictly forbidden, and E1 transitions in mirror nuclei must have equal strengths. It is also important to understand the mechanism of the isospin mixing in nuclei since this is used in the determination of the coupling constant from log(ft) values of superallowed Fermi decays, and may be also related to unitarity tests of the CKM matrix [33]. A possible way to study the isospin symmetry breaking is to observe experimental deviations from the two rules above. Observation of isospin forbidden E1 γ transitions in the even-even N=Z nuclei , as well as differences in E1 transitions in mirror nuclei may be used to determine the origins of the isospin symmetry breaking. Theoretical estimates, mainly limited to the ground state of even-even nuclei, show that the amount of isospin mixing increases with the mass number A, and, for given A, is maximum at N=Z. Forbidden E1 transitions have been observed only in the even-even nuclei ^{48}Cr and ^{64}Ge, in the latter the estimated minimum isospin mixing being 2.3(1.4)% [33], while irregular transitions have been observed in pairs of light mirrors, like ^{35}Ar-^{35}Cl [29]. Particularly important is the extension of such investigation to heavier N=Z systems. New powerful gamma detector arrays based on γ-ray tracking will play here a major role.

4. Increasing the sensitivity by means of gamma ray tracking

Gamma detector arrays of the present generation are built of Compton suppressed Ge detectors arranged in tightly packed spherical configurations. Despite the fact that these arrays are already composed of more than 100 detectors, their performance is limited to an efficiency of only about 10% due to the presence of the surrounding BGO shields. Recent advances in crystal segmentation technology and digital signal processing has opened the possibility to operate the detectors in a position sensitive mode. This enables to build a compact array solely out of Ge detectors omitting the BGO shields (the AGATA project). As it is expected from simulations it can have unprecedented features: an efficiency of up to 40% while maintaining a P/T-ratio of 60% . The commissioning of the first sub-array of AGATA, called the Demonstrator, is foreseen at LNL in year 2007.

5. Summary

Future perspectives at LNL are based on radioactive as well as on high intensity stable nuclear beams. Neutron rich nuclei far from stability can be investigated at medium spin using deep-inelastic and multinucleon transfer reactions. Here particularly powerful is the combination of large γ-ray detector array with large acceptance spectrometers like the CLARA-PRISMA detector at LNL. Results of an in-beam study, using the CLARA-PRISMA detector as well as the GASP γ-ray detector, are reported. High intensity stable beams with masses up to U are going to be accelerated at LNL by the new positive ion injector PIAVE. Future perspectives for more exotic systems will be based on new generation radioactive ion beam facilities like FAIR and EURISOL as well as on the intermediate radioactive ion beam facility like SPIRAL2 and SPES. The design of the SPES Isol facility, focused on the use of a direct uranium carbide target, is being to be completed.

6. Aknowledgment

I would like to thank A. Gadea, N. Marginean, E. Sahin, Y.H. Zhang and X.H. Zhou for providing most of the material here reported and to G. Duchene for discussing the results. Many thanks are due to all the colleagues of the PRISMA, CLARA groups for running the apparatus. This research has been supported by the European Commission through contract number HPRI-CT-1999-00078.

References

1. www.eurisol.org/site01/index.php
2. beta-beam.web.cern.ch/beta-beam/
3. A. Winter, Nucl. Phys. **A594**, 203 (1995).
4. SPES Thecnical Design for an Advanced Exotic Ion Beam Facility at LNL, LNL-INFN (REP) 181/02, June 2002.
5. G.A. Lalazissis *et al.*, Phys. Lett. **B418**, 7 (1998).
6. B. Blank and P.H. Regan, *Nuclear Physics News International 11 (1)*, 15 (2001).
7. J. Dobaczewski *et al.*, Phys. Scr. **T56**, 15 (1995).
8. R.C. Nayak, Phys. Rev. C **60**, 064305 (1999).
9. X. Campi *et al.*, Nucl. Phys. **A251**, 193 (1975).
10. N. Fukunishi *et al.*, Phys. Lett. **B296**, 279 (1992).
11. T. Otsuka *et al.*, Phys. Rev. Lett. **11**, 145 (1964).
12. T. Otsuka, Prog. Theor. Phys. Supp. **146**, 6 (2002).
13. H. Grawe, Springer Lect. Notes in Phys. **651**, 33 (2004).
14. R.B. Firestone, *Table of Isotopes*, 8th ed.(Wiley, 1996).
15. A. Gadea *et al.*, Eur. Phys. J. **A20**, 193 (2004).
16. A.M. Stefanini *et al.*, Nucl. Phys. **A701**, 217c (2002).
17. Y.H. Zhang *et al.*, Phys. Rev. C **70**, 024301 (2004).
18. D. Bazzacco, *Proceedings of the Intern. Conference on Nuclear Structure at High Angular Momentum*, Ottawa, 1992 Vol.II, p.376 Rep. N. AECL10613.
19. T. Otsuka *et al.*, Phys. Rev. Lett. **97**, 162501 (2006).
20. A.F. Lisetskiy *et al.*, Phys. Rev. C **70**, 044314 (2004).
21. G. de Angelis *et al.*, Phys. Lett. **B535**, 93 (2002).
22. G. de Angelis *et al.*, Phys. Lett. **B437**, 236 (1998).
23. N. Marginean *et al.*, Phys. Rev. C **63**, 031303 (2001).
24. J. Simpson, Z. Phys. **A358**, 139 (1997).
25. S. Lenzi *et al.*, Phys. Rev. Lett. **87**, 122501 (2001).
26. A.P. Zuker *et al.*, Phys. Rev. Lett. **89**, 142502 (2002).
27. M.A. Bentley *et al.*, Phys. Lett. **B437**, 243 (1998).
28. A. Gadea *et al.*, Phys. Rev. Lett. **97**, 152501 (2006).
29. J. Ekman *et al.*, Phys. Rev. Lett. **92**, 132501 (2004).
30. D.H. Wilkinson, *Isospin in Nuclear Physics, North Holland* (1969), p. 134.
31. M. Bentley, *Proceedings of the FINUSTAR Conference*, 12-16 September 2005, Kos, Greece.
32. M. Hasegawa *et al.*, Phys. Lett. **B617**, 150 (2005).
33. E. Farnea *et al.*, Phys. Lett. **B551**, 56 (2003).

EXPERIMENTAL NUCLEAR PHYSICS WITH INFN

A. BRACCO

Dipartimento di Fisica, Università di Milano, Via Celoria 16, 20133 Milano, Italy
E-mail:Bracco@mi.infn.it

The experimental activities in Nuclear Physics carried out by Italian researchers and financially supported by INFN are briefly reviewed. The experimental program covers the most important fields of the modern nuclear physics, namely the study of hadron dynamics, quark gluon plasma, nuclear structure and reaction dynamics and nuclear astrophysics. The experiments are performed at the four national laboratories, at CERN and several others laboratories abroad. In particular, LNL is mainly dedicated to nuclear structure, LNS is strongly involved on the study of equation of state, LNF has a program on hypernuclei and kaonic atoms and LNGS has a facility for measurements of cross sections of astrophysical interest. The large community working on the problem of quark gluon plasma is taking part in the ALICE experiment using the ultrarelativisc heavy ion collisions of LHC. Interdisciplinary researches are also supported. A brief outline of the research and of their present and future objectives is here given.

1. Introduction

The experimental program of INFN in nuclear physics aims at the investigation of the structure and dynamics of strongly interacting many body systems in the framework of fundamental interactions. In this context not only the properties of nuclei, but also several properties of hadrons are investigated. Probes with very different mass, energy and interaction properties are used at different laboratories in Italy and abroad. A wide spectrum of topics and problems is addressed ranging from the structure of the nucleon describes in terms of quarks and gluons to state and phase transition of nuclear matter; moreover it includes the investigation of nuclear properties under extreme conditions of temperature, angular momentum and isospin. About 507 (Full time equivalent) Italian Nuclear Physicists are working on these topics and according to the NuPECC classification the experiments are grouped in 4 major areas depending on the addressed physical problem. The areas are:

(i) Quarks and hadron dynamics
(ii) Phase transition of hadronic matter
(iii) Nuclear structure and reaction dynamics
(iv) Nuclear astrophysics and interdisciplinary researches

The resources for the different lines of research referred to the year 2007 are reported in Table 1. In the following sections some details of the ongoing research and on their perspectives are presented.

Table 1. Resources for the different areas funded by INFN for 2007.

Area	Researches	Experiments	Funds (%)
Quarks and hadron dynamics	136.9	11	31.9
Phase transition of hadronic matter	182.8	4	40.4
Nuclear Structure and reaction Dynamics	128.1	7	18.8
Nuclear Astrophysics and interdisciplinary research	53.4	6	8.9

2. Quark and hadron dynamics

One of the most exciting challenges of present day nuclear physics is to study the influence of the quark and gluon substructure on the properties and interactions of the hadrons and nuclei, i.e. to connect the QCD and low energy model descriptions. This requires an understanding of the region of non-perturbative QCD. This problem is approached by studying the properties of the internal structure of the nucleons, meson, baryonic resonances, kaonic atoms. In addition, the properties of nuclei in the presence of a strange quark are investigated in experiments measuring the formation and the decay of hypernuclei.

The internal structure of the free nucleons has been studied in great detail with HERMES using polarised lepton-polarised nucleon deep inelastic scattering. The exclusive measurements performed allows to disentangle the contributions of quarks, anti-quarks and gluons to the spin of the nucleon. The experiment, that will complete the data taking in 2007, has shown that the understanding of the nucleon spin requires consideration of the gluon contribution, and should take into account the strange pairs (see figure 1). In particular the decomposition in terms of different flavours of valence and sea quarks was obtained. Measurements of the deep virtual Compton scattering (DVCS) have given access to the Generalised Parton Distribution: the first experimental evidence of the nucleon transverse structure was obtained. The fragmentation function in the nuclear medium for different hadrons was also studied.

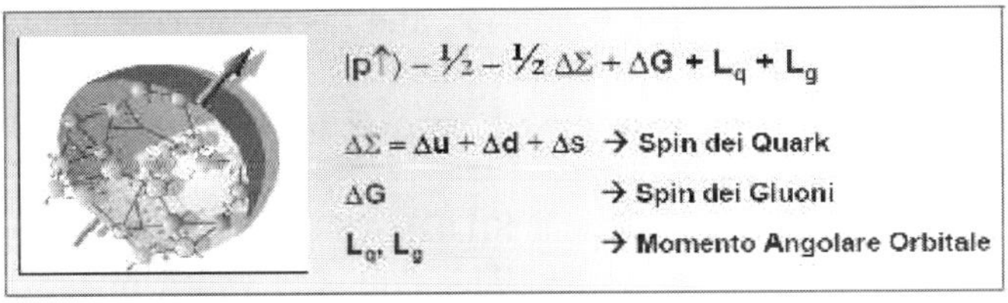

Fig. 1. The decomposition of the spin of the nucleon into the contributions of quarks (u,d,s) , gluons (G) , and orbital angular momentum L of quarks and gluons.

The experiments AIACE and LEDA at JLAB (USA) use the electron beams up to 6 GeV. The scientific program of AIACE concentrates on the study of baryon resonances in the meson electroproduction channel, the structure functions and the quark gluon structure. The unpolarised proton structure function F2 measured by the AIACE group revealed properties related to the constituent quark form factor and size. A new high statistics and precision esperiment had a great impact on the scientific community since did not confirm the previous controversial results on the exotic baryon with S=+1 pentaquark. A recent experiment is addressing the problem of understanding the behavior of the GDH integral at low Q^2. The experiment LEDA has studied hypernuclei in the mass region A<20 using the electroproduction mechanism.

The experiment CTT at MAMI proved the validity of the GDH rule on proton in the energy range 200–3000 MeV, result representing a test of the fundamental principles of the photon nucleon interaction theory. The more recent work is aiming at the mesurement of the magnetic moment of the Δ resonance. The detail study of the structure of the baronic resonances and mesons is also the main topic of the experiment GRAAL at ESFR using highly polarized photons and of PAINUC using π beams at DUBNA.

At LNF the DAΦNE produces monochromatic kaons from Φ decays. FINUDA the detector array covering the entire solid angle around one of the e^+e^- interaction point is the only one of this type to study the formation and decay spectroscopy of hypernuclei with high resolution. A first run of data taking was made in which were collected about one million events produced Λ hypernuclei on ^{6}Li, ^{7}Li, ^{12}C, ^{27}Al, ^{51}V targets. A very interesting and unexpected result was obtained, that of the possible existance of a deeply bound state KNN. The new experimental data taking presently underway will lead to a fb^{-1} of data. The results of the new data taking are expected to shed more light on the problem of deeply bound states and to provide details on the formation and decay process of hypernuclei. This will offer the possibility to test models describing the effects of strange quarks on nuclear structure and on t he Λ-N potential and therefore new calculations are very needed.

At LNF in 2007 the experiment SIDDHARTA will detect X-ray transitions from kaonic hydrogen atoms. The shift and width of the 1s level of kaonic hydrogen will be measured with unprecedented precision using X-ray triggerable detectors (silicon drift detectors) which will reduce considerably the backgroud. The goal is to determine the kaon nucleon scattering length. The possibility to determine the K mass at the 10 KeV level will be also investigated.

The PANDA and PAX experiments have as a major program the investigation of hadrons and nuclear physics problems at the new antiproton factory at GSI. Presently PANDA is concentrated on the realization of the TDR while the PAX collaboration is investigating the possibility of polarizing antiproton in order to design an experiment to measure the structure function transversity.

At LNF two letters of intents have been presented, one on the measurement of the nucleon form factors in the time like region and the other on the search of deeply bound states on ^{4}He. These experiments require major upgrades of DAΦNE complex in luminosity and energy which are presently under discussion and test.

3. Phase transitions of hadronic matter

Collisions of atomic nuclei at relativistic and ultrarelativistic energies are the means to study the phases of hadronic matter as a function of temperature and density of the system. At high density and temperature, hadronic matter may dissolve into the quark-gluon plasma via the associated phase transitions of quark deconfinement and of chiral symmetry restoration. The study of Ultrarelativistic Heavy Ions interactions was pursued vigorously by INFN with the IPER (NA50, NA60) experiment at CERN SPS. This activity will continue at the LHC at the largest energy ever produced with the ALICE detector whose construction has involved a large Italian participation. IPER studied, with a dimuon spectrometer, vector meson production in p-A and Pb-Pb and In-In interactions, confirming the anomalous J/Ψ suppression that it is hardly justified without accepting a phase transition of hadronic matter. In addition, the measurement of the low mass spectrum and of the D charmed meson production was performed. At GSI the dielectron spectrometer HADES has taken data on dielectron decay of vector mesons produced in relativistic heavy ion collisions. The aim is to search for a signal of chiral symmetry restoration.

Presently, the main engagement of INFN in this field is represented by the ALICE experiment in which italian physicists play a key role in the management (Chair of the collaboration board, deputy spokesperson and several members of the management board) and in the construction of important parts of the detector: ITS (Inner Tracking System), Time Of Flight (TOF), High Momentum Particle Identification (HMPID), Zero Degree Calorimeter (ZDC) and Dimuon Forward Spectrometer.

All these sub-systems developed innovative detectors and electronics and will to be ready for the first LHC operation with protons in 2007. For the sub-detectors many developments were entirely carried out in INFN laboratories. In particular the contribution of the italian groups for these detectors is briefly described below.

- *Inner Tracking System (ITS)*
  It is made of 6 layers using 3 different technologies, that of Pixel detectors of the first two layers, drift detectors for layers 3 and 4 and double sided strip detectors for layers 5 and 6. In addition to tracking the ITS allows to identify particles with momentum lower than 200 MeV/c.
- *Time of Flight System (TOF)*
  The detector is design to identify charged particles with momentum 0.5¡p¡2.5 GeV/c. It covers 160 m^2 , it has a time resolution of 50 ps and efficiency of 99 % and uses a 10 gap double stack Multigap RPC (resistive plate chambers). The design and construction are completely under Italian responsibility.
- *High momentum particle identification (HMPID)*
  This detector covers partially the surface of the barrel to identify the particles with momentum larger than 2.5 GeV/c. It constists of 7 modules of a proximity focusing RICH (Ring Imaging Cherenkov).
- *Muon spectrometer*
  The muon spectometer in the forward direction has the aim to detect heavy quarkonia through their decay into muons. It has RPC trigger chambers and tracking chambers.
- *ZDC*
  It is a zero degree calorimeter for protons and neutrons. It has also two electromagnetic calorimeters for the determination of the impact parameter of the collisions.

The computing activity advanced well with the realisation of the two Tier-2 prototype centres and a setting up of a data production and analysis environment based on the recently developed Alien tool. ALICE is preparing 4 Tier-2 centers and for two of them the construction is presently under way. The ALICE program on the quark-gluon plasma phase transition is not only relevant to test QCD but it has also a cosmological relevance because these very energetic collisions are expected to recreate the conditions as they existed about a microsecond after the Big-Bang. The different properties of the quark-gluon plasma will be investigated searching different signatures. Among these are the production of particles with heavy flavors and the jet-quenching mechanism.

For the jet-quenching problem the electromagnetic calorimeter, designed with the Italian contribution and mainly funded by DOE, will play a major role since the detection of the most energetic particles is crucial to understand this phenomenon.

Fig. 2. A pictorial view of the ALICE detector. Inside the large area magnet arount the interaction point there is the tracking system with silicon detectors and the drift chambers, the TOF and HMPD detectors, the TRD (transition radiation detector). In the front part, within a dipole magnet are the muon detectrs. In addition to detectors close to the interaction point there is the ZDC detector at 115 m which measures the centrality of the collisions.

4. Nuclear structure and reaction dynamics

The physics program within this field is carried out using heavy ion beams in the range 5-50 MeV/u. In particular, several different topics are approached depending both on the energy variation within this rather small interval and on the selected reaction mechanisms. These reaction mechanisms are illustrated pictorially in figure 3.

With heavy ion collisions at few MeV/u a significant progress has been made to in the understanding of Nuclear Structure which, at the same time, has generated a number of challenging new questions. These questions address the properties of the nucleus at the limits of excitation energy, spin, isospin and mass. The rapid development of experimental techniques leading to highly efficient detector arrays for gamma-rays, charged- and neutral particles made the investigation of central issues of Nuclear Structure possible. These include extreme nuclear shapes and their evolution with angular momentum as well as the influence of the thermal environment on collective modes at low and high excitation energy.

In addition, the isospin mixing in nuclei with N=Z, the neutron-proton pairing effects and the neutron rich nuclei have been investigated thanks to very exclusive measurements of gamma spectroscopy. In recent years the INFN experiments on nuclear spectroscopy have been assembled in the GAMMA collaboration. The program of GAMMA concerns measurements with existing instrumentation (using many detectors that originally were part of the EUROBALL array) and the R& D for the new generation of gamma spectroscopy arrays. In this context the RISING project at GSI makes use of radioactive beam "in-flight" while at LNL GASP and CLARA (that detects γ rays emitted by nuclei identified in mass and energy by the PRISMA spectrometer) operate with stable beams and concentrate on nuclei far from stability. In all these experiments the italian groups have a leading role. The Italian physicists have a leading role also in the European project AGATA for the construction of an array of segmented HPGe detector.

It should be noticed that the characterization of nuclei and of their structure with low and intermediate energy heavy ions rely on a continuing availability of accelerators providing high quality beams. Future developments are important to access regions so far unexplored that are relevant both to nuclear physics and astrophysics. These developments require the construction of radioactive beam facilities producing nuclei far from stability with high intensities. The project SPES at LNL is designed to provide neutron rich nuclei in the medium mass region giving the opportunity of an extensive program that together with those at GANIL and GSI will contribute to the understanding of the structure (shell and effective forces) in nuclei far from stability. For more details on this problem I refer to the contribution of G. De Angelis.

The other important topic involving an extensive experimental work is that of the study of different reaction mechanisms as tools to address some key questions. Collisions of atomic nuclei at low and intermediate energies are indeed the means to study thermalization properties, the fission process as a collective large amplitude mode probing nuclear viscosity, phases of nuclear matter. At low density and temperature, the nucleus may "multifragment" and evaporate into a cold gas of nucleons and light nuclei. Studies of dynamics and thermodynamics of excited nuclear matter have been mainly realised in the two national Laboratories LNS and LNL. At LNS the experiment ISOSPIN concerns the experimental program with the MEDEA + MULTICS and the CHIMERA arrays. The recent results are on the onset of cooperative processes, the emission time of intermediate mass fragments and the neck decay characterizing the very low density properties. The results strongly support a soft equation of state of nuclear matter and a momentum dependent mean field. In particular, the CHIMERA 4 charged particle detector is presently the most performing instrument producing high quality data on the multifragmentation studies in Heavy Ions collision at Fermi energy, while MEDEA is unique in the measurements of gamma-ray production an optimal probe for collective phenomena such as the giant dipole resonance. The most recent results on this topic are reported in the contribution of R. Alba in this volume.

At LNL the experiment Nucl-ex using the GARFIELD array focuses on the thermal properties of nuclei such as the determination of the heat capacity and of the problem of the damping of collective modes as a function of temperature. There is also within Nucl-ex an R& D program (FAZIA) for a new generation of charge particle arrays in view of experiments with radioactive beams.

The study of fission dynamics is carried out with the experiment FIESTA with the 8lp detectors at LNL while the neutron production for very heavy nuclei by the experiment N2P at LNL, Texas, and Yvaskyla.

An apparatus for the production of light EXOTIC beams using inverse kinematics reactions has been designed and set up at the LNL. The characterization of the produced ^{17}F beam has been made and measurements of proton break-up recently performed. In addition MAGNEX at LNS will perform measurements of nuclear structure with particle spectroscopy which is particularly suited for the study of resonances above the particle binding energy.

As already mentioned above the project for the future at LNL is SPES that aims to produce 10^{12-13} fission fragments per second using a proton

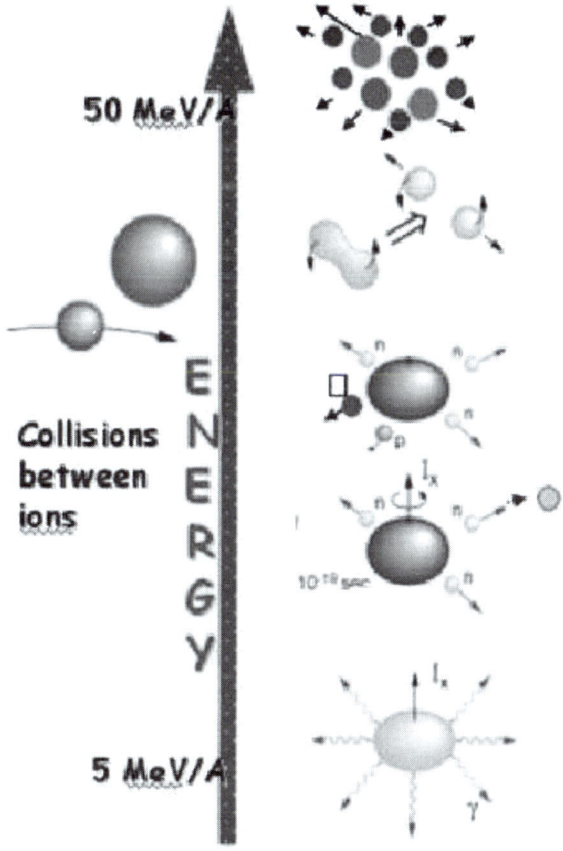

Fig. 3. Pictorial view of collisions among heavy ions in the energy range available at LNL (¡20 MeV/u) and LNS (¡50 MeV/u). 1) At energies around the Coulomb barrier both peripheral and central collisions are used to measure gamma spectroscopy (GASP and PRISMA-CLARA) at LNL and with particle spectroscopy with MAGNEX at LNS. 2) The study of the thermalization in fusion reaction is made with Garfield at LNL and 3) fission is studied at LNL with FIESTA and N2P. 4) Multifragmentation is investigated at LNS with charged particle production (CHIMERA) and gamma ray (MEDEA).

linear accelerator and a U target. In the meantime an extensive use of the ALPI–PIAVE complex is planned.

At LNS EXCYT has produced so far only the ^{8}Li beam and therefore the plan is to produce new exotic ion beams to be accelerated with Tandem and to make experiments with MAGNEX and CHIMERA. The FRIBS experiment, making separation with existing magnets has shown the possibility to produce in-flight radioactive beams and therefore more work will be made also in this connection.

5. Nuclear Astrophysics and Interdisciplinary research

The impact of Nuclear Physics on other fields of science is particularly strong for Astrophysics, applied physics and the study of Fundamental Interactions. High intensity low energy stable and radioactive ion beam accelerators are essential tools for the measurements of cross-sections relevant for astrophysical processes. The use of the antiprotons to study the fundamental interactions has also been very successful, while measurements of neutron capture cross sections have an impact on the waste transmutation problem.

Nuclear cross sections down to energies typical of the stellar processes have been experimentally studied. LUNA at LNGS realised the first measurement of the d(p, γ)^{3}He cross-section in the region of the Gamow peak of the Sun (down to 2.5 keV)and the first direct measurement of the ^{14}N(p,γ)^{15}O cross-section down to 135 keV. The result has halved the cross section value with respect to the one used in the Standard Solar Model. As a consequence, the estimate of the CNO solar neutrino flux has been decreased by about a factor 2 and that of the age of the Globular Clusters, the oldest components of the Galaxy, has been increased by 0.7-1 Gyr with respect to the current estimates. In the near future LUNA will measure the ^{25}Mg(p,γ)^{26}Al reaction which is of great importance both for nucleosynthesis and for γ ray astronomy. A new dedicated accelerator producing light ions at 4 MeV is planned for the future to study the abundancy of Carbon and Oxigen and heavier elements (see figure 4).

At Bochum the ERNA collaboration set up the apparatus to study the ^{12}C(α,γ)^{16}O reaction, obtaining a rejection factor of up to 107 for the beam and an acceptance of about 32 mrad and of 15 % in energy. The first measurements proved the effectiveness of the apparatus comprising also a jet target.

The ASFIN2 group at LNS developed and applied an indirect approach, the Trojan Horse Method, that allows extracting charged two-body cross-sections at sub-Coulomb energies, free of Coulomb suppression as well as of screening effects. The p-capture reactions on ^{6,7}Li was investigated by using the Trojan Horse method. Bare S(E)-factors were extracted together with the estimate of the screening potential. The method has been recently employed in an experiment a RIKEN with radioactive beams. The most recent results, including that on the reactions induced by ^{8}Li beams neces-

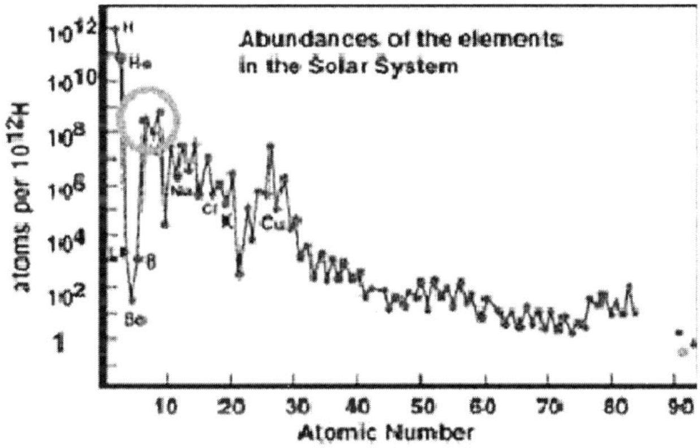

Fig. 4. The relative abundances of elements in the solar system. The abundancy of the C and O elements (in the circle) are investigated by ERNA and in the future at lower energy by LUNA at LNGS.

sary for the understanding of the inhomogeneous big-bang are reported in the contribution of R. Alba to this volume.

The experiment nTOF operating at CERN has provided data with unsurpassed resolution and with low background for the capture and fission cross-sections. In fact, important results on the capture reaction on 151Sm and 232Th were obtained: the first has implications both on Nuclear Astrophysics and on applications to nuclear waste transmutation, the second concerns an isotope involved in the Th/U fuel cycle for energy production in Accelerator Driven Systems. Several fission reactions have also been measured on ^{233,234,236}U, and in minor actinides, such as ^{237}Np, ^{241,243}Am and ^{245}Cm, interesting for the projects of nuclear waste transmutation. The high accuracy data collected at nTOF will result in an increased reliability of the cross-section data bases and will lead to a more refined study of the fission barrier.

ATHENA at the CERN AD is planning new measurements with a new set up to measure the spectroscopy of antihydrogen with high accuracy, spectroscopy interesting also to test gravity in antimatter. In addition VIP at LNGS will measure Pauli principle violation with high accuracy 10^{-15} by searching for X ray decay from an atomic transition in Cu forbidden by Pauli principle.

6. Conclusion

The Nuclear Physics activity supported by INFN and conducted by the italian community concerns all the major areas of the modern nuclear physics. A qualified and succesful participation is present in the programs both at national and international laboratories.

ALICE at LHC has a well planned program for the search of quark-gluon plasma. With the strong USA participation ALICE will be also equipped with an electromagnetic calorimeter well suited to measure JETs and understand the phenomenon of the Jet-quenching as a signature of quark-gluon plasma.

Concerning the field of hadron dynamics, the LNF is developing techniques to increase the luminosity and is preparing a proposal for an energy upgrade. There is also a strong involvment of italian physicists in the R& D for the realization of the PANDA e PAX projects at the facilty FAIR at GSI.

The two national laboratories LNL and LNS dedicated to nuclear physics at which sophisticated detectors are available will be engaged in experiments using the PIAVE-ALPI complex at LNL and the CS-EXCYT systems at LNS. With EXCYT particular attention will be devoted to astrophysical problems. The LNL laboratories have the long term project SPES aiming at producing radioactive ion beams for experiments on site and at technical developments within the longer term project EURISOL. The gamma spectroscopy will play a major role in the nuclear structure studies and the european collaboration has started the construction of the AGATA detector system to be use in campaigns at LNL, GANIL and GSI.

At LNGS the LUNA collaboration is preparing a project for the installation of a higher energy accelerator to be employed for measurements of the reactions close to the Gamow peak useful to understand the C/O ratio.

All these activies are in the Road-map of INFN, give future perspectives and the opportunity to INFN to maintain excellence in the field of Nuclear Physics.

AUTHOR INDEX

LIST OF PARTICIPANTS

AVOGADRO PAOLO
Università di Milano e INFN Milano

ANDREOZZI FRANCESCO
Università di Napoli Federico II

BACCA SONIA
GSI, Darmstadt

BALDO MARCELLO
INFN Catania

BARLETTA PAOLO
Università di Pisa

BENHAR OMAR
INFN Roma

BISCONTI CRISTIAN
Università di Lecce

BOMBACI IGNAZIO
Università di Pisa

BONACCORSO ANGELA
INFN Pisa

BONANNO LUCA
Università di Ferrara

CANTON LUCIANO
INFN Padova

CATARA FRANCESCO
Università di Catania

CO' GIAMPAOLO
Università di Lecce

CORAGGIO LUIGI
INFN Napoli

COVELLO ALDO
Università di Napoli Federico II

CRISTOFORETTI MARCO
Università di Trento

DI TORO MASSIMO
Università di Catania

FACCIOLI PIETRO
Università di Trento

GALATÀ GIUSEPPE
Università di Genova

GAMBACURTA DANILO
Università di Catania

GANDOLFI STEFANO
Università di Trento

GARGANO ANGELA
INFN Napoli

GIANNINI MAURO
Università di Genova

GIRLANDA LUCA
INFN Pisa

GIUSTI CARLOTTA
Università di Pavia

ITACO NUNZIO
Università di Napoli Federico II

KIEVSKY ALEJANDRO
INFN Pisa

LO IUDICE NICOLA
Università di Napoli

LOMBARDO UMBERTO
Università di Catania

LOVITCH LIONEL
Università di Pisa

MARCUCCI LAURA ELISA
Università di Pisa

MARTINI MARCO
Università di Lecce

MATERA FRANCESCO
Università di Firenze

MELDE THOMAS
Università di Graz

MEUCCI ANDREA
Università di Pavia

NARDI MARZIA
INFN Torino

NICOTRA ORAZIO
Università di Catania

PACATI FRANCO DAVIDE
Università di Pavia

PAGLIARA GIUSEPPE
Università di Ferrara

PARENTI IRENE
Università di Ferrara e INFN Ferrara

PASTORE ALESSANDRO
Università di Milano e INFN Milano

PISANO SILVIA
Università di Roma La Sapienza

PINCETTI MANUEL
Università di Pavia

PISENT GUALTIERO
Università di Padova

POLLS ARTURO
Università di Barcelona

PORRINO ANTONIO
Università di Napoli Federico II

RADICI MARCO
Università di Pavia

RATTI CLAUDIA
ECT*, Trento

ROSATI SERGIO
Università di Pisa

SALMÈ GIOVANNI
INFN Roma

SANTOPINTO ELENA
INFN Genova

SCHWAMB MICHAEL
Università di Trento e ECT*

TRAINI MARCO
Università Trento

VIGEZZI ENRICO
INFN Milano

VIVIANI MICHELE
INFN Pisa